Chemical Engineering Faculty Directory

Chemical Engineering Faculty Directory

2006–2007

Volume 55

A Publication of the
Chemical Engineering Education Projects Committee
of the
American Institute of Chemical Engineers

Edited by

S. Joe Qin and **J. Steven Swinnea**

Department of Chemical Engineering
University of Texas
Austin, Texas 78712

A JOHN WILEY & SONS, INC., PUBLICATION

A joint publication of the Chemical Engineering Education Projects Committee of the American Institute of Chemical Engineers and John Wiley & Sons, Inc.

Copyright © 2007 by American Institute of Chemical Engineers. All rights reserved.

Published by John Wiley & Sons, Inc., Hoboken, New Jersey.
Published simultaneously in Canada.

No part of this publication may be reproduced, stored in a retrieval system, or transmitted in any form or by any means, electronic, mechanical, photocopying, recording, scanning, or otherwise, except as permitted under Section 107 or 108 of the 1976 United States Copyright Act, without either the prior written permission of the Publisher, or authorization through payment of the appropriate per-copy fee to the Copyright Clearance Center, Inc., 222 Rosewood Drive, Danvers, MA 01923, (978) 750-8400, fax (978) 750-4470, or on the web at www.copyright.com. Requests to the Publisher for permission should be addressed to the Permissions Department, John Wiley & Sons, Inc., 111 River Street, Hoboken, NJ 07030, (201) 748-6011, fax (201) 748-6008, or online at http://www.wiley.com/go/permission.

Limit of Liability/Disclaimer of Warranty: While the publisher and author have used their best efforts in preparing this book, they make no representations or warranties with respect to the accuracy or completeness of the contents of this book and specifically disclaim any implied warranties of merchantability or fitness for a particular purpose. No warranty may be created or extended by sales representatives or written sales materials. The advice and strategies contained herein may not be suitable for your situation. You should consult with a professional where appropriate. Neither the publisher nor author shall be liable for any loss of profit or any other commercial damages, including but not limited to special, incidental, consequential, or other damages.

For general information on our other products and services or for technical support, please contact our Customer Care Department within the United States at (800) 762-2974, outside the United States at (317) 572-3993 or fax (317) 572-4002.

Wiley also publishes its books in a variety of electronic formats. Some content that appears in print may not be available in electronic format. For information about Wiley products, visit our web site at www.wiley.com.

Library of Congress Cataloging-in-Publication Data is available.

ISBN 978-0-470-14782-5

Printed in the United States of America.

10 9 8 7 6 5 4 3 2 1

Contents

United States
  Alabama
      University of Alabama 1
      University of Alabama in Huntsville 2
      Auburn University 3
      University of South Alabama 4
      Tuskegee University 4
  Arizona
      Arizona State University 5
      University of Arizona 6
  Arkansas
      University of Arkansas 7
  California
      University of California, Berkeley 8
      University of California, Davis 9
      University of California, Irvine 10
      University of California, Los Angeles 11
      University of California, Riverside 12
      University of California, San Diego 12
      University of California, Santa Barbara 13
      California Institute of Technology 14
      California State University, Long Beach 15
      California State Polytechnic University, Pomona 15
      San Jose State University 16
      University of Southern California 16
      Stanford University 17
  Colorado
      University of Colorado 18
      Colorado School of Mines 19
      Colorado State University 20
  Connecticut
      University of Connecticut 21
      University of New Haven 22
      Yale University 22
  Delaware
      University of Delaware 23
  District of Columbia
      Howard University 24
  Florida
      University of Florida 25
      Florida Institute of Technology 26
      Florida A&M University-Florida State University 27
      University of South Florida 28
  Georgia
      Georgia Institute of Technology 29
  Idaho
      University of Idaho 30
  Illinois
      University of Illinois at Chicago 31
      University of Illinois at Urbana-Champaign 32
      Illinois Institute of Technology 33
      Northwestern University 34

Indiana
University of Notre Dame	35
Purdue University	36
Rose-Hulman Institute of Technology	37
Tri-State University	37

Iowa
University of Iowa	38
Iowa State University of Science and Technology	39

Kansas
University of Kansas	40
Kansas State University	41

Kentucky
University of Kentucky	42
University of Louisville, J. B. Speed School of Engineering	43

Louisiana
Louisiana State University	44
Louisiana Tech University	45
McNeese State University	45
University of Louisiana at Lafayette	46
Tulane University	46

Maine
University of Maine	47

Maryland
Johns Hopkins University	48
University of Maryland, College Park	49
University of Maryland, Baltimore County	50

Massachusetts
University of Massachusetts Amherst	51
University of Massachusetts-Lowell	52
Massachusetts Institute of Technology	53
Northeastern University	54
Tufts University	55
Worcester Polytechnic Institute	56

Michigan
University of Michigan	57
Michigan State University	58
Michigan Technological University	59
Wayne State University	60
Western Michigan University	60

Minnesota
University of Minnesota at Duluth	61
University of Minnesota at Minneapolis St. Paul	61

Mississippi
University of Mississippi	63
Mississippi State University	63

Missouri
University of Missouri-Columbia	65
University of Missouri-Rolla	66
Washington University	67

Montana
Montana State University	68

Nebraska
University of Nebraska	68

Nevada
University of Nevada, Reno 69
New Hampshire
Dartmouth College 70
University of New Hampshire 70
New Jersey
New Jersey Institute of Technology 71
Princeton University 72
Rowan University 73
Rutgers, The State University of New Jersey 73
Stevens Institute of Technology 74
New Mexico
New Mexico Institute of Mining and Technology 75
University of New Mexico 76
New Mexico State University 77
New York
The City College of The City University of New York 78
Clarkson University 78
Columbia University 79
Cooper Union 79
Cornell University 80
Manhattan College 81
Polytechnic University 81
Rensselaer Polytechnic Institute 82
University of Rochester 83
State University of New York at Buffalo 84
Syracuse University 85
North Carolina
North Carolina Agricultural and Technical State University 86
North Carolina State University 87
North Dakota
University of North Dakota 88
Ohio
University of Akron 89
Case Western Reserve University 90
University of Cincinnati 91
Cleveland State University 92
University of Dayton 93
Ohio State University 94
Ohio University 95
University of Toledo 95
Youngstown State University 96
Oklahoma
University of Oklahoma 97
Oklahoma State University 98
University of Tulsa 99
Oregon
Oregon State University 100
Pennsylvania
Bucknell University 101
Carnegie Mellon University 102
Drexel University 103
Lafayette College 103
Lehigh University 104

University of Pennsylvania ... 105
Pennsylvania State University ... 106
University of Pittsburgh ... 107
Villanova University ... 108
Widener University ... 108

Puerto Rico
University of Puerto Rico ... 109

Rhode Island
Brown University ... 110
University of Rhode Island ... 111

South Carolina
Clemson University ... 112
University of South Carolina ... 113

South Dakota
South Dakota School of Mines and Technology ... 114

Tennessee
Christian Brothers University ... 115
The University of Tennessee at Chattanooga ... 115
University of Tennessee at Knoxville ... 116
Tennessee Technological University ... 117
Vanderbilt University ... 118

Texas
University of Houston ... 119
Lamar University ... 120
Prairie View A&M University ... 120
Rice University ... 121
Texas A&M University ... 122
Texas A&M University-Kingsville ... 123
University of Texas at Austin ... 123
Texas Tech University ... 125

Utah
Brigham Young University ... 126
University of Utah ... 127

Virginia
Hampton University ... 128
University of Virginia ... 128
Virginia Commonwealth University ... 129
Virginia Polytechnic Institute and State University/Virginia Tech ... 130

Washington
University of Washington ... 131
Washington State University ... 133

West Virginia
West Virginia University ... 134
West Virginia University Institute of Technology ... 135

Wisconsin
University of Wisconsin-Madison ... 136

Wyoming
University of Wyoming ... 137

Argentina
Instituto Tecnologico de Buenos Aires ... 138
Universidad Nacional de la Plata ... 138
Universidad Nacional del Litoral ... 138
Universidad Nacional del Salta ... 139

Universidad Nacional del Sur 140

Australia
University of Adelaide 140
Curtin University of Technology 141
University of Melbourne 141
Monash University 142
University of Newcastle 143
University of Queensland 143
Royal Melbourne Institute of Technology 144
University of Sydney 144

Bangladesh
Bangladesh University of Engineering and Technology 144

Belgium
University of Gent 145
Katholieke Uiniversiteit Leuven 145
University of Liege 145
Faculté Polytechnique de Mons 146

Brazil
Universidade Estadual de Campinas (UNICAMP) 146
Universidade Estadual de Maringa 147
Universidade Federal Do Rio De Janeiro 148
Universidade de Sao Paulo USP 149
Universidade Federal de Uberlandia 150

Canada
University of British Columbia 150
Dalhousie University 151
University of Alberta 151
University of Calgary 152
Ecole Polytechnique-University of Montreal 153
Lakehead University 154
Universite Laval 154
McGill University 154
McMaster University 155
University of New Brunswick 156
Queen's University 156
Royal Military College of Canada 157
Ryerson University 157
University of Saskatchewan 158
Universite de Sherbrooke 158
University of Toronto 159
University of Waterloo 160
University of Western Ontario 161

Chile
Pontificia Universidad Catolica de Chile 162
Universidad de Santiago de Chile 162

Peoples Republic of China
Beijing University of Chemical Technology 163
Tsinghua University 163

Colombia
    Universidad Industrial de Santander 164

Croatia
    University of Zagreb 165

Czech Republic
    University of Pardubice 165

Finland
    Abo Akademi University 166
    Helsinki University of Technology 166
    University of Oulu 166

France
    Ecole Nationale Superieure des Industries Chimiques (ENSIC) 167

Germany
    Aachen Technical University 168
    Technische Universitaet Braunschweig 168
    Technical University of Clausthal 168
    Darmstadt University of Technology 168
    University of Dortmund 169
    Universitat Erlangen-Nurnberg 169
    University of Hannover 170
    Universitat Kaiserslautern 171
    Universitat Karlsruhe 171
    Georg-Simon-Ohm Fachhochschule Nuernberg 171
    Fachhochschule Mannheim, University of Applied Sciences 172

Greece
    Aristotle University 172
    University of Patras 173

Hong Kong, SAR of China
    Hong Kong University of Science and Technology 173

Hungary
    Budapest University of Technology and Economics 174

India
    Annamalai University 174
    Birla Institute of Technology & Science (BITS) - Pilani 175
    Calcutta University 175
    Indian Institute of Science 175
    Indian Institute of Technology, Kanpur 176
    Indian Institute of Technology, Madras 176
    Indian Institute of Technology, Delhi 177
    Laxminarayan Institute of Technology, Nagpur University 177
    Manipal Institute of Technology 178
    Malaviya National Institute of Technology, Jaipur 178
    University of Mumbai 178
    National Institute of Technology Warangal 179
    Regional Engineering College, Affiliated with Bharathidasan University 179
    Indian Institute of Technology 179
    SDM College of Engineering and Technology 180

Siddaganga Institute of Technology, Affiliated with Viveswaraiah University 180
S. V. National Institute of Technology 180
Thapar Institute of Engineering and Technology 181
Vellore Institute of Technology 181

Indonesia
Institut Teknologi Bandung 181

Iran
Amirkabir University of Technology(Tehran Polytechnic) 182
Isfahan University of Technology 182
Sharif University of Technology 183

Ireland
Cork Institute of Technology 183
University College Cork 184
University College Dublin 184

Israel
Ben-Gurion University of the Negev 184
Technion-Israel Institute of Technology 185

Italy
Universita di Bologna 186
Universita di Cagliari 186
Universita de L'Aquila 187
Universita di Pisa 187
University of Salerno 188

Japan
Fukuoka University 189
Hiroshima University 189
Kanazawa University 189
Kansai University 190
Kyoto University 190
Kyushu University 191
Nagoya University 191
Niigata University 192
Osaka University 192
Seikei University 192
University of Tokyo 193
Tokyo Institute of Technology 193
Toyama University 194
Yamagata University 194
Yokohama National University 194

Republic of Korea
Chonnam National University 195
Keimyung University 195
Korea Advanced Institute of Science and Technology 195
Korea University 196
Pohang University of Science and Technology 196
Seoul National University 197

Kuwait
Kuwait University 197

Malaysia
- University of Malaya 198
- Universiti Sains Malaysia 198

Mexico
- Instituto Tecnologico y de Estudios Superiores de Monterrey 199
- Universidad Iberoamericana 199
- Universidad de las Americas, Puebla 199
- Universidad Autonoma Metropolitana-Iztapalapa 200

Netherlands
- University of Amsterdam 200
- Delft University of Technology 201
- Eindhoven University of Technology 201

New Zealand
- University of Auckland 202
- University of Canterbury 203

Norway
- Norwegian University of Science and Technology 203

Philippines
- Bicol University 204
- De La Salle University 204
- University of Santo Tomas 204

Poland
- Technical University of Lodz 205
- Technical University of Szczecin 205
- Warsaw University of Technology 205

Portugal
- Universidade do Porto 206

Qatar
- University of Qatar 207

Saudi Arabia
- King Fahd University of Petroleum & Minerals 207
- King Saud University 208

Singapore
- National University of Singapore 209

Slovenia
- University of Ljubljana 210
- University of Maribor 211

Republic of South Africa
- University of Cape Town 211
- University of KwaZulu-Natal 212
- North-West University, Potchefstroom Campus 212
- University of Pretoria 212
- University of the Witwatersrand, Johannesburg 213

Spain

University of Alcala de Henares Ingenieria Quimica 214
University of Alicante 214
Autonoma University of Barcelona 214
University of Cantabria 215
Universidad de Castilla-La Mancha 216
Complutense University of Madrid 216
University of Granada 217
University of Illes Balears 218
Universidad Politecnica de Madrid-ETSI Industriales 218
University of Murcia 218
University of Pais Vasco 219
Rovira i Virgili University 219
University of Salamanca 220
Universidad de Santiago de Compostela 220
Escuela Superior de Ingenieros Industriales de Sevilla 221
Universidad de Sevilla 221
Universitat de Valencia 222
University of Valladolid 223
University of Zaragoza 223

Sweden

University of Lund 224
Royal Institute of Technology 224

Switzerland

Swiss Federal Institute of Technology 225
Swiss Federal Institute of Technology, Ecole Polytechnique Federale de Lausanne 225

Taiwan

Chang Gung University 226
Feng-Chia University 226
National Central University 227
National Cheng Kung University 227
National Chung Cheng University 228
National Taiwan University of Science and Technology 228
National Taiwan University 229
National Tsing Hua University 229
Tamkang University 230
Tunghai University 230
Yuan Ze University 231

Thailand

Chulalongkorn University 231
Kasetsart University 232
Prince of Songkla University 232

Turkey

Ankara University 233
Bogazici University 233
University of Ege 234
Gazi University 234
Hacettepe University 235
Middle East Technical University 236
Yildiz Technical University 236

United Arab Emirates
United Arab Emirates University 236

United Kingdom
Aston University 237
University of Bath 237
University of Cambridge 238
University of Edinburgh 238
Heriot-Watt University 239
University of Leeds 239
University of London (Imperial College) 239
University of London (University College London) 240
Loughborough University 241
University of Manchester (formerly UMIST) 241
University of Newcastle Upon Tyne 243
University of Nottingham 243
Queen's University of Belfast 244
University of Sheffield 244
University of Strathclyde 245
University of Surrey 245
University of Teesside 246

Venezuela
Universidad Central de Venezuela 246

Omega Chi Epsilon   248

Index of U.S. Faculty   249

Index of Institutions   263

Preface

This edition of the directory now includes research interest areas for individual faculty. Areas are displayed as two letter codes. An explanation of the codes appears on the following page. The format of the present edition continues the changes set in place over the past few years. Most notably e-mail addresses for individuals appear where contributing institutions have provided them. Where possible U.S. entries begin at the top of a column. We hope that this provides a useful, readable document. We have also updated the Index of Schools (now the Contents that follows this section) and added a comprehensive index of all institutions at the end of the directory.

Production of this volume would not have been possible without the inspiration of John W. Eaton, who meticulously built the formatting and document processing scripts and gathered most of the original institution information.

Corrections and Updates

We would like to produce the most accurate and useful directory possible, but we cannot do this without your help. If you spot errors, or have suggestions for ways to improve the directory, or any other comments, we would like to hear from you. Send your corrections and comments via electronic mail to

    che-faculty-directory@che.utexas.edu

The most convenient method for updating or adding your institution's information can be found at the following URL

    http://www.che.utexas.edu/AIChE/directory/update.html

Updates will be added to the next edition. Please provide as much information as possible. We are especially interested in obtaining an e-mail address for each department.

Browsing the Directory on the WWW

Much of the information contained in this directory is also available via the World Wide Web at the URL

    http://www.che.utexas.edu/che-faculty/

The WWW version of the directory also contains electronic mail addresses and some links to university, department, and faculty home pages. Information at this site is nominally updated at the same time that the current directory is published.

S. Joe Qin and Steve Swinnea
Austin, Texas
September 2006

Key To Interest Areas

AM	Advanced Materials
BM	Biomedical Engineering
BT	Biotechnology
EN	Energy
EV	Environmental Engineering, Science & Sustainability
HT	Heat Transfer
ME	Microelectronics
MO	Modeling, Simulation & Theory
NT	Nanotechnology
PO	Polymers
PR	Process Systems, Engineering & Control
RE	Reactors, Kinetics & Catalysis
SE	Separations
SU	Surface & Interface Science
TH	Thermodynamics
TP	Transport Phenomena, Fluid Mechanics & Fluid Dynamics

United States

Alabama

University of Alabama

Chemical and Biological Engineering Department
Box 870203
Tuscaloosa, AL 35487-0203
Deliveries: A127 Bevill Building

University phone	(205) 348-4890
Department phone	(205) 348-6450
Department FAX	(205) 348-7558

Professors

April, Gary C. (Head) .... (205) 348-6452
gcapril@eng.ua.edu
EN,EV,MO
Arnold, David W. .... (205) 348-1741
darnold@eng.ua.edu
EN,PR
Clements, Wm. C. Jr. (Emeritus) ..... (205) 348-1726
wclements@eng.ua.edu
PR
Gupta, Arunava (Joint) .... (205) 348-3822
agupta@Mint.ua.edu
AM,ME,NT,SU
Lane, Alan M. .... (205) 348-1729
alane@eng.ua.edu
AM,EN,NT,RE
McKinley, Marvin D. (Emeritus) ...... (205) 348-6486
mmckinley@eng.ua.edu
EN,PR,SE
Weaver, Mark (Adjunct) .... (205) 348-7073
mweaver@eng.ua.edu
AM,EN,NT,SU
Wiest, John M. (Associate Dean) ...... (205) 348-1727
jwiest@coe.eng.ua.edu
AM,EN,MO,NT

Associate Professors

Brazel, Chris .... (205) 348-9738
cbrazel@eng.ua.edu
BM,BT
Carlson, Eric S. .... (205) 348-1581
ecarlson@eng.ua.edu
MO
Clark, Peter E. .... (205) 348-1682
pclark@eng.ua.edu
EN,PO
Johnson, Duane T. .... (205) 348-8402
djohnson@eng.ua.edu
AM,BT,EN,NT
Klein, Tonya M. .... (205) 348-9744
tklein@eng.ua.edu
AM,NT

Assistant Professors

Ritchie, Stephen M. .... (205) 348-2712
sritchie@eng.ua.edu
AM,EV,NT,SE
Turner, C. Heath .... (205) 348-1733
hturner@eng.ua.edu
AM,EN,MO,NT

Accredited by: ABET

Degrees granted 2005-2006:
  B.S.: 27   M.S.: 9   Ph.D.: 5

Graduate advisor: Alan M. Lane

Undergraduate advisor: David Arnold

Student organization: AIChE
Advisor: Peter Clark

Department reports to: Charles L. Karr, Dean of Engineering

Placement service: Career Center (205) 348-5848

United States

University of Alabama in Huntsville

Department of Chemical and Materials Engineering
EB 130
Huntsville, AL 35899
Deliveries: 301 Sparkman Dr.

University phone	(256) 824-6340
Department phone	(256) 824-6810
Department FAX	(256) 824-6839

Professors

Cerro, Ramon L. (Chair) .... (256) 824-7313
rlc@che.uah.edu
EN,EV,MO,NT,PR,SU

Chen, Chien-Pin .... (256) 824-6194
cchen@che.uah.edu
BM,EN,MO

Chittur, Krishnan K. .... (256) 824-3596
kchittur@che.uah.edu
BT,MO,SE

Smith, James E. Jr. .... (256) 824-6439
jesmith@eng.uah.edu
AM,EN,MO,PO,RE

Associate Professors

Banish, Michael .... (256) 824-6969
banishm@email.uah.edu
AM,BT,EN,ME,PO

Weimer, Jeffrey .... (256) 824-6954
jjweimer@matsci.uah.edu
AM,NT,PO

Assistant Professors

Taconi, Katherine .... (256) 824-3595
taconik@email.uah.edu
BM,BT,EN

Accredited by: ABET

Degrees granted 2005-2006:

B.S.: 9   M.S.: 5

Graduate advisor: C.P. Chen

Undergraduate advisor: Faculty

Student organization: Chemical Engineering Student's Club
Advisor: Katherine Taconi

Department reports to: Jorge Aunon, Dean, (256) 824-6474

Placement service: John R. Shrout (256) 824-6612, Director, Career Planning and Placement Office

Auburn University

Department of Chemical Engineering
212 Ross Hall
Auburn University, AL 36849-5127

University phone .... (334) 844-4000
Department phone .... (334) 844-4827
Department FAX .... (334) 844-2063

Professors

Chambers, Robert P. .... (334) 844-2054
chambro@eng.auburn.edu
Cullinan, Harry T. .... (334) 844-2016
culliht@eng.auburn.edu
Curtis, Christine W. .... (334) 844-5560
curticw@auburn.edu
Guin, James A. (Emeritus)
guinja@eng.auburn.edu
Gupta, Ram .... (334) 844-2013
gupta@auburn.edu
BM,NT,SE,TH
Hanley, Thomas .... (334) 844-7773
hanley@auburn.edu
BT,MO,PO,RE,TP
Krishnagopalan, Gopal A. .... (334) 844-2011
krishga@auburn.edu
EV,PR,RE
Lee, Y. Y. .... (334) 844-2019
leeyoon@auburn.edu
Maples, Glennon .... (334) 844-2041
maplegl@auburn.edu
Moore, O. C. (Emeritus)
Neuman, Ronald D. .... (334) 844-2017
neumard@auburn.edu
Roberts, Christopher (Chair) .... (334) 844-2036
robercr@auburn.edu
EN,NT,RE,TH
Tarrer, A. Ray .... (334) 844-2018
tarrear@auburn.edu
Tatarchuk, Bruce J. (Director) .... (334) 844-2023
tatarbj@auburn.edu

Associate Professors

Duke, Steve R. .... (334) 844-2087
dukeste@auburn.edu
AM,EN,NT,SE,TP
Hirth, Leo J. (Emeritus) .... (334) 844-2006
lhirth@eng.auburn.edu
Vives, Donald L. (Emeritus)

Assistant Professors

Ashurst, William (Bob) .... (334) 844-4827
ashurwr@eng.auburn.edu
Byrne, Mark E. .... (334) 844-4827
byrneme@auburn.edu
AM,BM,BT,NT,PO
Davis, Virginia A. .... (334) 844-2060
davisva@auburn.edu
AM,NT,PO
Eden, Mario .... (334) 844-4827
edenmar@auburn.edu
Josephson, Bill (Visiting) .... (334) 844-4827
Placek, Timothy D. .... (334) 844-2022
placetd@auburn.edu
EN,EV,MO,TP
Wang, Jin .... (334) 844-2020

Instructors

Mills, David R. .... (334) 844-2015
millsdr@auburn.edu

Other Faculty

Aksoy, Burak .... (334) 844-2016
aksoybu@auburn.edu
Cahela, Don .... (334) 844-2026
caheldr@auburn.edu
Hodges, Russel .... (334) 844-2031
hodgeru@auburn.edu
Kang, Choon-Hyoung .... (334) 844-2034
kangcho@auburn.edu
Liu, Juncheng .... (334) 844-2067
liujunc@auburn.edu
Pant, Kamal .... (334) 844-4827
You, Seong .... (334) 844-4827
Zhao, Jian .... (334) 844-4827
Zhao, Shihuai .... (334) 844-3111
Zhu, Wen-Hua .... (334) 844-2005
zhuwenh@auburn.edu

Accredited by: ABET, SACS

Degrees granted 2005-2006:
   B.S.: 39   M.S.: 4   Ph.D.: 7

Graduate advisor: Ram Gupta

Undergraduate advisor: Jennifer Harris (334) 844-2030

Student organization: AIChE; Omega Chi Epsilon; TAPPI, PIMA
Advisor: Yoon Y. Lee, Mario Eden; Gopal Krishnagopalan

Department reports to: Larry Benefield (334) 844-4326, Dean of Engineering

Placement service: Nancy Bernard, Coordinator (334) 844-4744

United States

University of South Alabama

Department of Chemical Engineering
EGLB 244
Mobile, AL 36688-0002

University phone	(251) 460-6101
Department phone	(251) 460-6160
Department FAX	(251) 461-1485

Professors

Askew, William C. .... (251) 460-7563
    waskew@jaguar1.usouthal.edu
Dhawan, Jagdish C. .... (251) 460-6160
    jdhawan@usouthal.edu
Harrison, B. Keith .... (251) 460-7461
    kharriso@usouthal.edu
Palanki, Srinivas (Chair) .... (251) 460-6160
Rodriguez, Harold V. (Emeritus)
Sylvester, Nicholas .... (251) 461-1432
    nsylvest@usouthal.edu

Assistant Professors

Misra, Manish .... (251) 460-6160
    manish@usouthal.edu
Morisani, Stephen J. .... (251) 460-7463
    smorisan@jaguar1.usouthal.edu

Accredited by: ABET

Degrees granted 2005-2006:
    B.S.: 21   M.S.: 9

Graduate advisor: Nicholas D. Sylvester

Undergraduate advisor: Srinivas Palanki

Student organization: AIChE
Advisor: J. C. Dhawan

Department reports to: Dr. John Steadman, Dean of Engineering

Placement service: Nicole Schultz, Coordinator of Co-op and Placement

Tuskegee University

Chemical Engineering Department
Tuskegee, AL 36088
Deliveries: Engineering Building, Room 513

University phone	(334) 727-8011
Department phone	(334) 727-8089
Department FAX	(334) 724-4188

Professors

Krishnagopalan, Jaya .... (334) 727-8795
    krishj@tuskegee.edu
Kwon, K. C. .... (334) 727-8976
    kwonk@tuskegee.edu
Nosa, Egiebor .... (334) 724-4265
    egibor@tuskegee.edu
Vahdat, Nader (Head) .... (334) 727-8978
    vahdatn@tuskegee.edu

Associate Professors

Josephson, William .... (334) 727-8996
    josepshon@tuskegee.edu

Assistant Professors

Floyd, Tamara .... (334) 727-8975
    floydt@tuskegee.edu
He, Qinghua .... (334) 727-4318
Naser, Jamil .... (334) 727-8050
    naserj@tuskegee.edu

Accredited by: ABET

Degrees granted 2005-2006:
    B.S.: 15

Undergraduate advisor: Nader Vahdat

Student organization: AIChE
Advisor: Tamara Floyd

Department reports to: Legand Burge (334) 727-8356, Dean

Placement service: Sara Stringer (334) 727-8294, Director of Cooperative Education and Placement

Arizona

Arizona State University

Department of Chemical Engineering
P.O. Box 876006
Tempe, AZ 85287-6006
Deliveries: Engineering CNTR G Wing #202

University phone .... (480) 965-9011
Department phone .... (480) 965-3313
Department FAX .... (480) 965-0037

Professors

Berman, Neil S. (Emeritus) .... (480) 965-4113
    Neil.Berman@asu.edu
Lin, Jerry Y. S. (Chair) .... (480) 965-7769
    Jerry.Lin@asu.edu
Raupp, Gregory B. (Associate Dean) .. (480) 965-8752
    raupp@asu.edu
Sater, Vernon E. (Emeritus) .... (480) 965-5529
    Vernon.Sater@asu.edu
Sierks, Michael .... (480) 965-2828
    sierks@asu.edu
Wang, Joseph .... (480) 727-0399
    Joseph.Wang@asu.edu

Associate Professors

Andino, Jean M.
Beckman, James R. (Associate Chair) .. (480) 965-4395
    jim.beckman@asu.edu
Burrows, Veronica .... (480) 965-4557
    burrows@asu.edu
Rivera, Daniel .... (480) 965-9476
    daniel.rivera@asu.edu

Assistant Professors

Allen, Jonathan .... (480) 965-4112
    Jonathan.O.Allen@asu.edu
Heys, Jeffrey .... (480) 965-0874
    Jeffrey.Heys@asu.edu
Rege, Kaushal
Vogt, Bryan

Accredited by: ABET

Degrees granted 2005-2006:
    B.S.: 30   M.S.: 6   Ph.D.: 2

Graduate advisor: Paul Grillos

Undergraduate advisor: Xavier Oaks

Student organization: AIChE
Advisor: James Beckman

Department reports to: Peter E. Crouch (480) 965-1722, Dean

Placement service: Ray Castillo, Director (480) 965-2350

United States

University of Arizona

Department of Chemical and Environmental Engineering
P.O. Box 210011
Tucson, AZ 85721
Deliveries: 1133 E. James E. Rogers Way, Room 108

University phone .... (520) 621-2211
Department phone .... (520) 621-6044
Department FAX .... (520) 621-6048

Professors

Arnold, Robert (Joint) .... (520) 621-2410
rga@engr.arizona.edu
EV
Bier, Milan (Emeritus) .... (520) 621-6763
Farrell, James (Joint) .... (520) 621-2465
farrellj@engr.arizona.edu
EV
Field, James .... (520) 626-5858
jimfield@email.arizona.edu
BT,EV
Guzman, Roberto .... (520) 621-6041
guzmanr@engr.arizona.edu
BT,NT,SE
Ogden, Kimberly (Joint) .... (520) 621-9484
ogden@email.arizona.edu
BT,TP
Peterson, Thomas W. (Dean) .... (520) 621-6594
peterson@erc.arizona.edu
Philipossian, Ara ....,........ (520) 621-6101
ara@engr.arizona.edu
AM,TP
Saez, Eduardo .... (520) 621-5369
esaez@engr.arizona.edu
EV,PO
Schrader, Glenn L. (Head) .... (520) 621-2591
schrader@email.arizona.edu
AM,ME,RE
Shadman, Farhang (Honorary) .... (520) 621-6052
shadman@erc.arizona.edu
AM
Sierka, Raymond (Emeritus) .... (520) 621-6586
sierka@engr.arizona.edu
EV
White, Don H. (Emeritus) .... (520) 621-6049
dwhite@engr.arizona.edu

Associate Professors

Baygents, James .... (520) 621-6043
baygents@email.arizona.edu
TP
Blowers, Paul .... (520) 626-5319
blowers@engr.arizona.edu
MO
Ela, Wendell (Joint) .... (520) 626-9323
wela@engr.arizona.edu
EV

Muscat, Anthony .... (520) 626-6580
muscat@erc.arizona.edu
ME,SU
Sierra Alvarez, Maria Reyes .... (520) 626-2896
rsierra@email.arizona.edu
BT,EV

Accredited by: ABET

Degrees granted 2005-2006:
    B.S.: 24   M.S.: 3   Ph.D.: 7

Graduate advisor: All Faculty

Undergraduate advisor: All Faculty

Student organization: AIChE
Advisor: Kimberly Ogden

Department reports to: Thomas W. Peterson (520) 621-6594, Dean

Placement service: Marie Rozenblit (520) 621-2719

Arkansas

University of Arkansas

Ralph E. Martin Department of Chemical Engineering
University of Arkansas
3202 Bell Engineering Center
Fayetteville, AR 72701
Deliveries: 3202 Bell Engineering Center

University phone .... (479) 575-2000
Department phone .... (479) 575-4951
Department FAX .... (479) 575-7926

Professors

Babcock, Robert E. .... (479) 575-5410
    rbabcoc@uark.edu
Beitle, Robert .... (479) 575-7566
    rbeitle@uark.edu
Clausen, E. C. .... (479) 575-5412
    eclause@uark.edu
Couper, James R. (Emeritus) .... (479) 575-5976
    jrc@engr.uark.edu
Cross, Robert (Research) .... (479) 575-5382
    racross@uark.edu
Havens, Jerry A. .... (479) 575-3857
    jhavens@uark.edu
King, Jerry .... (479) 575-5979
    jwking01@uark.edu
Penney, W. Roy .... (479) 575-5681
    wrp@engr.uark.edu
Silano, A. A. (Research) .... (479) 575-4502
Spicer, Thomas O. (Head) .... (479) 575-6516
    tos@uark.edu
Thatcher, Charles M. (Emeritus) ...... (479) 575-5382
Thoma, Greg .... (479) 575-7374
    gthoma@uark.edu
Turpin, Jim L. (Emeritus) .... (479) 575-5975
    jturpin@engr.uark.edu
Ulrich, Richard K. .... (479) 575-5645
    rulrich@uark.edu
Welker, J. Reed (Emeritus) .... (479) 575-6691
    rwelker@engr.uark.edu

Associate Professors

Ackerson, Michael D. .... (479) 575-5978
    Mike.Ackerson@processdyn.com

Instructors

Myers, Wm. A. .... (479) 575-5977
    wam@engr.uark.edu

Accredited by: ABET

Degrees granted 2005-2006:
    B.S.: 32   M.S.: 2   Ph.D.: 5

Graduate advisor: R. K. Ulrich

Undergraduate advisor: William A. Myers

Student organization: AIChE
Advisor: Ed Clausen

Department reports to: Dean of Engineering

Placement service: Career Services (479) 575-2805

United States

California

University of California, Berkeley

Chemical Engineering Department
201 Gilman Hall, MC# 1462
Berkeley, CA 94720-1462
Deliveries: B-84 Hildebrand Hall, Berkeley, CA 94720

University phone .... (510) 642-6000
Department phone .... (510) 642-2291
Department FAX .... (510) 642-4778

Professors

Balsara, Nitash P. .... (510) 643-8973
    nbalsara@cchem.berkeley.edu
Bell, Alexis T. (Chair) .... (510) 642-1536
    Bell@cchem.berkeley.edu
Blanch, Harvey W. .... (510) 642-1387
    Blanch@socrates.berkeley.edu
Cairns, Elton J. (Emeritus) .... (510) 642-8537
    EJcairns@lbl.gov
Clark, Douglas S. .... (510) 642-2408
    Clark@cchem.berkeley.edu
Frechet, Jean M. J. (Joint) .... (510) 643-3077
    frechet@cchem.berkeley.edu
Goren, Simon L. (Emeritus) .... (510) 642-2291
Graves, David B. .... (510) 642-2214
    graves@berkeley.edu
Hanson, Donald N. (Emeritus) .... (510) 642-2291
Iglesia, Enrique .... (510) 642-9673
    iglesia@cchem.berkeley.edu
Keasling, Jay D. .... (510) 642-4862
    keasling@socrates.berkeley.edu
King, C. Judson (Emeritus) .... (510) 643-3199
    cjking@berkeley.edu
Lynn, Scott (Emeritus) .... (510) 642-1634
    Lynn@cchem.berkeley.edu
Lyon, David N. (Emeritus) .... (510) 642-2291
    NBLX35A@prodigy.com
Maboudian, Roya .... (510) 643-7957
    maboudia@socrates.berkeley.edu
Muller, Susan J. .... (510) 642-4525
    muller2@socrates.berkeley.edu
Newman, John S. (Associate Chair) ... (510) 642-4063
    newman@newman.cchem.berkeley.edu
Petersen, Eugene E. (Emeritus) .... (510) 642-2291
    eepet@berkeley.edu
Prausnitz, John M. (Emeritus) .... (510) 642-3592
    prausnit@cchem.berkeley.edu
Radke, Clayton J. .... (510) 642-5204
    radke@cchem.berkeley.edu
Reimer, Jeffrey A. (Chair) .... (510) 642-8011
    reimer@socrates.berkeley.edu
Williams, Michael C. (Emeritus) ...... (510) 642-2291

Associate Professors

Schaffer, David .... (510) 643-5963
    schaffer@cchem.berkeley.edu

Assistant Professors

Chu, Jhih-Wei
    BM,BT,MO
Katz, Alexander .... (510) 643-3248
    katz@cchem.berkeley.edu
Segalman, Rachel A. .... (510) 642-7998
    segalman@berkeley.edu

Instructors

Plouffe, Paul .... (510) 642-8715
    Plouffe@cchem.berkeley.edu
Wallman, P. Henrik .... (510) 642-2295
    wallman@cchem.berkeley.edu

Other Faculty

Acrivos, Andreas (Adjunct) .... (510) 642-2291
Alexander, Keith .... (510) 642-4526
    kalexand@berkeley.edu
Grossberg, A. L. .... (510) 642-2291
Maiorella, Brian (Adjunct) .... (510) 642-2291
    bmaiorella@msn.com
Sternberg, Moshe (Adjunct) .... (510) 642-2291
    sternber@socrates.berkeley.edu
Zones, Stacey (Adjunct) .... (510) 642-2291

Accredited by: ABET

Degrees granted 2005-2006:
    B.S.: 88   M.S.: 4   Ph.D.: 18

Graduate advisor: John S. Newman

Undergraduate advisor: Sandra Rehling

Student organization: AIChE
Advisor: Alexander Katz

Department reports to: Dean Charles Harris, College of Chemistry

Placement service: L. Hernandez-Lewis (510) 642-1162

University of California, Davis

Chemical Engineering and Materials Science
One Shields Avenue
Davis, CA 95616-5294
Deliveries: 1334 Bainer Hall

Department phone .... (530) 752-0400
Department FAX .... (530) 752-1031

Professors
Asta, Mark .... (530) 752-5132
mdasta@ucdavis.edu
AM,MO,NT
Bell, Richard L. (Emeritus) .... (530) 752-0400
Boulton, Roger .... (530) 752-0900
rbboulton@ucdavis.edu
BT,PR,RE
Browning, Nigel .... (530) 754-5358
nbrowning@ucdavis.edu
AM
Dungan, Stephanie R. .... (530) 752-5447
srdungan@ucdavis.edu
PO,SU
Gates, Bruce C. .... (530) 752-3953
bcgates@ucdavis.edu
RE
Gibeling, Jeffery C. (Dean) .... (530) 752-7037
jcgibeling@ucdavis.edu
AM
Groza, Joanna R. .... (530) 752-8825
jrgroza@ucdavis.edu
AM
Higgins, Brian G. .... (530) 752-8780
bghiggins@ucdavis.edu
MO
Howitt, David G. .... (530) 752-1164
dghowitt@ucdavis.edu
AM
Jackman, Alan P. (Emeritus) .... (530) 752-8777
apjackman@ucdavis.edu
BT,EV
Lavernia, Enrique (Dean) .... (530) 752-0400
ejlavernia@ucdavis.edu
AM
Longo, Marjorie L. .... (530) 754-6348
mllongo@ucdavis.edu
BM,NT,PO,SU
McCoy, Benjamin J. (Emeritus) .... (530) 752-8773
bjmccoy@ucdavis.edu
MO,RE,SE
McDonald, Karen A. (Associate Dean) . (530) 752-8314
kamcdonald@ucdavis.edu
BT
Mukherjee, Amiya K. .... (530) 752-1776
akmukherjee@ucdavis.edu
AM
Munir, Zuhair A. .... (530) 752-6348
zamunir@ucdavis.edu
AM
Navrotsky, Alexandra .... (530) 752-3292
anavrotsky@ucdavis.edu
Palazoglu, Ahmet N. .... (530) 752-8774
anpalazoglu@ucdavis.edu
MO,PR
Phillips, Ronald J. .... (530) 752-2803
rjphillips@ucdavis.edu
MO,PO
Powell, Robert L. (Chair) .... (530) 752-8779
rlpowell@ucdavis.edu
MO,NT,PO
Risbud, Subhash .... (530) 752-0474
shrisbud@ucdavis.edu
AM
Ryu, Dewey .... (530) 752-8954
ddyryu@ucdavis.edu
BT
Shackelford, James F. .... (530) 752-4030
jfshackelford@ucdavis.edu
AM
Smith, J.M. (Emeritus) .... (530) 752-8773
jmsmith@ucdavis.edu
RE
Stroeve, Pieter .... (530) 752-8778
pstroeve@ucdavis.edu
BT,ME,NT,PO,SU
Whitaker, Stephen (Emeritus) .... (530) 752-0400
swhitaker@ucdavis.edu
MO

Associate Professors
Block, David E. .... (530) 752-6046
deblock@ucdavis.edu
BT,PR
Faller, Roland .... (530) 752-5389
rfaller@ucdavis.edu
AM,MO,PO
Kuhl, Tonya L. .... (530) 754-5911
tlkuhl@ucdavis.edu
NT,PO,SU
Schoenung, Julie M. .... (530) 752-5840
jmschoenung@ucdavis.edu
AM

Assistant Professors
El-Farra, Nael .... (530) 754-6919
nhelfarra@ucdavis.edu
MO,PR

Accredited by: ABET

Degrees granted 2005-2006:
    B.S.: 20   M.S.: 2   Ph.D.: 8

Graduate advisor: Pieter Stroeve

Student organization: AIChE; TMS
Advisor: Tonya Kuhl

Department reports to: Enrique Lavernia, Dean Engr.

Placement service: Pam Swartwood, Coordinator

University of California, Irvine

Chemical Engineering and Materials Science
303 Engineering Tower, Room 916
Irvine, CA 92697-2575

Department phone .... (949) 824-3426
Department FAX .... (949) 824-2541

Professors

Allbritton, Nancy L. (Joint) .... (949) 824-9137
nlallbri@uci.edu
BM,BT,SE
DaSilva, Nancy A. .... (949) 824-8288
ndasilva@uci.edu
BM,BT,EN,EV
Earthman, James C. .... (949) 824-5018
earthman@uci.edu
AM,BM,EN,NT
George, Steven C. (Joint) .... (949) 824-3941
scgeorge@uci.edu
BM,BT,MO
Grant, Stanley (Chair) .... (949) 824-7320
sbgrant@uci.edu
Hatfield, G. Wesley (Joint) .... (949) 824-5344
wes.hatfield@uci.edu
Hong, Juan .... (949) 824-8278
jhong@uci.edu
Lim, Henry C. .... (949) 824-3785
hclim@uci.edu
BT,PR,RE
Madou, Marc (Joint) .... (949) 824-6585
mmadou@uci.edu
Mecartney, Martha .... (949) 824-2919
martham@uci.edu
AM
Mohamed, Farghalli .... (949) 824-5807
famohame@uci.edu
AM,MO,NT
Rangel, Roger H. (Joint) .... (949) 824-4033
rhrangel@uci.edu
Shea, Kenneth (Joint) .... (949) 824-5844
kjshea@uci.edu
AM,BT,NT,PO,SE
Shi, Frank .... (949) 824-5362
fgshi@uci.edu
AM,EN,ME
Yee, Albert F. .... (949) 824-9887
afyee@uci.edu
AM,BM,ME,NT,PO,SU

Associate Professors

Lu, Jia Grace .... (949) 824-8714
jglu@uci.edu
AM,ME,NT
Sun, Lizhi (Joint) .... (949)824-8670
lsun@uci.edu
AM,BM,BT,MO,NT
Venugopalan, Vasan .... (949) 824-5802
vvenugop@uci.edu
BM,BT,HT,TP

Assistant Professors

Jeon, Noo Li (Joint) .... (949) 824-9032
njeon@uci.edu
BM
Mumm, Daniel .... (949) 824-3858
mumm@uci.edu
Putnam, Andrew .... (949) 824-1243
aputnam@uci.edu
Ragan, Regina .... (949) 824-6830
rragan@uci.edu
Wang, Szu-Wen .... (949) 824-2383
wangsw@uci.edu

Accredited by: ABET

Degrees granted 2005-2006:
    B.S.: 20   M.S.: 9   Ph.D.: 5

Graduate advisor: Vasan Venugopalan

Undergraduate advisor: Frank Shi

Student organization: AIChE

Department reports to: Nicolaos Alexopoulos, Dean

Placement service: Kathryn Van Ness, Director

University of California, Los Angeles

Department of Chemical & Biomolecular Engineering
5531 Boelter Hall
Box 951592
Los Angeles, CA 90095-1592
Deliveries: 420 Westwood Plaza

University phone .... (310) 825-4321
Department phone .... (310) 825-2046
Department FAX .... (310) 206-4107

Professors

Christofides, Panagiotis D. .... (310) 794-1015
　　　　　　　　　　　　　　　pdc@seas.ucla.edu
Cohen, Yoram .... (310) 825-8766
　　　　　　　　　　　　　　　yoram@ucla.edu
Davis, James .... (310) 206-0011
　　　　　　　　　　　　　　　jdavis@conet.ucla.edu
Friedlander, Sheldon K. .... (310) 825-2206
　　　　　　　　　　　　　　　skf@ucla.edu
Hicks, Robert F. .... (310) 206-6865
　　　　　　　　　　　　　　　rhicks@ucla.edu
Ignarro, L. J. .... (310) 825-5159
　　　　　　　　　　　　　　　lignarro@mednet.ucla.edu
Knuth, Eldon L. (Emeritus) .... (310) 825-8485
　　　　　　　　　　　　　　　elknuth@ucla.edu
Liao, James C. .... (310) 825-1656
　　　　　　　　　　　　　　　liaoj@seas.ucla.edu
Manousiouthakis, Vasilios (Chair) ..... (310) 206-0300
　　　　　　　　　　　　　　　vasilios@ucla.edu
Monbouquette, Harold G. .... (310) 825-8946
　　　　　　　　　　　　　　　harold@seas.ucla.edu
Nobe, Ken (Emeritus) .... (310) 825-2447
　　　　　　　　　　　　　　　nobe@seas.ucla.edu
Senkan, Selim M. .... (310) 206-4106
　　　　　　　　　　　　　　　senkan@seas.ucla.edu
Van Vorst, W. D. (Emeritus) .... (310) 825-2816
　　　　　　　　　　　　　　　wvanvors@ucla.edu
Wazzan, A. R. (Emeritus) .... (310) 206-1598
　　　　　　　　　　　　　　　wazzan@seas.ucla.edu

Associate Professors

Chang, Jane P. .... (310) 206-7980
　　　　　　　　　　　　　　　jpchang@seas.ucla.edu

Assistant Professors

Orkoulas, Gerassimos .... (310) 267-0169
　　　　　　　　　　　　　　　makis@seas.ucla.edu
Segura, Tatiana .... (310) 825-2046
Tang, Yi .... (310) 825-0375
　　　　　　　　　　　　　　　yitang@ucla.edu

Accredited by: ABET

Degrees granted 2004-2005:

　　B.S.: 44　　M.S.: 8　　Ph.D.: 18

Graduate advisor: Panagiotis Christofides

Undergraduate advisor: Harold G. Monbouquette

Student organization: AIChE
Advisor: Vasilios Manousiouthakis

Department reports to: Dean, School of Engineering and Applied Sci.

Placement service: William Beard

University of California, Riverside

Department of Chemical and Environmental Engineering
Riverside CA 92521

Department phone .... (951) 827-6460
Department FAX .... (951) 827-5696

Professors

Chen, Wilfred .... (951) 827-2473
wilfred@engr.ucr.edu
BM,BT,EV,NT
Deshusses, Marc A. (Chair) .... (951) 827-2477
mdeshuss@engr.ucr.edu
BM,BT,EN,EV,NT
Haddon, Robert C. (Joint) .... (951) 827-2044
haddon@engr.ucr.edu
AM,BM,EN,ME,NT,PO
Matsumoto, Mark R. (Associate Dean) . (951) 827-5318
matsumot@engr.ucr.edu
EV
Mulchandani, Ashok .... (951) 827-6419
adani@engr.ucr.edu
AM,BM,BT,EN,EV,NT
Norbeck, Joseph M. .... (951) 827-2262
Joe_Norbeck@cert.ucr.edu
EN,EV
Wyman, Charles E. .... (951) 781-5703
cewyman@engr.ucr.edu
BM,BT,EN,EV
Yan, Yushan .... (951) 827-2068
yushan@engr.ucr.edu
AM,EN,EV,ME,NT,PO

Associate Professors

Cocker, David R. .... (951) 827-2408
dcocker@cert.ucr.edu
EV
Wu, Jianzhong .... (951) 827-2413
jwu@engr.ucr.edu
AM,BM,EN,MO,SU,TH

Assistant Professors

Myung, Nosang V. .... (951) 827-7710
Myung@engr.ucr.edu
AM,BM,BT,EN,EV,ME,NT
Walker, Sharon L. .... (951) 827-6094
swalker@engr.ucr.edu
EV

Accredited by: ABET

Degrees granted 2005-2006:

        B.S.: 17   Ph.D.: 8

Graduate advisor: Jianzhong Wu

Student organization: AIChE
Advisor: Nosang Myung

Department reports to: Reza Abbaschian

University of California, San Diego

Chemical Enginnering Program
Mechanical and Aerospace Engineering Department
9500 Gilman Drive
La Jolla, CA 92093-0411
Deliveries: 371 Engineering Building II

University phone................. (858) 534-2000
Department phone .... (858) 534-3174
Department FAX .... (858) 534-4543

Professors

Chau, Pao C. .... (858) 534-6935
Miller, David R. .... (858) 534-3182
dmiller@ucsd.edu
Pozrikidis, Constantine .... (858) 534-6530
costas@ucsd.edu
Seshadri, K. .... (858) 534-4876
seshadri@ucsd.edu
Talbot, Jan B. (Director) .... (858) 534-3176
jtalbot@ucsd.edu

Associate Professors

Herz, Richard K. .... (858) 534-6540
herz@ucsd.edu

Accredited by: ABET

Graduate advisor: Michelle Vavra

Undergraduate advisor: Gerri Johnson

Student organization: AIChE
Advisor: Pao Chau

Department reports to: Dean of Engineering (858) 534-4575

Placement service: Marci Swain (858) 534-0141

University of California, Santa Barbara

Department of Chemical Engineering
University of California at Santa Barbara
Santa Barbara, CA 93106-5080
Deliveries: Engineering II, Room 3357

University phone .... (805) 893-8000
Department phone .... (805) 893-3412
Department FAX .... (805) 893-4731

Professors

Banerjee, Sanjoy .... (805) 893-3456
   banerjee@anemone.ucsb.edu
Chmelka, Bradley F. (Associate Chair) . (805) 893-3673
   bradc@engineering.ucsb.edu
   AM,NT,PO,RE,SU
Doherty, Michael F. .... (805) 893-5309
   mfd@engineering.ucsb.edu
   PR,SE
Doyle, Frank .... (805) 893-8133
   doyle@engineering.ucsb.edu
   BM,BT,MO,PR
Fredrickson, Glenn H. .... (805) 893-8308
   ghf@engineering.ucsb.edu
   AM,MO,NT,PO,SU
Israelachvili, Jacob .... (805) 893-8407
   jacob@engineering.ucsb.edu
   SU
Kramer, Edward J. .... (805) 893-4999
   edkramer@mrl.ucsb.edu
   AM,NT,PO
Leal, L. Gary (Chair) .... (805) 893-8510
   lgl20@engineering.ucsb.edu
   MO,PO,TP
Lucas, Glenn E. (Exec. Vice Chancellor) (805) 893-4069
   gene@engineering.ucsb.edu
McFarland, Eric .... (805) 893-4343
   mcfar@engineering.ucsb.edu
   RE
Mellichamp, Duncan A. (Emeritus) .... (805) 893-2821
   dmell@engineering.ucsb.edu
   PR
Rinker, Robert G. (Emeritus) .... (805) 893-2610
   rinker@engineering.ucsb.edu
Sandall, Orville C. (Emeritus) .... (805) 893-2908
   sandall@engineering.ucsb.edu
Scott, Susannah L. .... (805) 893-5606
   sscott@engineering.ucsb.edu
   AM,EN,EV,NT,PO,RE,SU
Seborg, Dale E. (Associate Chair) ..... (805) 893-3352
   seborg@engineering.ucsb.edu
   PR
Theofanous, Theofanis .... (805) 893-4900
   theo@tcrss.ucsb.edu
Tirrell, Matthew (Dean) .... (805) 893-3141
   tirrell@engineering.ucsb.edu
   AM,NT,PO
Zasadzinski, Joseph .... (805) 893-4769
   gorilla@engineering.ucsb.edu
   BM,NT,SU

Assistant Professors

Daugherty, Patrick .... (805) 893-2610
   psd@engineering.ucsb.edu
   BM,BT
Mitragotri, Samir .... (805) 893-7532
   samir@engineering.ucsb.edu
Shell, Scott
   BT,MO
Squires, Todd .... (805) 893-7383
   squires@engineering.ucsb.edu
   MO,PO,SU

Accredited by: ABET

Graduate advisor: Jacob Israelachvili

Undergraduate advisor: Dale Seborg

Student organization: AIChE
Advisor: Frank Doyle

Department reports to: Matthew Tirrell, (805) 893-3141, Dean

Placement service: Carol A. Geer (805) 893-4419

California Institute of Technology

Chemical Engineering Department
Mail Stop 210-41
Pasadena, CA 91125-4100
Deliveries: 391 S. Holliston Ave.

University phone	(626) 395-6811
Department phone	(626) 395-4115
Department FAX	(626) 568-8743

Professors

Arnold, Frances H. .... (626) 395-4162
frances@cheme.caltech.edu
BT,EN,RE
Brady, John F. .... (626) 395-4183
jfb@cheme.caltech.edu
AM,MO,NT,SU
Davis, Mark E. .... (626) 395-4251
mdavis@cheme.caltech.edu
AM,BM,EN,RE
Flagan, Richard C. (Head) .... (626) 395-4383
flagan@cheme.caltech.edu
AM,EV,NT,PO,RE,SE
Gavalas, George R. (Emeritus) .... (626) 395-4152
gavalas@cheme.caltech.edu
EN,RE
Haile, Sossina .... (626) 395-2958
smhaile@caltech.edu
AM,EN,RE
Kornfield, Julia A. .... (626) 395-4138
jak@cheme.caltech.edu
AM,BM,BT,NT,PO,SU
Seinfeld, John H. .... (626) 395-4635
seinfeld@caltech.edu
EV,MO
Tirrell, David A. (Chair) .... (626) 395-3140
tirrell@caltech.edu
AM,BM,BT,PO
Tschoegl, Nicholas W. (Emeritus)
Wang, Zhen-Gang .... (626) 395-4647
zgw@cheme.caltech.edu
MO,PO,SU

Associate Professors

Giapis, Konstantinos P. .... (626) 395-4180
giapis@cheme.caltech.edu
AM,ME,MO,NT,SU

Assistant Professors

Asthagiri, Anand R. .... (626) 395-8130
anand@cheme.caltech.edu
BM,BT,MO
Smolke, Christina D. .... (626) 395-2460
smolke@cheme.caltech.edu
BM,BT,NT

Other Faculty

Wagner, Eric S.

Accredited by: ABET

Degrees granted 2005-2006:
B.S.: 15   M.S.: 11   Ph.D.: 11

Graduate advisor: Konstantinos P. Giapis

Student organization: AIChE
Advisor: Christina D. Smolke

Department reports to: David A. Tirrell, Chairman, Div. of Chem. and Chemical Engineering

Placement service: Gerald Houser, Director, Career Development

California State University, Long Beach

Department of Chemical Engineering
1250 Bellflower Blvd., VEC 136
Long Beach, CA 90840

University phone	(562) 985-4111
University FAX	(562) 985-7561
Department phone	(562) 985-4909
Department FAX	(562) 985-7561

Professors

Hile, Lloyd (Emeritus) .... (562) 985-7532
    lhile@csulb.edu
Jang, Larry (Chair) .... (562) 985-7533
    jang@csulb.edu
Kavianian, Hamid .... (562) 985-2260
    hrkche@cox.net
Naimpally, Ashok .... (562) 985-1508
    avnaimpa@csulb.edu
Tsai, Shirley .... (562) 985-7534
    sctsai@csulb.edu

Accredited by: ABET

Degrees granted 2005-2006:
    B.S.: 22   M.S.: 8

Graduate advisor: Ashok Naimpally

Undergraduate advisor: Larry Jang

Student organization: AIChE
Advisor: Larry Jang

Department reports to: Michael Mahoney, (562) 985-5121, Dean of College of Engineering

Placement service: Ed Morton (562) 985-4151, Director

California State Polytechnic University, Pomona

Chemical and Materials Engineering
3801 West Temple Avenue
Pomona, CA 91768-4069

University phone	(909) 869-2600
Department phone	(909) 869-2626
Department FAX	(909) 869-6920

Professors

Aldrich, J. Winthrop (Emeritus) ...... (909) 869-2626
    jwaldrich@csupomona.edu
Bray, Robert S. (Emeritus) .... (909) 869-2626
Caenepeel, Christopher L. (Chair) ..... (909) 869-2626
    clcaenepeel@csupomona.edu
Hacker, Barbara .... (909) 869-2629
    bahacker@csupomona.edu
Harris, William M. (Emeritus) .... (909) 869-2626
Hohmann, Edward C. (Dean) .... (909) 869-2600
    echohmann@csupomona.edu
Nguyen, Thuan K. .... (909) 869-2631
    tknguyen@csupomona.edu
Ontiveros, Cordelia .... (909) 869-2626
    contiveros@calstate.edu
Pang, K. Hing (Emeritus) .... (909) 869-2626
    khpang@csupomona.edu
Ravi, Vilupanur .... (909) 869-2627
    vravi@csupomona.edu
Scott, Garland E. (Emeritus) .... (909) 869-2626
    gescott@csupomona.edu
Sheng, Henry P. (Emeritus) .... (909) 869-2626
Stoll, A. George (Emeritus) .... (909) 869-2626
    agstoll@csupomona.edu
Tassoney, Joseph P. (Emeritus) .... (909) 869-2626
Tomlinson, John L. (Emeritus) .... (909) 869-2626

Associate Professors

Dong, Winny .... (909) 869-2634
    winnydong@csupomona.edu
Le, Lloyd
    lle@csupomona.edu

Accredited by: ABET

Degrees granted 2005-2006:
    B.S.: 17

Undergraduate advisor: Chris Caenepeel, clcaenepeel@csupomona.edu

Student organization: AIChE
Advisor: C.L. Caenepeel (909)-869-2626

Department reports to: Edward H. Hohmann (909) 869-2600, Dean

Placement service: Stan Hebert, 909-869-2337, sphebert@csupomona.edu, Director

San Jose State University

Department of Chemical and Materials Engineering
Engineering 385
One Washington Square
San Jose, CA 95192-0082

University phone	(408) 924-1000
Department phone	(408) 924-4000
Department FAX	(408) 924-4057

Professors

Allen, Emily (Chair) .... (408) 924-4010
elallen@email.sjsu.edu
ME,NT
Jennings, Michael B. (Director) .... (408) 924-3926
jennimi@email.sjsu.edu
EV,PR
McNeil, Melanie .... (408) 924-3873
mcneil@sjsuvm1.sjsu.edu
BT,EV,NT,RE

Associate Professors

Komives, Claire .... (408) 924-4002
ckomives@email.sjsu.edu
BT,SE
Young, Gregory L. (Associate Chair) .. (408) 924-3945
glyoung@email.sjsu.edu
EN,HT,ME,NT,TP

Accredited by: ABET

Degrees granted 2005-2006:
   B.S.: 8   M.S.: 4

Graduate advisor: M. McNeil

Undergraduate advisor: Claire Komives

Student organization: AIChE
Advisor: Claire Komives

Department reports to: Belle Wei (408) 924-3800, Dean of Engineering

Placement service: Lina Melkonian (408) 924-6033, Director

University of Southern California

Mork Family Department of Chemical Engineering
and Materials Science
University Park, HED 216
Los Angeles, CA 90089-1211
Deliveries: 925 Bloom Walk, HED 216

University phone	(213) 740-2311
Department phone	(213) 740-2225
Department FAX	(213) 740-8053

Professors

Dougherty, Elmer L. (Emeritus) .... (213) 740-0310
Ershaghi, Iraj .... (213) 740-0321
ershaghi@usc.edu
Kalia, Rajiv .... (213) 821-2658
rkalia@usc.edu
Madhukar, Anupam .... (213) 740-4323
madhukar@usc.edu
Mansfeld, Florian .... (213) 740-3016
mansfeld@usc.edu
Nutt, Steven .... (213) 740-1634
nutt@usc.edu
Sahimi, Muhammad .... (213) 740-2064
moe@usc.edu
Salovey, Ronald (Emeritus) .... (213) 740-2225
salove@usc.edu
Tsotsis, Theodore T. (Chair) .... (213) 740-2069
tsotsis@usc.edu
Vashishta, Priya .... (213) 821-2663
priyav@usc.edu
Yortsos, Yannis C. (Dean) .... (213) 740-0617
yortsos@usc.edu

Associate Professors

Chang, W. Victor .... (213) 740-2072
wenji@usc.edu
Goo, Edward .... (213) 740-4426
ekgoo@usc.edu
Roberts, Richard .... (213) 821-4132
richard.roberts@usc.edu
Shing, Katherine S. .... (213) 740-2068
shing@usc.edu

Assistant Professors

Jessen, Kristian .... (213) 740-7320
EN,EV,MO,TP
Kezirian, Michael (Adjunct)
kezirian@usc.edu
Konkar, Atul .... (213) 821-2965
konkar@usc.edu
Lee, C. Ted .... (213) 740-2066
tedlee@usc.edu
Wang, Pin .... (213) 740-0780
pinwang@usc.edu

Accredited by: ABET

Degrees granted 2005-2006:
   B.S.: 18   M.S.: 26   Ph.D.: 5

Graduate advisor: Dr. Katherine Shing

Student organization: AIChE
Advisor: Dr. Katherine Shing

Department reports to: Yannis C. Yortsos, (213) 740-0617, Dean of Engineering

Placement service: Eileen Kohan, STU 111, Los Angeles, CA 90089-4897

Stanford University

Department of Chemical Engineering
Stauffer III, Room 113
381 North-South Axis
Stanford, CA 94305-5025

University phone .... (650) 723-2300
Department phone .... (650) 723-4906
Department FAX .... (650) 723-9780

Professors

Bent, Stacey (Associate Chair) .... (650) 723-0385
    sbent@stanford.edu
Boudart, Michel (Emeritus) .... (650) 723-4906
    mboudart@stanford.edu
Frank, Curtis W. .... (650) 723-4573
    curt.frank@stanford.edu
Fuller, Gerald G. .... (650) 723-9243
    ggf@stanford.edu
    AM,PO,TP
Khosla, Chaitan S. (Chair) .... (650) 723-6538
    khosla@stanford.edu
    BT,EN,RE
Robertson, Channing R. (Associate Dean) (650) 723-3936
    chanbo@stanford.edu
Shaqfeh, Eric S. G. .... (650) 723-3764
    esgs@stanford.edu
Swartz, James .... (650) 723-5398
    jswartz@stanford.edu

Associate Professors

Bao, Zhenan .... (650) 723-2419
    zbao@stanford.edu
    EN,NT,PO,SU

Assistant Professors

Kao, Camilla .... (650) 736-0547
Musgrave, Charles .... (650) 725-9176
    chasm@stanford.edu
    AM,EN,ME,MO,NT,RE,SU
Spakowitz, Andrew .... (650) 736-8733
    ajspakow@stanford.edu

Accredited by: ABET

Degrees granted 2005-2006:
    B.S.: 19   M.S.: 38   Ph.D.: 14

Graduate advisor: Eric Shaqfeh, Zhenan Bao

Undergraduate advisor: James Swartz

Student organization: AIChE
Advisor: Zhenan Bao

Department reports to: James D. Plummer, Dean

Placement service: Lance Choy, Director, Career Development Center

Colorado

University of Colorado

Department of Chemical and Biological Engineering
UCB 424
Boulder, CO 80309-0424

University phone .... (303) 492-0111
Department phone .... (303) 492-7471
Department FAX .... (303) 492-4341

Professors
Anseth, Kristi S. .... (303) 492-3147
kristi.anseth@colorado.edu
AM,BM,PO
Bowman, Christopher N. (Chair) ...... (303) 492-3247
christopher.bowman@colorado.edu
AM,BM,PO,RE
Clough, David E. .... (303) 492-6638
david.clough@colorado.edu
EN,PR
Davis, Robert H. (Dean) .... (303) 492-7314
robert.davis@colorado.edu
MO,SE,TP
Falconer, John L. (Associate Chair) ... (303) 492-8005
john.falconer@colorado.edu
AM,RE,SE
George, Steven .... (303) 492-3398
steven.george@colorado.edu
AM,ME,NT
Noble, Richard D. .... (303) 492-6100
nobler@colorado.edu
AM,MO,SE
Ramirez, W. Fred .... (303) 492-8660
fred.ramirez@colorado.edu
EN,EV,PR
Randolph, Theodore W. .... (303) 492-4776
theodore.randolph@colorado.edu
BT,RE,TH
Sani, Robert L. .... (303) 492-5517
robert.sani@colorado.edu
MO,TP
Schwartz, Daniel .... (303) 735-0240
daniel.schwartz@colorado.edu
NT,TP
Stansbury, Jeff .... (303) 724-1044
jeffrey.stansbury@UCHSC.edu
BM,BT,PO
Weimer, Alan W. .... (303) 492-3759
alan.weimer@colorado.edu
EN,EV,NT,RE,TP

Associate Professors
Gin, Douglas .... (303) 492-7640
douglas.gin@colorado.edu
AM,NT,PO
Hrenya, Christine M. .... (303) 492-7689
hrenya@colorado.edu
MO,TP
Kompala, Dhinakar S. .... (303) 492-6350
dhinakar.kompala@colorado.edu
BT,RE

Assistant Professors
Bryant, Stephanie .... (303) 735-6714
stephanie.bryant@colorado.edu
AM,BM,BT,PO
Gill, Ryan .... (303) 492-2627
rtg@colorado.edu
BM,BT,EN
Mahoney, Melissa .... (303) 492-3573
melissa.mahoney@colorado.edu
AM,BM,BT,PO
Medlin, J. William .... (303) 492-2418
Will.Medlin@colorado.edu
MO,RE,SU

Instructors
deGrazia, Janet .... (303) 735-4763
deGrazia@colorado.edu
Louie, Beverly .... (303) 492-4967
beverly.louie@colorado.edu

Accredited by: ABET

Degrees granted 2005-2006:
    B.S.: 45   M.S.: 4   Ph.D.: 13

Graduate advisor: Theodore Randolph

Undergraduate advisor: Janet deGrazia

Student organization: AIChE
Advisor: Alan Weimer

Department reports to: (303) 492-7006, Dean of Engineering and Applied Sciences

Placement service: Jodi Schneiderman

Colorado School of Mines

Chemical Engineering Department
451 Alderson Hall
Golden, CO 80401
Deliveries: Warehouse, 900 18th Street

University phone	(303) 273-3000
Department phone	(303) 273-3720
Department FAX	(303) 273-3730

Professors

Baldwin, Robert M. (Emeritus) 303-273-3720
rbaldwin@pi.ac.ae
Bunge, Annette L. (Emeritus) .... (303) 273-3722
abunge@mines.edu
Dean, Anthony M. .... (303) 273-3643
amdean@mines.edu
Dorgan, John R. .... (303) 273-3539
jdorgan@mines.edu
Ely, James F. (Head) .... (303) 273-3885
jely@mines.edu
Evans, Robert (Research) 303-929-0030
evans@mines.edu
Gary, James H. (Emeritus) .... (303) 273-3720
Golden, John O. (Emeritus) .... (303) 273-3720
jgolden@mines.edu
Kidnay, Arthur J. (Emeritus) .... (303) 384-2215
akidnay@mines.edu
Knecht, Robert D. (Research) .... (303) 273-3592
rknecht@mines.edu
Marr, David W. M. .... (303) 273-3008
dmarr@mines.edu
McKinnon, J. Thomas .... (303) 273-3098
jmckinno@mines.edu
Miller, Ronald L. .... (303) 273-3892
rlmiller@mines.edu
Sloan, E. Dendy Jr. .... (303) 273-3723
esloan@mines.edu
Way, J. Douglas .... (303) 273-3519
dway@mines.edu
Yesavage, Victor F. (Emeritus) .... (303) 273-3725
vyesavag@mines.edu

Associate Professors

Abbud-Madrid, Angel (Research) 303-384-2300
aabudma@mines.edu
Carstensen, Hans-Heinrich (Research) ... 303-384-2312
hcarsten@mines.edu
Herring, Andy .... (303) 384-2082
aherring@mines.edu
Kiselev, Sergei B. (Research) .... (303) 273-3190
skiselev@mines.edu
Koh, Carolyn .... (303) 273-3237
ckoh@mines.edu
Miller, Kelly (Research) .... (303) 273-3951
ktmiller@mines.edu
Riedel, Edward (Research) 303-384-2406
eriedel@mines.edu

Thoen, Paul (Research) 303-273-3193
pthoen@mines.edu
Wolden, Colin A. .... (303) 273-3544
cwolden@mines.edu
Wu, David T. .... (303) 384-2024
dwu@mines.edu

Assistant Professors

Agarwal, Sumit .... (303) 273-3720
sagarwal@mines.edu
Jechura, John (Adjunct) 303-273-3614
jjechura@mines.edu
Liberatore, Matthew W. .... (303) 273-3720
mliberat@mines.edu
Murray, Glen (Research) 303-273-3873
gmurray@mines.edu
Oakey, John (Research) 303-273-3172
joakey@mines.edu
Romonchuk, Wayne (Research) 303-384-2204
wromonch@mines.edu
Shin, Eun-Jae (Research) 303-384-2046
eshin@mines.edu
Stempfer, Berthe (Research) 303-384-2204
bstempfe@mines.edu
Vestal, Charles (Adjunct) 303-273-3614
cvestal@mines.edu

Other Faculty

Gardner, Tracy Q. 303-273-3846
tgardner@mines.edu
Perschetti, John 303-273-3724
jpersiche@mines.edu

Accredited by: ABET

Degrees granted 2005-2006:
    B.S.: 49   M.S.: 10   Ph.D.: 1

Graduate advisor: Colin Wolden

Undergraduate advisor: James F. Ely

Student organization: AIChE
Advisor: James F. Ely

Department reports to: Nigel Middleton (303) 273-3327, Executive Vice President for Academic Affairs

Placement service: Anna Hanley, CSM Placement Office

United States

Colorado State University

Department of Chemical and Biological Engineering
1370 Campus Delivery
100 Glover Building
Fort Collins, CO 80523-1370

University phone	(970) 491-1101
Department phone	(970) 491-5252
Department FAX	(970) 491-7369

Professors

Belfiore, Laurence A. .... (970) 491-5395
belfiore@engr.colostate.edu
AM,NT,PO

Dandy, David S. (Head) .... (970) 491-7437
dandy@colostate.edu
AM,BM,ME,MO,NT

Linden, James C. .... (970) 491-6122
linden@engr.colostate.edu
BT

Murphy, Vincent G. (Emeritus) .... (970) 491-1791
vince@engr.colostate.edu

Reardon, Kenneth F. (Associate Head) . (970) 491-6505
reardon@engr.colostate.edu
BM,BT,RE

Watson, A. Ted .... (970) 491-5253
atw@engr.colostate.edu
BM,MO,NT

Associate Professors

Wickramasinghe, Ranil .... (970) 491-5276
wickram@engr.colostate.edu
BM,BT,SE

Assistant Professors

Bailey, Travis .... (970) 491-4648
tsbailey@engr.colostate.edu
AM,ME,NT,PO

Kipper, Matt J. .... (970) 491-5253
mkipper@engr.colostate.edu
AM,BM,NT,PO

Reisfeld, Brad .... (970) 491-1019
breisfel@lamar.colostate.edu
BM,BT,MO

Wang, Qiang (David) .... (970) 491-2763
qwang@engr.colostate.edu
MO,NT,PO

Instructors

Perkins, Tracy .... (970) 491-5175
tperkins@engr.colostate.edu
MO,PR

Smith, T. Gordon .... (970) 491-2227
tgsmith@engr.colostate.edu

Accredited by: ABET

Degrees granted 2005-2006:
 B.S.: 32   M.S.: 9   Ph.D.: 6

Graduate advisor: Ranil Wickramasinghe

Undergraduate advisor: Kenneth F. Reardon

Student organization: AIChE
Advisor: T. Gordon Smith

Department reports to: Dr. Sandra Woods, Dean, College of Engineering

Placement service: Ann Malen, Director, The Career Center (970) 491-5707, Career Center

Connecticut

University of Connecticut

Chemical, Materials & Biomolecular Engineering Department
Chemical Engineering Program
191 Auditorium Road, Unit 3222
Storrs, CT 06269-3222
Deliveries: Engineering II Building, Room 204

University phone .... (860) 486-2000
Department phone .... (860) 486-4019
Department FAX .... (860) 486-2959

Professors

Achenie, Luke E. K. .... (860) 486-2756
    achenie@engr.uconn.edu
Cooper, Douglas J. .... (860) 486-4092
    cooper@engr.uconn.edu
Erkey, Can .... (860) 486-4601
    cerkey@engr.uconn.edu
Shaw, Montgomery T. (Head) .... (860) 486-3980
    shawmt@uconnvm.uconn.edu
Weiss, Robert A. .... (860) 486-4698
    rweiss@mail.ims.uconn.edu

Associate Professors

Parnas, Richard S. .... (860) 486-9060
    rparnas@mail.ims.uconn.edu

Assistant Professors

Lei, Yu .... (860) 486-4554
    ylei@engr.uconn.edu
Srivastava, Ranjan .... (860) 486-2802
    srivasta@engr.uconn.edu
Wang, Yong .... (860) 486-4072
    yongwang@engr.uconn.edu
Wilhite, Benjamin .... (860) 486-3689
    bwilhite@engr.uconn.edu
Zhu, Lei .... (860) 486-8708
    lzhu@ims.uconn.edu

Other Faculty

Anderson, Thomas F. (Emeritus) ...... (860) 486-2473
    anderson@engr.uconn.edu
Bell, James P. (Emeritus) .... (860) 486-4019
    jbell@mail.ims.uconn.edu
Bennett, Carroll O. (Emeritus) .... (860) 486-4019
Coughlin, Robert W. (Emeritus) ...... (860) 486-4489
    coughlin@engr.uconn.edu
Cutlip, Michael (Emeritus) .... (860) 486-0321
    mcutlip@uconnvm.uconn.edu
DiBenedetto, Anthony T. (Emeritus) .. (860) 486-4019
    adibened@mail.ims.uconn.edu
Howard, G. Michael (Emeritus) .... (860) 486-2479
    howard@uconnvm.uconn.edu
Kunz, H. Russell .... (860) 486-5389
    russkunz@engr.uconn.edu
Sundstrom, Donald W. (Emeritus) .... (860) 486-4019

Accredited by: ABET

Degrees granted 2005-2006:
    B.S.: 34   M.S.: 6   Ph.D.: 12

Graduate advisor: Lei Zhu

Student organization: AIChE
Advisor: Ranjan Srivastava

Department reports to: Erling Smith (860) 486-2221, Dean of Engineering

Placement service: Cynthia Jones, Director (860) 486-3013

University of New Haven

Department of Chemistry and Chemical Engineering
University of New Haven
300 Boston Post Road
West Haven, CT 06516
Deliveries: Receiving Section
University phone .... (203) 932-7000
Department phone .... (203) 932-7404
Department FAX .... (203) 931-6077

Professors
Collura, Michael A. .... (203) 932-7149
    mcollura@newhaven.edu
Desio, Peter J. (Emeritus)
    pdesio@newhaven.edu
Saliby, Michael J. .... (203) 932-7169
    msaliby@newhaven.edu
Wheeler, George L. (Emeritus)
    georgelwheeler2@comcast.com

Associate Professors
Gow, Arthur S. III .... (203) 932-7173
    agow@newhaven.edu
Harding, W. David (Chair) .... (203) 932-7438
    dharding@newhaven.edu
Schwartz, Pauline M. .... (203) 932-7170
    pschwartz@newhaven.edu

Assistant Professors
Luzik, Eddie .... (203) 932-7006
    eluzik@newhaven.edu
Savage, Nancy O.
    nsavage@newhaven.edu

Instructors
Del Valle, Eddie .... (203) 932-4577
    edelvalle@newhaven.edu

Accredited by: ABET

Degrees granted 2005-2006:
    B.S.: 6

Undergraduate advisor: W. David Harding

Student organization: AIChE
Advisor: W. David Harding

Department reports to: Dr. Barry Farbrother, Dean, Tagliatela School of Engineering (203) 932-7167

Placement service: Kathryn Link (203) 932-7334

Yale University

Chemical Engineering Department
P. O. Box 208286
New Haven, CT 06520
Deliveries: 9 Hillhouse Avenue, New Haven, CT 06511
Department phone .... (203) 432-2218
Department FAX .... (203) 432-4387

Professors
Altman, Eric .... (203) 432-4375
    eric.altman@yale.edu
Benoit, Gaboury (Joint) .... (203) 432-5139
    gaboury.benoit@yale.edu
Elimelech, Menachem (Chair) .... (203) 432-2789
    menachem.elimelech@yale.edu
Fenn, John B. (Emeritus) .... (203) 432-2222
Firoozabadi, Abbas (Adjunct) .... (203) 432-4379
    abbas.firoozabadi@yale.edu
Graedel, Thomas E. (Joint) .... (203) 432-9733
    thomas.graedel@yale.edu
Haller, Gary L. .... (203) 432-4378
    gary.haller@yale.edu
Loewenberg, Michael .... (203) 432-4334
    michael.loewenberg@yale.edu
Pfefferle, Lisa D. .... (203) 432-4377
    pfefferle@biomed.med.yale.edu
Pignatello, Joseph (Adjunct) .... (203) 432-2222
    joseph.pignatello@po.state/ct.us
Rosner, Daniel E. .... (203) 432-4391
    daniel.rosner@yale.edu
Saiers, James E. (Joint) .... (203) 432-5121
    james.saiers@yale.edu
Saltzman, Mark (Joint) .... (203) 432-4262
    mark.saltzman@yale.edu
Zilm, Kurt (Joint) .... (203) 432-3956
    kurt.zilm@yale.edu

Associate Professors
Khalil, Yehia F. (Adjunct) .... (203) 440-0348
    yehia.khalil@yale.edu
Van Tassel, Paul .... (203) 432-7983
    paul.vantassel@yale.edu

Assistant Professors
Mitch, William .... (203) 432-4386
    william.mitch@yale.edu
Peccia, Jordan .... (203) 432-4385
    jordan.peccia@yale.edu

Accredited by: ABET

Degrees granted 2005-2006:
    B.S.: 6   M.S.: 1   Ph.D.: 6

Department reports to: Bruce Carmichael (203) 432-4448, Associate Provost, for Science & Technology

Placement service: Philip Jones, (203) 432-0802

Delaware

University of Delaware

Department of Chemical Engineering
Newark, DE 19716-3110
Deliveries: Colburn Laboratory, 150 Academy Street

Department phone .... (302) 831-2543
Department FAX .... (302) 831-1048

Professors

Barteau, Mark A. (Chair) .... (302) 831-8905
barteau@udel.edu
AM,EN,EV,MO,NT,RE,SU
Beris, Antony N. .... (302) 831-8018
beris@udel.edu
AM,BM,EN,EV,MO,NT,PO,TH,TP
Bischoff, Kenneth B. (Emeritus)
Buttrey, Douglas J. .... (302) 831-2034
dbuttrey@udel.edu
AM,EN,NT,RE,TH
Chen, Jingguang .... (302) 831-0642
jgchen@udel.edu
EN,NT,RE,SU
Dhurjati, Prasad .... (302) 831-2879
dhurjati@udel.edu
BM,BT,MO,PR
Kaler, Eric W. (Dean) .... (302) 831-3553
kaler@udel.edu
AM,BT,NT,SU,TH
Lauterbach, Jochen A. .... (302) 831-6327
lauterba@udel.edu
AM,EN,NT,PO,RE,SU
Lenhoff, Abraham M. .... (302) 831-8989
lenhoff@udel.edu
BT,MO,NT,SE,SU,TH,TP
Lobo, Raul F. .... (302) 831-1261
lobo@udel.edu
AM,EN,EV,RE
Ogunnaike, Babatunde A. .... (302) 831-4504
ogunnaik@udel.edu
BM,PR
Olson, Jon H. (Emeritus) .... (302) 831-8472
olson@udel.edu
Russell, T. W. Fraser .... (302) 831-1714
twfr@udel.edu
EN,TP
Sandler, Stanley I. .... (302) 831-2945
sandler@udel.edu
BT,EN,EV,SE,TH
Schultz, Jerold M. (Emeritus) .... (302) 831-8145
schultz@che.udel.edu
Vlachos, Dionisios .... (302) 831-2830
vlachos@udel.edu
AM,BT,EN,MO,NT,RE,SE,TP
Wagner, Norman J. .... (302) 831-8079
wagnernj@udel.edu
NT,PO,SU,TH,TP
Wool, Richard P. .... (302) 831-3312
wool@udel.edu
AM,EN,EV,MO,PO,SU

Associate Professors

Robinson, Anne Skaja .... (302) 831-0557
skaja@udel.edu
BM,BT
Shine, Annette D. .... (302) 831-2010
shine@udel.edu
AM,BM,NT,PO,TP

Assistant Professors

Epps, Thomas H. .... (302) 831-0215
thepps@udel.edu
AM,NT,PO,TH
Furst, Eric M. .... (302) 831-0102
furst@udel.edu
BM,NT,PO,TP
Roberts, Christopher J. .... (302) 831-0838
cjr@udel.edu
BT,MO,TH
Sullivan, Millicent M. O. .... (302) 831-8072
msulliva@udel.edu
BM,BT,NT,PO,TP
Willis, Brian G. .... (302) 831-6856
bgwillis@udel.edu
ME,NT,SU

Other Faculty

Butera, Robert J.
Diemer, Russell (Adjunct)
Etchells, Arthur W. III
Grant, James J. III
Grasselli, Robert (Adjunct)
Grenville, Richard K.
LaRoche, Richard K.
Lyons, James E.
Manogue, William H.
Mulholland, Kenneth L.
Richards, John (Adjunct)
Schure, Mark R.
Schwaber, James S.
Short, David .... (302) 831-2399
shortd@udel.edu
Tilton, James N.
Uebler, E. Alan

Accredited by: ABET

Degrees granted 2005-2006:
    B.S.: 52   M.S.: 4   Ph.D.: 14

Graduate advisor: Jingguang Chen and Antony Beris

Student organization: AIChE
Advisor: Raul F. Lobo

Department reports to: Eric W. Kaler, Dean Engr.

Placement service: David J. Berilla, Assoc. Director

District of Columbia

Howard University

Chemical Engineering Department
2300 6th Street, NW
Lewis K. Downing Hall, Room 1009
Washington, DC 20059

University phone	(202) 806-6100
Department phone	(202) 806-6624
Department FAX	(202) 806-4635

Professors

Aluko, Mobolaji E. .... (202) 806-4793
    alukome@gmail.com
    EN,ME,MO

Cannon, Joseph N. .... (202) 806-6626
    jcannon@howard.edu
    HT,TP

Chawla, Ramesh C. (Chair) .... (202) 806-6617
    rchawla@howard.edu
    EV,RE,SE

Lutz, Robert J. (Visiting) .... (301) 435-1944
    rlutz@Box-r.nih.gov
    BM,TP

Mitchell, James .... (202) 806-9086
    jwm@msrce.howard.edu
    AM,NT,SU

Rao, M. Gopala (Emeritus) .... (202) 806-4796
    mgrao@comcast.com
    EN,EV,PO,SE

Tharakan, John P. .... (202) 806-4811
    jtharakan@howard.edu
    BM,BT,EV,RE

Associate Professors

Collins, William E. .... (202) 806-4595
    wcollins@howard.edu
    BM,PO

Assistant Professors

Ekechukwu, Kenneth (Adjunct) .... (301) 585-9569
    ela@prodicy.net
    MO,PR

Ganley, Jason .... (202) 806-4796
    jganley@howard.edu
    EN,PR

Accredited by: ABET

Degrees granted 2005-2006:
    B.S.: 17   M.S.: 3

Graduate advisor: Dr. Ramesh C. Chawla

Student organization: AIChE
Advisor: Dr. Joseph N. Cannon

Department reports to: James H. Johnson, Jr., Dean, College of Engineering, Architecture & Computer Science

Placement service: Samuel Hall, Office of Placement and Career Planning

Florida

University of Florida

P. O. Box 116005
Chemical Engineering Department
Gainesville, FL 32611-6005
Deliveries: Museum Road, Bldg. 723

University phone .... (352) 392-3261
Department phone .... (352) 392-0881
Department FAX .... (352) 392-9513

Professors

Anderson, Timothy J. (Associate Dean) (352) 392-0946
   tim@ufl.edu
Block, Seymour S. (Emeritus) .... (352) 392-9102
   block@che.ufl.edu
Crisalle, Oscar D. .... (352) 392-5120
   crisalle@che.ufl.edu
Curtis, Jennifer (Chair) .... (352) 392-0882
   jcurtis@che.ufl.edu
Dickinson, Richard .... (352) 392-0898
   dickinso@che.ufl.edu
Hoflund, Gar B. .... (352) 392-9104
   garho@hotmail.com
Johns, Lewis E. .... (352) 392-0881
   chemical@che.ufl.edu
Ladd, Anthony J.C. .... (352) 392-6509
   ladd@che.ufl.edu
Narayanan, Ranga .... (352) 392-9103
   ranga@che.ufl.edu
Orazem, Mark E. .... (352) 392-6207
   meo@che.ufl.edu
Park, Chang-Won .... (352) 392-6205
   park@che.ufl.edu
Ren, Fan .... (352) 392-4727
   ren@che.ufl.edu
Shah, Dinesh O. (Emeritus) .... (352) 392-0877
   shah@che.ufl.edu
Svoronos, Spyros A. .... (352) 392-9101
   svoronos@ufl.edu

Associate Professors

Chauhan, Anuj .... (352) 392-2592
   chauhan@che.ufl.edu
Tseng, Yiider .... (352) 392-0862
   ytseng@che.ufl.edu
Weaver, Jason .... (352) 392-0869
   weaver@che.ufl.edu

Assistant Professors

Asthagiri, Aravind .... (352) 392-0868
   aasthagiri@che.ufl.edu
Butler, Jason .... (352) 392-2591
   butler@che.ufl.edu
Jiang, Peng .... (352) 392-0881
   pjiang@che.ufl.edu
Kopelevich, Dmitry .... (352) 392-4422
   dkopelevich@che.ufl.edu
Lele, Tanmay .... (352) 392-0881
   tlele@che.ufl.edu
Narang, Atul .... (352) 392-0028
   narang@che.ufl.edu
Vasenkov, Sergey .... (352) 392-0315
   svasenkov@che.ufl.edu
Ziegler, Kirk .... (352) 392-3412
   kziegler@che.ufl.edu

Accredited by: ABET

Degrees granted 2005-2006:
   B.S.: 44   M.S.: 8   Ph.D.: 11

Graduate advisor: Richard Dickinson

Undergraduate advisor: Spyros A. Svoronos

Student organization: AIChE
Advisor: Aravind R. Asthagiri

Department reports to: Pramod Khargonekar (352) 392-6000, Dean of Engineering

Placement service: Wayne Wallace, Director

Florida Institute of Technology

Chemical Engineering Department
150 West University Boulevard
Melbourne, FL 32901-6975
Deliveries: Building 540

University phone	(321) 674-8000
University FAX	(321) 984-8461
Department phone	(321) 674-8068
Department FAX	(321) 674-7565

Associate Professors

Jennings, Paul A. (Head) .... (321) 674-7561
jennings@fit.edu
BT,EV

Tomadakis, Manolis M. .... (321) 674-7243
tomadaki@fit.edu
SE

Whitlow, Jonathan E. .... (321) 674-7354
whitlow@fit.edu
MO

Assistant Professors

Brenner, James R. .... (321) 674-7560
jbrenner@fit.edu
AM,NT

Pozo de Fernandez, Maria E. .... (321) 674-8838
mpozo@fit.edu
PO

Other Faculty

Barile, Ron (Adjunct) .... (321) 674-8068
rbarile@fit.edu

Dutta, Subhash (Adjunct) .... (321) 674-8068
sdutta@fit.edu

Accredited by: ABET, SACS

Degrees granted 2005-2006:
  B.S.: 9   M.S.: 3

Graduate advisor: Paul A. Jennings

Undergraduate advisor: All Faculty

Student organization: AIChE
Advisor: Maria E. Pozo de Fernandez

Department reports to: Dr. Thomas Waite (321) 674-8020, Dean of College of Engineering

Placement service: Dona Gaynor, Career Services and Cooperative Education

Florida A&M University-Florida State University

Department of Chemical and Biomedical Engineering
Room A131, FAMU-FSU College of Engineering
2525 Pottsdamer Street
Tallahassee, FL 32310-6046
Deliveries: 2525 Pottsdamer Street, Room A131 CEB

University phone .... (850) 644-2525
Department phone .... (850) 410-6149
Department FAX .... (850) 410-6150

Professors
Alamo, Rufina .... (850) 410-6376
                  alamo@eng.fsu.edu
                  AM,PO
Locke, Bruce R. (Chair) .... (850) 410-6165
                  locke@eng.fsu.edu
                  BT,EV,MO,RE,SE,TP
Palanki, Srinivas .... (850) 410-6164
                  palanki@eng.fsu.edu
                  MO,PR,RE
Schrieber, Loren (Adjunct) .... (850) 410-6684
                  schreiber@eng.fsu.edu
                  HT,SE,TP

Associate Professors
Chella, Ravindran .... (850) 410-6170
                  rchella@eng.fsu.edu
                  AM,MO,PO,TP
Kalu, Eric .... (850) 410-6148
                  ekalu@eng.fsu.edu
                  EN,RE
Ma, Teng .... (850) 410-6149
                  teng@eng.fsu.edu
                  BM,BT
Telotte, John C. .... (850) 410-6168
                  telotte@eng.fsu.edu
                  RE,SE,TH
Vinals, Jorge (Adjunct) .... (850) 410-6149
                  vinals@eng.fsu.edu
                  MO,SU

Assistant Professors
Chen, Kevin .... (850) 410-6684
                  kevinchen@eng.fsu.edu
                  BM,BT
Grant, Samuel .... (850) 410-6158
                  grantsa@eng.fsu.edu
                  BM,BT
Kwon, Soonjo .... (850) 410-6411
                  skwon@eng.fsu.edu
                  BM,BT
Ramakrishnan, Subramanian .... (850) 410-6159
                  ramakrishnan@eng.fsu.edu
                  AM,PO,RE,SU,TP

Other Faculty
Finney, Wright C. .... (850) 410-6309
                  finney@eng.fsu.edu
                  BM,EV

Accredited by: ABET, SACS

Degrees granted 2005-2006:
    B.S.: 28   M.S.: 6   Ph.D.: 5

Graduate advisor: Teng Ma

Undergraduate advisor: Wright C. Finney

Student organization: AIChE
Advisor: Wright C. Finney

Department reports to: Ching-Jen Chen, Dean, (850) 410-6439

Placement service: Todd Kramer, (850)410-6388

University of South Florida

Chemical Engineering Department
4202 East Fowler Avenue
ENB 118
Tampa, FL 33620-5350

University phone	(813) 974-2011
Department phone	(813) 974-3997
Department FAX	(813) 974-3651

Professors

Bhethanabotla, Venkat R. .... (813) 974-2116
bethana@eng.usf.edu
AM,BM,ME,MO,NT,SU
Busot, J. Carlos (Emeritus) .... (813) 974-2141
busot@eng.usf.edu
Campbell, Scott W. .... (813) 974-3970
campbell@eng.usf.edu
EV,ME,MO
Gilbert, Richard .... (813) 974-2139
gilbert@eng.usf.edu
AM,BM,ME,NT
Goswami, Yogi 813 974 0956
goswami@eng.usf.edu
EN,EV
Joseph, Babu (Chair) .... (813) 974-0692
joseph@eng.usf.edu
EN,MO,PR
Lee, William E. III .... (813) 974-2136
lee@eng.usf.edu
BM
Llewellyn, J. Anthony .... (813) 974-1780
tony@eng.usf.edu
Smith, Carlos A. (Associate Dean) .... (813) 974-3780
csmith@eng.usf.edu
PR
Sunol, Aydin K. .... (813) 974-3566
sunol@eng.usf.edu
AM,EN,EV,MO,PO,PR,SE,SU

Associate Professors

Jaroszeski, Mark .... (813) 974-4662
mjarosze@eng.usf.edu
BM

Assistant Professors

Alcantar, Norma .... (813) 974-3997
alcantar@eng.usf.edu
AM,BM,EV,NT,PO,SU
Toomey, Ryan .... (813) 974-9164
rtoomey@eng.usf.edu
AM,NT,PO,SU
VanAuker, Michael .... (813) 974-3186
vanauker@eng.usf.edu
BM

Accredited by: ABET

Degrees granted 2005-2006:
   B.S.: 34   M.S.: 4   Ph.D.: 4

Graduate advisor: Dr. Wolan

Undergraduate advisor: Dr. Campbell

Student organization: AIChE/OCE/BMES
Advisor: Dr. Carlos Smith

Department reports to: Dr. Louis Martin-Vega, Dean of Engineering

Placement service: Drema K. Howard (813) 974-2171, Director

Georgia

Georgia Institute of Technology

School of Chemical & Biomolecular Engineering
Atlanta, GA 30332-0100
Deliveries: 311 Ferst Drive

University phone .... (404) 894-2000
Department phone .... (404) 894-2865
Department FAX .... (404) 894-2866

Professors

Allen, Mark G. (Joint) .... (404) 894-9419
mark.allen@ece.gatech.edu
ME
Banerjee, Sujit .... (404) 894-9709
sujit.banerjee@che.gatech.edu
EV,RE
Bidstrup Allen, Sue Ann (Associate Chair) (404) 894-2872
sue.bidstrup@che.gatech.edu
ME,PO
Bommarius, Andreas .... (404) 385-1334
andreas.bommarius@che.gatech.edu
BT
Eckert, Charles A. .... (404) 894-7070
charles.eckert@che.gatech.edu
SE,TH
Empie, Howard L. .... (404) 894-9704
howard.empie@che.gatech.edu
RE
Ernst, William R. (Emeritus) .... (404) 894-2889
william.ernst@che.gatech.edu
Frederick, William .... (404) 894-2082
jim.frederick@ipst.gatech.edu
RE
Fuller, Thomas (Joint) .... (770) 528-7075
tom.fuller@chbe.gatech.edu
EN
Hess, Dennis W. .... (404) 894-5922
dennis.hess@che.gatech.edu
ME
Hsieh, Jeffery .... (404) 894-3556
jeffery.hsieh@che.gatech.edu
RE
Kohl, Paul .... (404) 894-2893
paul.kohl@che.gatech.edu
ME
Koros, William J. .... (404) 385-2684
william.koros@che.gatech.edu
PO
Lee, Jay .... (404) 385-2148
jay.lee@che.gatech.edu
PR
Liotta, Charles L. (Joint) .... (404) 894-4048
charles.liotta@carnegie.gatech.edu
RE
McIntire, Larry (Joint) .... (404) 894-5057
larry.mcintire@bme.gatech.edu
BM
Muzzy, John D. .... (404) 894-2882
john.muzzy@che.gatech.edu
PO
Nerem, Robert M. (Joint) .... (404) 894-2768
robert.nerem@ibb.gatech.edu
BM
Prausnitz, Mark R. .... (404) 894-5135
mark.prausnitz@che.gatech.edu
BM
Rousseau, Ronald W. (Chair) .... (404) 894-2867
ronald.rousseau@che.gatech.edu
SE
Sambanis, Athanassios .... (404) 894-2869
athanassios.sambanis@che.gatech.edu
BM
Samuels, Robert J. (Emeritus) .... (404) 894-2885
robert.samuels@che.gatech.edu
Schork, F. Joseph .... (404) 894-3274
joseph.schork@che.gatech.edu
PO
Skelland, A. H. Peter (Emeritus) ...... (404) 894-2889
anthony.skelland@che.gatech.edu
Sommerfeld, Jude T. (Emeritus) ...... (404) 894-2873
jude.sommerfeld@che.gatech.edu
Stancell, Arnold (Emeritus) .... (404) 894-0316
arnold.stancell@che.gatech.edu
Teja, Amyn (Associate Chair) .... (404) 894-3098
amyn.teja@che.gatech.edu
SE,TH
Winnick, Jack (Emeritus) .... (404) 894-2839
jack.winnick@che.gatech.edu
Yoganathan, Ajit P. (Joint) .... (404) 894-7063
ajit.yoganathan@bme.gatech.edu

Associate Professors

Agrawal, Pradeep K. (Associate Chair) (404) 894-2826
pradeep.agrawal@che.gatech.edu
RE
Behrens, Sven .... (404) 894-1838
sven.behrens@chbe.gatech.edu
PO
Chaikof, Elliot L. (Adjunct) .... (404) 727-8413
elliot.chaikof@che.gatech.edu
BM
Chen, Rachel .... (404) 894-1255
rachel.chen@chbe.gatech.edu
BT
Deng, Yulin .... (404) 894-5759
yulin.deng@che.gatech.edu
NT,PO
Forney, Larry J. .... (404) 894-2825
larry.forney@che.gatech.edu
EV,TP
Henderson, Clifford .... (404) 385-0525
cliff.henderson@che.gatech.edu
ME,PO

United States

Jones, Christopher .... (404) 385-1683
chris.jones@che.gatech.edu
PO,RE
Ludovice, Peter .... (404) 894-1835
pete.ludovice@chbe.gatech.edu
MO,PO
Meredith, J. Carson .... (404) 385-2151
carson.meredith@chbe.gatech.edu
MO,PO
Realff, Matthew J. .... (404) 894-1834
matthew.realff@che.gatech.edu
PR
Roberts, Ronnie S. (Emeritus) .... (404) 894-2889
ronnie.roberts@che.gatech.edu
Tedder, Dan W. .... (404) 894-2856
daniel.tedder@che.gatech.edu
EN

Assistant Professors

Breedveld, L. Victor .... (404) 894-5134
victor.breedveld@che.gatech.edu
TP
Gallivan, Martha .... (404) 894-2878
martha.gallivan@che.gatech.edu
MO,PR
Lu, Hang .... (404) 894-8473
hang.lu@chbe.gatech.edu
BM,BT
Nair, Sankar .... (404) 894-4826
sankar.nair@che.gatech.edu
AM,NT
Nenes, Athanasios (Joint) .... (404) 894-9225
thanos.nenes@che.gatech.edu
EV

Other Faculty

Chance, Ronald (Joint) .... (404) 894-1838
ron.chance@chbe.gatech.edu
NT
Iisa, M. Kristina .... (404) 894-0810
kristina.iisa@ipst.gatech.edu
RE
Mohalley-Snedeker, Jacqueline .... (404) 894-8471
jacqueline.mohalley@chbe.gatech.edu

Accredited by: ABET

Degrees granted 2005-2006:
B.S.: 73   M.S.: 23   Ph.D.: 26

Graduate advisor: Amyn S. Teja

Undergraduate advisor: Pradeep Agrawal

Student organization: AIChE
Advisor: Clifford Henderson

Department reports to: Don P. Giddens, Dean of Engineering

Placement service: Ralph Mobley, Director

Idaho

University of Idaho

Chemical Engineering Department
PO Box 441021
6th and Urquhart
Moscow, ID 83844-1021

Department phone .... (208) 885-6793
Department FAX .... (208) 885-7462

Professors

Admassu, Wudneh .... (208) 885-8918
wadmassu@uidaho.edu
BT,EV,PO
Edwards, Louis L. Jr. .... (208) 885-6793
jkidd@uidaho.edu
MO,PR
Korus, Roger A. (Chair) .... (208) 885-6005
rkorus@uidaho.edu
BT,EV,RE
Park, Jin Y. .... (208) 885-6970
jinpark@uidaho.edu
RE
Scheldorf, Jay J. (Emeritus) .... (208) 885-7282
von Braun, Margrit C. (Associate Dean) (208) 885-7838
vonbraun@uidaho.edu
EV

Associate Professors

Drown, David C. .... (208) 885-7848
ddrown@uidaho.edu
EN,MO,PR

Assistant Professors

Aston, Eric .... (208) 885-6953
aston@uidaho.edu
AM,ME,NT,SE,SU
Thomas, Aaron .... (208) 885-7652
amthomas@uidaho.edu
NT,SE
Utgikar, Vivek .... (208) 282-7720
vutgikar@if.uidaho.edu
BT,EN

Accredited by: ABET

Degrees granted 2005-2006:
B.S.: 14   M.S.: 3   Ph.D.: 2

Graduate advisor: Eric Aston/Aaron Thomas

Undergraduate advisor: Wudneh Admassu

Student organization: AIChE
Advisor: David C. Drown

Department reports to: Charles Petersen, Dean

Placement service: Daniel A. Blanco, Career Services Center

Illinois

University of Illinois at Chicago

Chemical Engineering Department
810 S. Clinton M/C 110
Chicago, IL 60607

University phone	(312) 996-7000
Department phone	(312) 996-3424
Department FAX	(312) 996-0808

Professors

Jameson, Cynthia J. (Adjunct) .... (312) 996-3424
cjjames@uic.edu
BM,NT,SU
Kiefer, John (Emeritus) .... (312) 996-5711
kiefer@uic.edu
Mansoori, G. Ali .... (312) 996-5592
mansoori@uic.edu
Murad, Sohail (Head) .... (312) 996-5593
Murad@uic.edu
Nemeth, Laszlo (Adjunct) .... (312) 996-3424
lnemeth@uic.edu
AM,RE,SE
Regalbuto, John .... (312) 996-0288
JRR@uic.edu
Saxena, Satish C. (Emeritus)
saxena@uic.edu
Takoudis, Christos .... (312) 996-3424
Takoudis@uic.edu
Turian, Raffi .... (312) 996-8734
Turian@uic.edu

Associate Professors

Linninger, Andreas .... (312) 996-2581
Linninge@uic.edu
Nitsche, Ludwig .... (312) 996-3469
LCN@uic.edu
Szepe, Stephen (Emeritus) .... (312) 996-2342
SSzepe@uic.edu

Assistant Professors

Meyer, Randall
Rjm@uic.edu

Other Faculty

Funk, Edward (Adjunct) .... (312) 355-5149
ewf@ewfconsulting.net
Oroskar, Anil (Adjunct) .... (312) 996-3424
anil@orochem.com

Accredited by: ABET

Degrees granted 2005-2006:
    B.S.: 17   M.S.: 6   Ph.D.: 3

Graduate advisor: Lewis Wedgewood

Undergraduate advisor: L.C. Nitsche

Student organization: AIChE
Advisor: L.C. Nitsche

Department reports to: Prith Banerjee (312) 996-2400, Dean, College of Engineering

Placement service: Andres Garza (312) 996-2300, Director

University of Illinois at Urbana-Champaign

Department of Chemical and Biomolecular Engineering
114 Roger Adams Lab, Box C-3
600 South Mathews Avenue
Urbana, IL 61801-3602

University phone	(217) 333-1000
Department phone	(217) 333-3640
Department FAX	(217) 333-5052

Professors

Alkire, Richard C. .... (217) 333-0063
r-alkire@uiuc.edu
AM,ME,MO,NT,SU

Braatz, Richard D. .... (217) 333-5073
braatz@uiuc.edu
AM,BM,BT,ME,MO,NT,PR,SE

Hammack, William S. .... (217) 244-4146
hammack@netbox.com

Hanratty, Thomas J. (Emeritus) ...... (217) 333-1318
thanratt@uiuc.edu

Higdon, Jonathan J. L. .... (217) 333-1479
jhigdon@uiuc.edu
EN,MO

Leckband, Deborah E. .... (217) 244-9214
leckband@uiuc.edu
BM,BT,SU

Masel, Richard I. .... (217) 244-2819
masel@scs.uiuc.edu
EN,EV,RE,SU

McHugh, Anthony J. (Emeritus)

Pack, Daniel W. .... (217) 244-2816
dpack@uiuc.edu
BM,BT,NT

Sahinidis, Nikolaos V. .... (217) 244-1304
nikos@uiuc.edu
MO,PR

Seebauer, Edmund G. (Head) .... (217) 333-4402
eseebaue@uiuc.edu
ME,NT

Zhao, Huimin .... (217) 333-2631
zhao5@uiuc.edu
BM,BT,EN,EV,RE

Zukoski, Charles F. .... (217) 333-0034
czukoski@uiuc.edu
AM,BT,NT,PO,SU

Assistant Professors

Kenis, Paul J. A. .... (217) 265-0523
kenis@uiuc.edu
BM,BT,EN,EV,NT,RE,SU

Rao, Christopher V. .... (217) 244-2247
chris@scs.uiuc.edu
BT,MO,PR

Strano, Michael S. .... (217) 333-3634
strano@uiuc.edu

Instructors

Miletic, Marina .... (217) 244-6730
marina@uiuc.edu

Accredited by: ABET

Degrees granted 2005-2006:
B.S.: 52   M.S.: 12   Ph.D.: 3

Undergraduate advisor: N. V. Sahinidis

Student organization: AIChE
Advisor: M. Miletic

Department reports to: Sarah Mangelsdorf, Dean, College of Liberal Arts and Sciences

Placement service: Debe Deeb Williams, (217) 333-1050, 105 Noyes Lab

Illinois Institute of Technology

Department of Chemical and Environmental Engineering
10 W. 33rd St., Perlstein Hall 127
Chicago, IL 60616-3793
Deliveries: Perlstein Hall 112

University phone .... (312) 567-3000
Department phone .... (312) 567-3040
Department FAX .... (312) 567-8874

Professors

Abbasian, Javad (Associate Chair) .... (312) 567-3047
    abbasian@iit.edu
Arastoopour, Hamid (Dean) .... (312) 567-3034
    arastoopour@iit.edu
Bernstein, Barry .... (312) 567-3166
    bernsteinb@iit.edu
Cinar, Ali (Dean) .... (312) 567-3042
    cinar@iit.edu
Gidaspow, Dimitri .... (312) 567-3045
    gidaspow@iit.edu
Linden, Henry .... (312) 567-3095
    linden@iit.edu
Moschandreas, Demetrios .... (312) 567-3532
    moschandreas@iit.edu
Myerson, Allan .... (312) 567-3101
    myerson@iit.edu
Noll, Kenneth .... (312) 567-3536
    noll@iit.edu
Parulekar, Satish J. .... (312) 567-3044
    parulekar@iit.edu
Schieber, Jay D. .... (312) 567-3046
    schieber@iit.edu
Selman, J. Robert .... (312) 567-3037
    selman@iit.edu
Teymour, Fouad A. (Chair) .... (312) 567-8947
    teymour@iit.edu
Venerus, David C. (Associate Chair) ... (312) 567-5177
    venerus@iit.edu
Wasan, Darsh T. .... (312) 567-3001
    wasan@iit.edu

Associate Professors

Anderson, Paul .... (312) 567-3531
    andersonp@iit.edu
Pagilla, Krishna .... (312) 567-5717
    pagilla@iit.edu
Prakash, Jai .... (312) 567-3639
    prakash@iit.edu

Assistant Professors

Chmielewski, Donald .... (312) 567-3537
    chmielewski@iit.edu
Gidalevitz, David .... (312) 567-3534
    gidalevitz@iit.edu
Perez-Luna, Victor .... (312) 567-3963
    perezluna@iit.edu
Ramani, Vijay .... (312) 567-3064
    ramani@iit.edu

Instructors

Lindahl, Harold (Adjunct) .... (312) 567-5115
    lindahl@iit.edu

Other Faculty

Aderangi, Nader .... (312) 567-8874
    aderangi@iit.edu
Al-Hallaj, Said .... (312) 567-5118
    alhallaj@iit.edu
Hatziavramidis, Dimitri .... (312) 567-3010
    hatziavramidis@iit.edu
Nikolov, Alex .... (312) 567-5980
    nikolov@iit.edu

Accredited by: ABET

Degrees granted 2004-2005:
    B.S.: 18   M.S.: 51   Ph.D.: 9

Graduate advisor: Dr. David Venerus (312) 567-5177

Undergraduate advisor: Dr. Javad Abbasian (312) 567-3047

Student organization: AIChE
Advisor: Dr. Don Chmielewski (312) 567-3537

Department reports to: Hamid Arastoopour (312) 567-3034, Dean of Armour College of Engineering

Placement service: Director (312) 567-6800

Northwestern University

Department of Chemical and Biological Engineering
2145 Sheridan Road, Room E136
Evanston, IL 60208-3120

University phone	(847) 491-3741
Department phone	(847) 491-7398
Department FAX	(847) 491-3728

Professors

Bankoff, S. George (Emeritus) .... (847) 491-5267
gbankoff@northwestern.edu
Barron, Annelise E. .... (847) 491-2778
a-barron@northwestern.edu
Broadbelt, Linda J. .... (847) 491-5351
broadbelt@northwestern.edu
Burghardt, Wesley R. (Chair) .... (847) 467-1401
w-burghardt@northwestern.edu
Butt, John B. (Emeritus)
Carr, Stephen H. (Associate Dean) .... (847) 491-7379
s-carr@northwestern.edu
Cohen, William C. (Emeritus)
bill-cohen@northwestern.edu
Crist, Buckley Jr. .... (847) 491-3279
b-crist@northwestern.edu
Dranoff, Joshua S. .... (847) 491-5252
j-dranoff@northwestern.edu
Goldstick, Thomas K. (Emeritus) ..... (847) 491-5518
t-goldstick@northwestern.edu
Kung, Harold H. .... (847) 491-7492
hkung@northwestern.edu
Miller, William M. .... (847) 491-4828
wmmiller@northwestern.edu
Mockros, Lyle F. (Emeritus) .... (847) 491-3172
lmockros@northwestern.edu
Ottino, Julio M. (Dean) .... (847) 491-3558
ottino@chem-eng.northwestern.edu
Papoutsakis, E. Terry .... (847) 491-7455
e-paps@northwestern.edu
Snurr, Randall Q. .... (847) 467-2977
snurr@northwestern.edu
Stevens, William F. (Emeritus)

Associate Professors

Amaral, Luis A. N. .... (847) 491-1750
amaral@northwestern.edu
Brown, George M. (Emeritus)
Shea, Lonnie D. .... (847) 491-7043
l-shea@northwestern.edu

Assistant Professors

Grzybowski, Bartosz A. .... (847) 491-3024
grzybor@northwestern.edu
Hatzimanikatis, Vassily .... (847) 491-5357
vassily@northwestern.edu

Other Faculty

Graessley, William W. (Adjunct) ...... (847) 491-7398
Haug, Warren (Adjunct) .... (847) 467-5712
w-haug@northwestern.edu
Kung, Mayfair .... (847) 491-5085
m-kung@northwestern.edu

Accredited by: ABET

Degrees granted 2005-2006:

Ph.D.: 21

Student organization: AIChE
Advisor: Vassily Hatzimanikatis

Department reports to: Julio Ottino, Dean, MEAS

Placement service: William J. Banis, Director

Indiana

University of Notre Dame

Department of Chemical & Biomolecular Engineering
Notre Dame, IN 46556-5637
Deliveries: 182 Fitzpatrick Hall
University phone .... (574) 631-5000
Department phone .... (574) 631-5580
Department FAX .... (574) 631-8366

Professors

Bohn, Paul W.
    Paul.W.Bohn.6@nd.edu
    AM,NT,SE
Brennecke, Joan F. .... (574) 631-5847
    jfb@nd.edu
    EN,EV,SE
Chang, Hsueh-Chia .... (574) 631-5697
    chang2@nd.edu
    BM,MO,NT
Kantor, Jeffrey C. .... (574) 631-5797
    Kantor.1@nd.edu
    EN,MO,PR
Leighton, David T. Jr. .... (574) 631-6698
    dtl@nd.edu
    BM,BT,EN,MO,SE
Maginn, Edward J. .... (574) 631-5687
    ed@nd.edu
    AM,EN,MO,SE
McCready, Mark J. (Chair) .... (574) 631-7146
    mjm@nd.edu
    BT,EN,RE
McGinn, Paul J. .... (574) 631-6151
    mcginn.1@nd.edu
    AM,EN,NT,RE
Miller, Albert E. .... (574) 631-8307
    miller.3@nd.edu
    AM,BT,EN,ME,NT,RE,SU
Schmitz, Roger A. (Emeritus) .... (574) 631-7798
    schmitz.1@nd.edu
    EN,MO
Stadtherr, Mark A. .... (574) 631-9318
    markst@nd.edu
    EN,EV,MO,PR,SE
Strieder, William C. .... (574) 631-5648
    strieder.1@nd.edu
    AM,BT,EN,MO,RE,SU
Wolf, Eduardo E. .... (574) 631-5897
    wolf.1@nd.edu
    EN,NT,RE,SU

Associate Professors

Hill, Davide A. .... (574) 631-8487
    hill.1@nd.edu
    AM,EN,ME,MO,PO
Schneider, William F. .... (574) 631-9754
    wschneider@nd.edu
    EN,MO,RE,SU

Assistant Professors

Palmer, Andre F. .... (574) 631-4776
    apalmer@nd.edu
    BM,BT,NT
Zhu, Yingxi Elaine .... (574) 631-2667
    yzhu3@nd.edu
    BM,NT,PO,SU

Other Faculty

Chen, Zilin .... (574) 631-5749
    Zilin.Chen.88@nd.edu
    BM,SE
Kamat, Prashant (Joint) .... (574) 631-5411
    pkamat@nd.edu
    AM,EN,EV,NT,SU
Mukasyan, Alexander .... (574) 631-9825
    amoukasi@nd.edu
    AM,EN,NT,RE
Saddawi, Salma .... (574) 631-3324
    Saddawi.1@nd.edu
    BT,EN,EV,RE

Accredited by: ABET

Degrees granted 2005-2006:
    B.S.: 30   M.S.: 9   Ph.D.: 15

Graduate advisor: Mark A. Stadtherr

Student organization: AIChE
Advisor: William F. Schneider

Department reports to: Frank Incropera, McCloskey Dean of Engineering

Placement service: Tricia Ford (219) 631-3315, 248 Flanner Hall

Purdue University

School of Chemical Engineering
Forney Hall of Chemical Engineering
480 Stadium Mall Drive, 104
West Lafayette, IN 47907-2100

Department phone .... (765) 494-4050
Department FAX .... (765) 494-0805

Professors

Agrawal, Rakesh .... (765) 494-2257
agrawalr@purdue.edu
EN,NT,PR,SE
Albright, Lyle F. (Emeritus) .... (765) 494-4087
albright@ecn.purdue.edu
Andres, Ronald P. (Emeritus) .... (765) 494-4047
ronald@ecn.purdue.edu
Basaran, Osman .... (765) 494-4061
obasaran@ecn.purdue.edu
Beaudoin, Stephen .... (765) 494-7944
sbeaudoi@ecn.purdue.edu
AM,BM,BT,EV,ME,MO,NT,SU
Blau, Gary E. (Visiting) .... (765) 494-9472
blau@ecn.purdue.edu
Caruthers, James M. .... (765) 494-6625
caruther@ecn.purdue.edu
AM,MO,PO,PR,RE
Chao, K. C. (Emeritus) .... (765) 494-4050
kwang@ecn.purdue.edu
Delgass, W. Nicholas .... (765) 494-4059
delgass@ecn.purdue.edu
EN,MO,NT,PO,RE,SU
Emery, Alden H. (Emeritus) .... (765) 494-4050
emery@ecn.purdue.edu
Franses, Elias I. .... (765) 494-4078
franses@ecn.purdue.edu
BM,NT,SE,SU
Greenkorn, Robert A. (Emeritus) ...... (765) 494-4051
greenkor@ecn.purdue.edu
Hannemann, Robert (Visiting) .... (765) 494-4079
hanneman@ecn.purdue.edu
Harris, Michael T. .... (765) 494-0963
mtharris@ecn.purdue.edu
Houze, R. Neal .... (765) 494-4076
houze@ecn.purdue.edu
Kim, Sangtae .... (765) 494-5692
kim55@ecn.purdue.edu
BM,BT,MO,NT,PR
Pekny, Joseph F. .... (765) 494-7901
pekny@ecn.purdue.edu
MO,PR
Pipes, Byron .... (765) 494-5767
bpipes@purdue.edu
AM,NT,PO
Ramkrishna, D. (Associate Head) ..... (765) 494-4066
ramkrish@ecn.purdue.edu
BM,BT
Reklaitis, Gintaras V. .... (765) 494-9662
reklaiti@ecn.purdue.edu
EN,MO,PR
Ribeiro, Fabio .... (765) 494-7799
fabio@ecn.purdue.edu
EN,NT,RE,SU
Squires, Robert G. (Emeritus) .... (765) 494-4093
squires@ecn.purdue.edu
Tsao, George T. .... (765) 494-4068
tsaogt@ecn.purdue.edu
Varma, Arvind (Head) .... (765) 494-4904
avarma@purdue.edu
AM,EN,RE
Venkatasubramanian, Venkat .... (765) 494-0734
venkat@ecn.purdue.edu
AM,MO,PR
Wang, Linda Nien-Hwa .... (765) 494-4081
wangn@ecn.purdue.edu
BT,EN,SE
Wankat, Phillip C. .... (765) 494-0814
wankat@ecn.purdue.edu
SE

Associate Professors

Corti, David S. .... (765) 496-6064
dscorti@ecn.purdue.edu
MO,SU
Lee, Gil U. .... (765) 494-0492
gl@ecn.purdue.edu
Liu, Julie .... (765) 494-4050
Morgan, John A. .... (765) 494-4088
jamorgan@ecn.purdue.edu
BT,RE
Thomson, Kendall T. .... (765) 496-6706
thomsonk@ecn.purdue.edu

Assistant Professors

Baertsch, Chelsey D. .... (765) 496-7826
baertsch@ecn.purdue.edu
Hillhouse, Hugh W. .... (765) 496-6056
hugh@ecn.purdue.edu
Won, You-Yeon .... (765) 494-4077
yywon@ecn.purdue.edu
AM,BM,BT,NT,PO,SU

Other Faculty

Okos, M. .... (765) 494-1211
okos@ecn.purdue.edu

Accredited by: ABET

Degrees granted 2005-2006:
    B.S.: 81   M.S.: 11   Ph.D.: 18

Graduate advisor: Osman Basaran

Student organization: AIChE
Advisor: John A. Morgan

Department reports to: Leah Jamieson, Interim Dean

Placement service: Center for Career Opportunities

Rose-Hulman Institute of Technology

Chemical Engineering Department
5500 Wabash Ave.
Terre Haute, IN 47803-3999

University phone .... (812) 877-1511
Department phone .... (812) 877-8430
Department FAX .... (812) 877-8992

Professors
Abegg, Carl F. (Emeritus) .... (812) 877-8430
  Carl.F.Abegg@rose-hulman.edu
Artigue, Ronald S. .... (812) 877-8369
  Ronald.S.Artigue@rose-hulman.edu
  BT
Bowden, Warren W. (Emeritus) .... (812) 877-8430
  Trebol1125@aol.com
Carlson, Alfred .... (812) 877-8423
  Alfred.Carlson@rose-hulman.edu
  BT
Caskey, Jerry A. (Emeritus) .... (812) 877-8430
  Jerry.A.Caskey@rose-hulman.edu
Hariri, M. Hossein (Head) .... (812) 877-8381
  M.H.Hariri@rose-hulman.edu
  EN,EV,SE
Leipziger, Stuart (Emeritus) .... (812) 877-8430
  Stuart.Leipziger@rose-hulman.edu
Moore, Noel E. (Emeritus) .... (812) 877-8430
  Noel.E.Moore@rose-hulman.edu

Associate Professors
Anklam, Mark R. .... (812) 877-8098
  Mark.R.Anklam@rose-hulman.edu
  BT,RE,SU
Coronell, Daniel G. .... (812) 877-8419
  Daniel.G.Coronell@rose-hulman.edu
  MO,RE
Miller, David C. .... (812) 877-8506
  David.Miller@rose-hulman.edu
  EV,MO,PO,PR
Serbezov, Atanas .... (812) 877-8097
  Atanas.Serbezov@rose-hulman.edu
  EN,SE

Assistant Professors
McClellan, Scott J. .... (812) 877-8599
  mcclell1@rose-hulman.edu
  BT,PO
Sauer, Sharon G. .... (812) 877-8527
  Sharon.G.Sauer@rose-hulman.edu
  MO

Accredited by: ABET

Degrees granted 2005-2006:
  B.S.: 38   M.S.: 2

Graduate advisor: David Miller

Student organization: AIChE
Advisor: Atanas Serbezov

Department reports to: Art Western, Dean of Faculty

Placement service: Jan Ford (812) 877-8338, Associate Director

Tri-State University

McKetta Department of Chemical and Bioprocess Engineering
1 University Ave.
Angola, IN 46703-0307

University phone .... (260) 665-4100
Department phone .... (260) 665-4815
Department FAX .... (260) 665-4814

Associate Professors
Finley, David .... (260) 665-4224
  finleyd@tristate.edu
  EV
Salim, Majid .... (260) 665-4223
  salimm@tristate.edu
  PO
Wagner, John E. .... (260) 665-4226
  wagnerj@tristate.edu
  BT

Assistant Professors
Hersel, Allen (Chair) .... (260) 665-4252
  hersela@tristate.edu
  BT,SE

Accredited by: ABET

Degrees granted 2005-2006:
  B.S.: 12

Undergraduate advisor: Dr. Majid Salim/Dr. John Wagner/Dr. Allen Hersel

Student organization: AIChE
Advisor: Dr. Majid Salim

Department reports to: Dr. Allen Hersel (260) 665-4252, Chair, Chemical Engineering Dept.

Placement service: Ms. Debbie Roemke (219) 665-4123, Asst. Director of Placement

United States

Iowa

University of Iowa

Chemical and Biochemical Engineering Department
College of Engineering
Iowa City, IA 52242-1527
Deliveries: 4133 Seamans Center

University phone .... (319) 335-3500
Department phone .... (319) 335-1400
Department FAX .... (319) 335-1415

Professors

Beddow, John K. (Emeritus) .... (319) 335-1400
Carmichael, Gregory R. .... (319) 335-1399
               gcarmich@icaen.uiowa.edu
Murhammer, David W. .... (319) 335-1228
               murham@icaen.uiowa.edu
Rethwisch, David G. .... (319) 335-1413
               drethwis@icaen.uiowa.edu
Scranton, Alec .... (319) 335-1400
               alec-scranton@uiowa.edu
Subramanian, Venkiteswaran (Joint) ... (319) 335-5813
               manisubr@engineering.uiowa.edu
               RE
Wiencek, John (Chair) .... (319) 353-2377
               wiencek@icaen.uiowa.edu

Associate Professors

Peeples, Tonya .... (319) 335-2251
               peeples@icaen.uiowa.edu

Assistant Professors

Guymon, C. Allan .... (319) 335-5015
               allan-guymon@uiowa.edu
Jessop, Julie .... (319) 335-0681
               julie-jessop@uiowa.edu
Stanier, Charles .... (319) 335-1399
               charles-stanier@uiowa.edu

Instructors

Aurand, Gary .... (319) 384-0970
               gary-aurand@uiowa.edu
Butler, Audrey .... (319) 384-0538
               audrey-butler@uiowa.edu

Accredited by: ABET

Degrees granted 2005-2006:
    B.S.: 18   M.S.: 1   Ph.D.: 6

Graduate advisor: Allan Guymon

Student organization: AIChE
Advisor: David W. Murhammer

Department reports to: P. Barry Butler, (319) 335-5766, Dean of Engineering

Placement service: Engineering Career Services (319) 335-5774

Iowa State University of Science and Technology

Chemical and Biological Engineering Department
2114 Sweeney Hall
Ames, IA 50011-2230

University phone .... (515) 294-4111
Department phone .... (515) 294-7642
Department FAX .... (515) 294-2689

Professors

Brown, Robert C. (Joint) .... (515) 294-3759
    rcbrown@iastate.edu
Burnet, George (Emeritus) .... (515) 294-8670
    gxb@iastate.edu
Doraiswamy, L. K. (Emeritus) .... (515) 294-4117
    dorai@iastate.edu
    RE
Fox, Rodney O. .... (515) 294-9104
    rofox@iastate.edu
    EN,HT,MO,NT,PO,RE,TP
Glatz, Charles E. .... (515) 294-8472
    cglatz@iastate.edu
    BT,SE,TP
Hebert, Kurt R. .... (515) 294-6763
    krhebert@iastate.edu
    TP
Hill, James C. (Chair) .... (515) 294-4959
    jchill@iastate.edu
    TP
Jolls, Kenneth R. .... (515) 294-5222
    jolls@iastate.edu
    TH,TP
Kushner, Mark (Dean) .... (515) 294-9988
    mjk@iastate.edu
    AM,BT,ME,MO,PO,RE,TP
Mallapragada, Surya K. .... (515) 294-7407
    suryakm@iastate.edu
    AM,BM,NT,PO,SU
Porter, Marc D. (Adjunct) .... (515) 294-6433
    porter@ameslab.gov
Reilly, Peter J. .... (515) 294-5968
    reilly@iastate.edu
    BT,MO
Schrader, Glenn L. (Adjunct)
    schrader@iastate.edu
    RE
Seagrave, Richard C. (Emeritus) ...... (515) 294-0518
    seagrave@iastate.edu
    BM,TH
Shanks, Jackie V. .... (515) 294-4828
    jshanks@iastate.edu
    BT
Ulrichson, Dean L. (Emeritus)
    dlulrich@iastate.edu
Wheelock, Thomas D. (Emeritus) ..... (515) 294-5226
    wheel@iastate.edu
    EN,RE,SE,SU
Youngquist, Gordon R. (Emeritus)
    gyoungqu@iastate.edu

Associate Professors

Hanneman, Larry F. (Adjunct) .... (515) 294-0253
    lfhannem@iastate.edu
Hillier, Andrew .... (515) 294-3678
    hillier@iastate.edu
    RE,SU
Narasimhan, Balaji .... (515) 294-8019
    nbalaji@iastate.edu
    AM,BM,NT,PO
Rollins, Derrick K. .... (515) 294-5516
    drollins@iastate.edu
    AM,BM,BT,MO,PR
Shanks, Brent H. .... (515) 294-1895
    bshanks@iastate.edu
    BT,EN,RE
Vigil, R. Dennis .... (515) 294-6438
    vigil@iastate.edu
    RE,TP

Assistant Professors

Clapp, Aaron .... (515) 294-9514
    clapp@iastate.edu
    BM,BT,NT,SU
Cochran, Eric W. .... (515) 294-0625
    ecochran@iastate.edu
    AM,EN,MO,NT,PO,TH
Lamm, Monica .... (515) 294-6533
    mhlamm@iastate.edu
    AM,MO,TH

Instructors

Loveland, Stephanie .... (515) 294-3024
    prairie@iastate.edu
    PR
Stiehl, Cory .... (515) 294-5825
    cstiehl@iastate.edu

Accredited by: ABET

Degrees granted 2005-2006:
    B.S.: 50   M.S.: 2   Ph.D.: 12

Graduate advisor: Balaji Narasimhan

Undergraduate advisor: R. Dennis Vigil & Brenda Kutz

Student organization: AIChE
Advisor: Kenneth R. Jolls

Department reports to: Mark R. Kushner (515) 294-5933; Dean

Placement service: Larry F. Hanneman, (515) 294-0253; Director, Engineering Career Services

Kansas

University of Kansas

Department of Chemical and Petroleum Engineering
Learned Hall, Room 4132
1530 W. 15th St.
Lawrence, KS 66045-7609

University phone . . . . (785) 864-2700
Department phone . . . . (785) 864-4965
Department FAX . . . . (785) 864-4967

Professors

Bishop, Kenneth A. (Emeritus) . . . . (785) 864-2918
kbishop@ku.edu
Gehrke, Stevin H. (785) 864-4956
shgehrke@ku.edu
Green, Don W. . . . . (785) 864-2911
dgreen@ku.edu
Locke, Carl E. Jr. (Emeritus) . . . . (785) 864-2929
lok@ku.edu
Maloney, James O. (Emeritus) . . . . (785) 864-4942
Mesler, Russell B. (Emeritus) . . . . (785) 843-0863
Nguyen, Trung Van . . . . (785) 864-3938
cptvn@ku.edu
Preston, Floyd W. (Emeritus) . . . . (785) 843-6212
petlep@ku.edu
Rosson, Harold F. (Emeritus) . . . . (785) 331-0747
Subramaniam, Bala . . . . (785) 330-4490
bsubramaniam@ku.edu
Swift, George W. (Emeritus) . . . . (785) 842-2751
cbswift@ku.edu
Vossoughi, Shapour . . . . (785) 864-2902
shapour@ku.edu
Walas, Stanley M. (Emeritus) . . . . (785) 843-0547
Weatherley, Laurence R. (Chair) . . . . . . (785) 864-3553
lweather@ku.edu
Willhite, G. Paul . . . . (785) 864-2906
willhite@ku.edu

Associate Professors

Camarda, Kyle V. . . . . (785) 864-2908
camarda@ku.edu
Howat, Colin S. . . . . (785) 218-3718
cshowat@ku.edu
Liang, Jenn-Tai . . . . (785) 864-2669
jtliang@ku.edu
Nordheden, Karen J. . . . . (785) 864-8820
nordhed@ku.edu
Ostermann, Russell D. (Associate Chair) (785) 864-2907
ostermann@ku.edu
Southard, Marylee Z. . . . . (785) 864-3868
marylee@ku.edu
Williams, Susan M. . . . . (785) 864-2919
smwilliams@ku.edu

Assistant Professors

Berkland, Cory J. . . . . (785) 864-4949
berkland@ku.edu
Detamore, Michael S. . . . . (785) 864-4943
detamore@ku.edu
Guzman, Javier . . . . (785) 864-4920
Laurence, Jennifer S. . . . . (785) 864-4949
laurencj@ku.edu
Scurto, Aaron M. . . . . (785) 864-4947
ascurto@ku.edu

Instructors

Howat, Julie . . . . (785) 218-3260
jnhowat@ku.edu

Accredited by: ABET

Degrees granted 2005-2006:
   B.S.: 27   Ph.D.: 1

Graduate advisor: Trung V. Nguyen

Undergraduate advisor: All faculty

Student organization: AIChE
Advisor: Colin S. Howat

Department reports to: Stuart R. Bell, Dean, School of Engineering, (785) 864-3881

Placement service: Cathy Schwabauer, Coordinator, (785) 864-3891

Kansas State University

Department of Chemical Engineering
105 Durland
Manhattan, KS 66506-5102

University phone .... (785) 532-6011
Department phone .... (785) 532-5584
Department FAX .... (785) 532-7372

Professors

Akins, Richard G. (Emeritus) .... (785) 532-5584
    akins@ksu.edu
Czermak, -Ing. Peter (Adjunct)
    peter.czermak@tg.fh-giessen.de
    AM,BM,BT,EN,EV,MO,NT,PR,RE,SE
Edgar, James H. .... (785) 532-4320
    edgarjh@ksu.edu
    AM,ME,NT
Erickson, Larry E. .... (785) 532-4313
    lerick@ksu.edu
    AM,BT,EN,EV,MO,NT,PR,RE
Fan, L. T. .... (785) 532-4327
    fan@ksu.edu
    BT,EV,MO,PR,RE
Glasgow, Larry A. .... (785) 532-4314
    glasgow@ksu.edu
    EV,SE,TP
King, Terry S. (Adjunct) .... (765) 285-1333
    tsking@ksu.edu
    AM,EN,NT,RE
Kyle, Benjamin G. (Emeritus) .... (785) 532-5584
    che@ksu.edu
Matthews, John C. (Emeritus) .... (785) 532-5584
    che@ksu.edu
Pfromm, Peter .... (785) 532-4312
    pfromm@ksu.edu
    BT,RE,SE
Rezac, Mary (Head) .... (785) 532-5584
    rezac@ksu.edu
    BT,PO,SE
Schlup, John R. .... (785) 532-4319
    jrsch@ksu.edu
    AM,BT,PO,PR,RE
Walawender, Walter P. .... (785) 532-4318
    walawen@ksu.edu

Associate Professors

Hohn, Keith L. .... (785) 532-4315
    hohn@ksu.edu
    AM,EN,NT,RE

Assistant Professors

Anthony, Jennifer .... (785) 532-4321
    anthonyj@ksu.edu
    AM,EV,SE,TH
Walton, Krista .... (785) 532-5584
    kwalton@ksu.edu
    AM,MO,NT,SE

Instructors

Castro, Sigifredo .... (785) 532-4262
    scastro@ksu.edu
    BT,EV,PO,SE

Accredited by: ABET

Degrees granted 2005-2006:
    B.S.: 19   M.S.: 6   Ph.D.: 2

Graduate advisor: James H. Edgar

Undergraduate advisor: John R. Schlup

Student organization: AIChE
Advisor: Walter P. Walawender

Department reports to: Richard R. Gallagher, Interim Dean of Engineering

Placement service: Tom Hollinberger, Assistant Director (785) 532-1685

Kentucky

University of Kentucky

Department of Chemical and Materials Engineering
177 F. Paul Anderson Tower
Lexington, KY 40506-0046

University phone	(859) 257-9000
Department phone	(859) 257-8028
Department FAX	(859) 323-1929

Professors

Anderson, Kimberly .... (859) 257-2300x232
kanderson@engr.uky.edu
Bhattacharyya, Dibakar .... (859) 257-2794
db@engr.uky.edu
Gillis, Peter P. (Emeritus) .... (859) 257-8884
Grulke, Eric A. .... (859) 257-2300x289
egrulke@engr.uky.edu
Hamrin, Charles E. Jr. (Emeritus) .... (859) 257-4959
hamrin@engr.uky.edu
Kermode, Richard I. .... (859) 257-2823
Morris, James G. (Emeritus) .... (859) 257-8090
Okazaki, Kenji (Emeritus) .... (859) 257-1307
Ray, Asit K. .... (859) 257-7990
akray@engr.uky.edu
Reucroft, Philip (Emeritus) .... (859) 257-8723
reuc@engr.uky.edu
Schrodt, J. Thomas (Emeritus) .... (859) 257-3969
che207@ukcc.uky.edu
Tsang, Tate T. H. (Chair) .... (859) 257-8059
tsang@engr.uky.edu

Associate Professors

Kalika, Douglass .... (859) 257-5507
kalika@engr.uky.edu
Knutson, Barbara L. .... (859) 257-5715
bknutson@engr.uky.edu
Rankin, Stephen .... (859) 257-9799
srankin@engr.uky.edu
Silverstein, David .... (270) 534-3132
silverdl@engr.uky.edu
Smart, Jimmy .... (270) 534-3119
jsmart@engr.uky.edu

Assistant Professors

Balk, John ....(859)257-4582
balk@engr.uky.edu
Dunbar, Paul .... (270) 534-3105
pdunbar@engr.uky.edu
Dziubla, Thomas ....(859)257-4063
dziubla@engr.uky.edu
Eitel, Richard .... (859)-257-5076
reitel@engr.uky.edu
Hilt, J. Zachary ....(859)257-9844
hilt@engr.uky.edu
Hinds, Bruce .... (859) 257-5507
bhind0@engr.uky.edu
Lee-DeSautels, Rhonda .... (270) 534-3122
rlee@engr.uky.edu
Nychka, John ....(859)257-0039
nychka@engr.uky.edu
Yang, Fuqian .... (859) 257-2994
fyang0@engr.uky.edu
Zhai, Tongguang .... (859) 257-4958
tzhai@engr.uky.edu

Other Faculty

Huffman, Gerald .... (859) 257-4029
cffls@pop.uky.edu
Huggins, Frank .... (859) 257-4045
frank@funky.cffls.uky.edu
Shah, Naresh .... (859) 257-5119
naresh@pop.uky.edu

Accredited by: ABET

Degrees granted 2005-2006:
    B.S.: 24   M.S.: 2   Ph.D.: 4

Graduate advisor: Dr. Barbara Knutson

Undergraduate advisor: Dr. Kim Anderson

Student organization: AIChE
Advisor: Dr. D. Bhattacharyya

Department reports to: Thomas W. Lester, Dean of Engineering

University of Louisville, J. B. Speed School of Engineering

Chemical Engineering Department
Ernst Hall 106
Louisville, KY 40292
Deliveries: 2301 South Third Street, Louisville, KY 40208

University phone	(502) 852-5555
Department phone	(502) 852-6347
Department FAX	(502) 852-6355

Professors

Collins, Dermot J. .... (502) 852-6737
dermot@louisville.edu
EV
Deshpande, Pradeep B. (Emeritus) .... (502) 852-0885
pbdesh01@louisville.edu
PR
Fleischman, Marvin (Emeritus) .... (502) 852-6347
m0flei01@louisville.edu
EV
Gerhard, Earl R. (Emeritus)
Harper, Dean O. (Emeritus) .... (502) 852-6347
doharp01@louisville.edu
PO
Kang, Kyung A. .... (502) 852-2094
kyung.kang@louisville.edu
BM,BT,NT
Laukhuf, W. L. S. (Associate Chair) ... (502) 852-6350
scrap@louisville.edu
PR,SE
Plank, Charles A. (Emeritus)
SE,SU
Ralston, Patricia A. S. .... (502) 852-0479
parals01@louisville.edu
MO,PR
Spencer, Hugh T. (Emeritus)
EV
Starr, Thomas L. (Associate Dean) .... (502) 852-1073
tom.starr@louisville.edu
AM,NT
Sunkara, Mahendra K. .... (502) 852-1558
mksunk01@louisville.edu
AM,EN,ME,NT,SU
Watters, James C. (Chair) .... (502) 852-0802
jim.watters@louisville.edu
EN,EV,PO

Assistant Professors

Berson, R. Eric .... (502) 852-1567
rebers01@louisville.edu
BT,EN
Willing, Gerold A. .... (502) 852-7860
G0Will05@Louisville.edu
AM,EN,NT,PO,SU

Accredited by: ABET

Degrees granted 2005-2006:
B.S.: 17   M.S.: 13   Ph.D.: 5

Graduate advisor: Kyung A. Kang

Undergraduate advisor: James C. Watters

Student organization: AIChE
Advisor: James C. Watters

Department reports to: Mickey R. Wilhelm, Dean (502) 852-6281

Placement service: Penny Hoerter (502) 852-0348

Louisiana

Louisiana State University

Chemical Engineering Department, Rm. 110
Baton Rouge, LA 70803-7303

University phone	(225) 578-5011
Department phone	(225) 578-1426
Department FAX	(225) 578-1476

Professors

Corripio, Armando B. (Emeritus) ..... (225) 578-3061
corripio@lsu.edu
Dooley, Kerry M. .... (225) 578-3063
dooley@lsu.edu
Griffin, Gregory L. .... (225) 578-3064
griffin@lsu.edu
Groves, Frank R. (Emeritus) .... (225) 578-3060
Harrison, Douglas P. (Emeritus) ..... (225) 578-3066
harrison@lsu.edu
Hjortso, Martin .... (225) 578-3058
hjortso@lsu.edu
Knopf, F. Carl .... (225) 578-1426
knopf@lsu.edu
McLaughlin, Edward (Emeritus) .... (225) 578-3546
Pike, Ralph W. .... (225) 578-3428
pike@lsu.edu
Reible, Danny (Emeritus)
Romagnoli, Jose A. .... (225) 578-1377
jose@lsu.edu
Sterling, Arthur (Emeritus) .... (225) 578-3057
sterling@lsu.edu
Thibodeaux, Louis J. .... (225) 578-3055
thibod@lsu.edu
Valsaraj, Kalliat T. (Chair) .... (225) 578-6522
valsaraj@lsu.edu

Associate Professors

Flake, John C. .... (225) 578-1426
Podlaha, Elizabeth J. .... (225) 578-3056
podlaha@lsu.edu
Spivey, James J. .... (225) 578-3690
jjspivey@lsu.edu
Thompson, Karsten E. .... (225) 578-3069
karsten@lsu.edu
Wetzel, David M. .... (225) 578-3071
dwetzel@lsu.edu
Wornat, Mary J. .... (225) 578-7509
mjwornat@lsu.edu

Assistant Professors

Henry, James .... (225) 578-1750
jehenry@lsu.edu

Instructors

Cygan, Margaret .... (225) 578-3062
cygan@lsu.edu
Lashover, Jacob .... (225) 578-3072
jacklash@lsu.edu

Mowrey, Daniel B. .... (225) 578-1321
dbmowrey@lsu.edu
Toups, Harry .... (225) 578-3068
hjtoups@lsu.edu

Accredited by: ABET

Degrees granted 2005-2006:
    B.S.: 54   M.S.: 5   Ph.D.: 3
Graduate advisor: Mary J. Wornat, Elizabeth Podlaha

Undergraduate advisor: Karsten E. Thompson

Student organization: AIChE
Advisor: James E. Henry

Department reports to: Zaki Bassiouni (225) 578-5731, Dean of Engineering

Placement service: Mary Feduccia (225) 578-2162

Louisiana Tech University

Chemical Engineering Department
P. O. Box 10348 T.S.
Ruston, LA 71272
Deliveries: 600 W. Arizona St., Bogard Hall, Room 257

University phone .... (318) 257-0211
University FAX 318-257-2928
Department phone .... (318) 257-2483
Department FAX .... (318) 257-2562

Associate Professors

Palmer, James (Chair) .... (318) 257-2885
    jpalmer@coes.latech.edu
    BT,EN,NT,RE,SE,TP

Assistant Professors

Gold, Scott .... (318) 257-5148
    sgold@latech.edu
    AM,EN,NT,SU
Mainardi, Daniela .... (318) 257-4714
    mainardi@latech.edu
    MO

Instructors

Behbahani, Ahmed (Visiting) .... (318) 257-3075
    sbehbaha@coes.latech.edu

Accredited by: ABET

Degrees granted 2005-2006:
    B.S.: 11   M.S.: 9   Ph.D.: 1

Graduate advisor: James Palmer

Undergraduate advisor: James Palmer

Student organization: AIChE
Advisor: Scott Gold

Department reports to: Stan Napper (318) 257-4647, Dean of Engineering and Science

Placement service: Cheryl B. Myers (318) 257-4336, Dean of Student Services
Louisiana Tech University, Ruston, LA 71272

McNeese State University

Engineering Department
P.O. Box 91735
Lake Charles, LA 70609-1735
Deliveries: 143 Drew Hall

University phone .... (337) 475-5000
Department phone .... (337) 475-5874
Department FAX .... (337) 475-5286

Professors

Barkat, Omar .... (337) 475-5817
    obarkat@mail.mcneese.edu
Griffith, D. John .... (337) 475-5858
    griffith@mail.mcneese.edu
    PR,TH

Associate Professors

Robinson, Richard L. .... (337) 475-5250
    rrobin@mail.mcneese.edu

Assistant Professors

Sullivan, Jonathan .... (337) 475-5859
    jsullivan@mail.mcneese.edu
    AM

Accredited by: ABET (general engineering curriculum)

Student organization: AIChE
Advisor: Dr. Jonathon Sullivan

Department reports to: Nikos Kiritsis, Dean, College of Engineering and Technology

Placement service: Kathy Bond

University of Louisiana at Lafayette

Chemical Engineering Department
P.O. Box 44130
Lafayette, LA 70504-4130
Deliveries: Madison Hall, Room 217A, 131 Rex Street

University phone	(337) 482-1000
Department phone	(337) 482-6562
Department FAX	(337) 482-1220

Professors

Farshad, Fred F. .... (337) 482-5862
    TP
Garber, James D. (Head) .... (337) 482-6151
    garber@louisiana.edu
    AM,MO
Misra, Devesh .... (337) 482-6430
    AM,NT
Ponter, Anthony B. .... (337) 482-5350
    SU
Reinhardt, James R. .... (337) 482-5351
    MO
Zappi, Mark E. (Dean) .... (337) 482-6685
    BT

Assistant Professors

Chirdon, William .... (337) 482-6564
    AM,HT,PO

Accredited by: ABET

Degrees granted 2005-2006:
    B.S.: 16   M.S.: 12

Graduate advisor: Devesh Misra

Student organization: AIChE
Advisor: J.R. Reinhardt

Department reports to: Mark E. Zappi, (337) 482-6685, Dean of Engineering

Placement service: Kimberly Billeaudeau (337) 262-5300

Tulane University

Department of Chemical and Biomolecular Engineering
New Orleans, LA 70118
Deliveries: 300 Lindy Boggs Building

University phone	(504) 862-8000
Department phone	(504) 865-5772
Department FAX	(504) 865-6744

Professors

De Kee, Daniel .... (504) 865-5620
    ddekee@tulane.edu
Gonzalez, Richard (Emeritus) .... (504) 865-5741
    richard.gonzalez@tulane.edu
John, Vijay T. (Chair) .... (504) 865-5883
    vj@tulane.edu
Law, Victor J. .... (504) 865-5773
    law@tulane.edu
Lu, Yunfeng .... (504) 865-5827
    ylu@tulane.edu
Mitchell, Brian S. .... (504) 862-8257
    brian@tulane.edu
O'Connor, Kim C. .... (504) 865-5740
    koc@tulane.edu
Papadopoulos, Kyriakos D. .... (504) 865-5826
    kyriakos@tulane.edu

Associate Professors

Sullivan, Samuel L. Jr. (Emeritus) ..... (504) 865-5772

Assistant Professors

Ashbaugh, Henry S. .... (504) 862-8258
    hanka@tulane.edu
Godbey, W T. .... (504) 865-5872
    godbey@tulane.edu

Accredited by: ABET

Degrees granted 2005-2006:
    B.S.: 15   M.S.: 4   Ph.D.: 2

Graduate advisor: Kim O'Connor

Undergraduate advisor: John Prindle, Jr.

Student organization: AIChE
Advisor: John Prindle, Jr.

Department reports to: Nicholas J. Altiero (504) 865-5764, Dean of Engineering

Placement service: John Prindle, Jr. (504) 865-5774

Maine

University of Maine

Chemical and Biological Engineering Department
5737 Jenness Hall
Orono, ME 04469-5737
Deliveries: 5737 Jenness Hall

University phone .... (207) 581-1110
Department phone .... (207) 581-2276
Department FAX .... (207) 581-2323

Professors

Bousfield, Douglas W. .... (207) 581-2300
  bousfld@maine.edu
  SU,TP
Donahue, Darrell .... (207) 581-2728
  darrell_donahue@umit.maine.edu
  BM,MO,PR
Genco, Joseph M. .... (207) 581-2284
  genco@maine.edu
Mumme, Kenneth (Emeritus)
  mumme@maine.edu
Pendse, Hemant (Chair) .... (207) 581-2290
  pendse@maine.edu
  EN,RE
Ruthven, Douglas M. .... (207) 581-2283
  druthven@umche.maine.edu
Thompson, Edward V. (Emeritus)
  ethompson@umche.maine.edu
van Heiningen, Adriaan .... (207) 581-2278
  avanheiningen@umche.maine.edu
  MO

Associate Professors

Co, Albert .... (207) 581-2282
  albertco@maine.edu
  PO,TP
DeSisto, William .... (207) 581-2291
  wdesisto@umche.maine.edu
  AM,NT
Hwalek, John J. .... (207) 581-2302
  hwalek@maine.edu
  HT,PR

Assistant Professors

Ashworth, Sharon (Research) .... (207) 581-2277
  sashworth@umche.maine.edu
  BT
Mason, Michael .... (207) 581-2344
  mmason@umche.maine.edu
  NT
Millard, Paul .... (207) 581-2265
  paul.millard@maine.edu
  BT
Neivandt, David .... (207) 581-2288
  dneivandt@umche.maine.edu

Nohe, Anja .... (207) 581-2270
  anohe@umche.maine.edu
  BM,BT,NT
Wheeler, Clayton .... (207) 581-2280
  cwheeler@umche.maine.edu

Accredited by: ABET

Degrees granted 2005-2006:
  B.S.: 25   M.S.: 1   Ph.D.: 2

Student organization: AIChE
Advisor: John J. Hwalek

Department reports to: Dana Humphrey, Interim Dean of Engineering

Placement service: Patricia Counihan (207) 581-1359

Maryland

Johns Hopkins University

Department of Chemical and Biomolecular Engineering
Johns Hopkins University
3400 N. Charles Street
221 Maryland Hall
Baltimore, MD 21218

University phone .... (410) 516-8000
Department phone .... (410) 516-7170
Department FAX .... (410) 516-5510

Professors

Betenbaugh, Michael J. .... (410) 516-5461
beten@jhu.edu
BM,BT
Donohue, Marc D. (Associate Dean) ... (410) 516-7761
mdd@jhu.edu
EN,MO,SE,SU
Katz, Joseph L. .... (410) 516-8484
jlk@jhu.edu
AM,EN,NT,SU
Stebe, Kathleen J. (Chair) .... (410) 516-7769
kjs@jhu.edu
MO,NT,SU
Wirtz, Denis .... (410) 516-7006
wirtz@jhu.edu
BM,BT,NT,PO

Associate Professors

Hanes, Justin .... (410) 516-3484
hanes@jhu.edu
AM,BM,BT,NT,PO
Konstantopoulos, Konstantinos .... (410) 516-6290
konst_k@jhu.edu
BM,BT

Assistant Professors

Asthagiri, Dilip .... (410) 516-3475
dilipa@jhu.edu
MO
Drazer, German M. .... (410) 516-0170
drazer@jhu.edu
BT,MO,NT,SE,SU
Frechette, Joelle .... (410) 516-0113
jfrechette@jhu.edu
AM,NT,SU
Gracias, David .... (410) 516-5284
dgracias@jhu.edu
AM,BM,EN,ME,NT,PO,SU
Gray, Jeffrey .... (410) 516-5313
jgray@jhu.edu
BM,BT,MO,SU
Ostermeier, Marc .... (410) 516-7144
oster@jhu.edu
BM,BT,NT

Other Faculty

Aranovich, Gregory (Research) .... (410) 516-7079
aranovich@jhu.edu
Erlebacher, Jonah (Joint) .... (410) 516-6077
Jonah.Erlebacher@jhu.edu
Gaddy, Glen (Research) .... (410) 516-8747
ggaddy@jhu.edu
Hoh, Jan (Joint) .... (410) 614-3795
jhoh@jhmi.edu
Karweit, Michael (Research) .... (410) 516-8415
mjk@jhu.edu
Krambeck, Fredrick (Research) .... (410) 516-7158
fjkrambeck@jhu.edu
Lai, Eva (Research)
chengineer@yahoo.com
Leong, Kam W. (Joint) .... (410) 614-5644
kleong@bme.jhu.edu
Paulaitis, Michael (Research) .... (410) 516-6604
michaelp@jhu.edu
Searson, Peter (Joint) .... (410) 516-8774
searson@jhu.edu

Accredited by: ABET

Degrees granted 2005-2006:
    B.S.: 20   M.S.: 6   Ph.D.: 10

Graduate advisor: Konstantinos Konstantopoulos

Undergraduate advisor: Marc Ostermeier

Student organization: AIChE
Advisor: Jeffrey Gray

Department reports to: Nicholas Jones, Dean of Engineering

Placement service: Patricia Matteo, Director, Career Planning and Development

University of Maryland, College Park

Chemical and Biomolecular Engineering Department
College Park, MD 20742-2111
Deliveries: Chemical & Nuclear Engineering Building, Room 2113

University phone .... (301) 405-1000
Department phone .... (301) 405-1935
Department FAX .... (301) 405-0523

Professors

Anisimov, Mikhail A. .... (301) 405-8049
anisimov@eng.umd.edu
Barbari, Timothy A. .... (301) 405-2983
barbari@eng.umd.edu
Bentley, William E. .... (301) 405-1906
bentley@eng.umd.edu
Calabrese, Richard V. .... (301) 405-1908
rvc@eng.umd.edu
Choi, Kyu Yong .... (301) 405-1907
choi@eng.umd.edu
Gentry, James W. (Emeritus) .... (301) 405-1915
gentry@eng.umd.edu
Gomezplata, Albert (Emeritus)
Greer, Sandra C. .... (301) 405-1895
sgreer@umd.edu
McAvoy, Thomas J. (Emeritus) .... (301) 405-1939
mcavoy@eng.umd.edu
Regan, Thomas M. (Emeritus) .... (301) 405-1936
tregan@eng.umd.edu
Sengers, Jan V. (Emeritus) .... (301) 405-4805
js45@umail.umd.edu
Smith, Theodore G. (Emeritus) .... (301) 405-1918
tgsmith@eng.umd.edu
Weigand, William A. .... (301) 405-1916
weigand@eng.umd.edu

Associate Professors

Adomaitis, Raymond A. (Chair) ...... (301) 405-2969
adomaiti@umd.edu
Ehrman, Sheryl H. .... (301) 405-1917
sehrman@eng.umd.edu
Kofinas, Peter .... (301) 405-7335
kofinas@eng.umd.edu
Wang, Nam Sun .... (301) 405-1910
nsw@eng.umd.edu
Zafiriou, Evanghelos .... (301) 405-6625
zafiriou@isr.umd.edu

Assistant Professors

Aranda-Espinoza, Helim .... (301) 405-8250
helim@umd.edu
Dimitrakopoulos, Panagiotis .... (301) 405-8166
dimitrak@eng.umd.edu
Fisher, John P. .... (301) 405-7475
jpfisher@eng.umd.edu
Klapa, Maria I. .... (301) 405-1320
mklapa@eng.umd.edu
Raghavan, Srinivasa R. .... (301) 405-8164
sraghava@eng.umd.edu

Other Faculty

DiMarzio, Edmund A. (Adjunct)
Ranade, Madhav B. (Adjunct)
Wesson, Rosemarie D. (Adjunct)
Yang, Arthur J.-M. (Adjunct)

Accredited by: ABET

Degrees granted 2004-2005:
    B.S.: 19   M.S.: 8   Ph.D.: 4

Graduate advisor: Sheryl H. Ehrman

Undergraduate advisor: Nam S. Wang

Student organization: AIChE
Advisor: Raymond A. Adomaitis

Department reports to: Nariman Farvardin (301) 405-3868, Dean of Engineering

Placement service: Heidi Sauber (301) 405-3863, Director of Engineering Career Services

University of Maryland, Baltimore County

Department of Chemical and Biochemical Engineering
1000 Hilltop Circle
ECS 101
Baltimore, MD 21250
Deliveries: UMBC/CBE

University phone .... (410) 455-1000
Department phone .... (410) 455-3400
Department FAX .... (410) 455-1049

Professors

Bayles, Taryn .... (410) 455-3428
tbayles@umbc.edu
TP
Frey, Douglas D. .... (410) 455-3400
dfrey1@umbc.edu
BT,SE
Moreira, Antonio R. .... (410) 455-6576
moreira@umbc.edu
BT
Payne, Gregory F. .... (410) 455-3413
payne@research.umbc.edu
BT,SE
Rao, Govind .... (410) 455-3415
grao@umbc.edu
BM,BT,RE

Associate Professors

Good, Theresa .... (410) 455-3405
tgood@umbc.edu
BM,NT
Marten, Mark .... (410) 455-3439
marten@umbc.edu
BT
Ross, Julia M. (Chair) .... (410) 455-3414
jross@umbc.edu
BM,TP

Assistant Professors

Castellanos, Mariajose .... (410) 455-8151
mariaca@umbc.edu
MO
Leach, Jennie .... (410) 544-3408
jleach@umbc.edu
AM,BM,NT

Accredited by: ABET

Degrees granted 2005-2006:
 B.S.: 30   M.S.: 7   Ph.D.: 2

Graduate advisor: Theresa Good

Undergraduate advisor: Julia Ross and Taryn Bayles

Student organization: AIChE
Advisor: Theresa Good

Department reports to: Warren R. DeVries, Ph.D., Dean of Engineering (410) 455-3270

Placement service: Betty J. Glascoe (410) 455-2069, Director

Massachusetts

University of Massachusetts Amherst

159 Goessmann Lab
Chemical Engineering Department
Amherst, MA 01003-9303
Deliveries: 686 North Pleasant Street

University phone .... (413) 545-0111
Department phone .... (413) 545-2507
Department FAX .... (413) 545-1647

Professors

Conner, W. Curtis Jr. .... (413) 545-0316
    wconner@ecs.umass.edu
Douglas, James M. (Emeritus) .... (413) 545-2252
    cheweb@ecs.umass.edu
Henson, Michael A. .... (413) 545-3481
    henson@ecs.umass.edu
Hoagland, David A. (Adjunct) .... (413) 577-1513
    dah@neurotica.pse.umass.edu
Laurence, Robert L. (Emeritus) .... (413) 545-0470
    rlaurence@ecs.umass.edu
Malone, Michael F. (Dean) .... (413) 545-2359
    mmalone@ecs.umass.edu
Maroudas, Dimitrios .... (413) 545-3617
    maroudas@ecs.umass.edu
Monson, Peter A. .... (413) 545-0661
    monson@ecs.umass.edu
Mountziaris, T.J. (Head) .... (413) 545-2359
    tjm@ecs.umass.edu
Santore, Maria M. (Adjunct) .... (413) 577-1417
    santore@mail.pse.umass.edu
Watkins, James J. (Adjunct) .... (413) 545-2569
    watkins@ecs.umass.edu
Westmoreland, Phillip R. .... (413) 545-1750
    westm@ecs.umass.edu
Winter, H. Henning .... (413) 545-0922
    winter@ecs.umass.edu

Associate Professors

Auerbach, Scott M. (Adjunct) .... (413) 545-1240
    auerbach@chem.umass.edu
Bhatia, Surita R. .... (413) 545-0096
    sbhatia@ecs.umass.edu
Ford, David M. .... (413) 577-0134
    ford@ecs.umass.edu
Roberts, Susan C. .... (413) 545-1660
    sroberts@ecs.umass.edu

Assistant Professors

Davis, Jeffrey M. .... (413) 545-2916
    jmdavis@ecs.umass.edu
Forbes, Neil S. .... (413) 577-0132
    forbes@ecs.umass.edu
Huber, George W. .... (413) 545-2507
    huber@ecs.umass.edu
Sun, Lianhong .... (413) 545-6143
    lsun@ecs.umass.edu

Accredited by: ABET

Degrees granted 2005-2006:
    B.S.: 24   M.S.: 4   Ph.D.: 7

Graduate advisor: Dimitrios Maroudas

Undergraduate advisor: Surita Bhatia

Student organization: AIChE
Advisor: Neil S. Forbes

Department reports to: Michael F. Malone (413) 545-0300, Dean of Engineering

Placement service: (413) 545-0614, Cheryl Brooks

University of Massachusetts-Lowell

Department of Chemical Engineering
1 University Avenue
Lowell, MA 01854
Deliveries: One University Avenue

University phone	(978) 934-4000
Department phone	(978) 934-3171
Department FAX	(978) 934-3047

Professors

Bonner, Francis J. .... (978) 934-3154
Francis_Bonner@uml.edu
Brown, Gilbert Jay .... (978) 934-3166
Gilbert_Brown@uml.edu
Donatelli, Alfred Anthony (Chair) ..... (978) 934-3156
Alfred_Donatelli@uml.edu
Higgins, Charles J. (Emeritus) .... (978) 934-3171
Phelps, James Parkhurst (Emeritus) ... (978) 934-3171
Sama, Dominick Anthony (Emeritus) .. (978) 934-3171
Dominick_Sama@uml.edu
Sheff, James Robert .... (978) 934-3169
James_Sheff@uml.edu
Vasilos, Thomas (Emeritus) .... (978) 934-3171
Thomas_Vasilos@uml.edu
Vedula, Krishna .... (978) 934-2737
Krishna_Vedula@uml.edu
Walkinshaw, John W. .... (978) 934-3159
John_Walkinshaw@uml.edu
White, John Robert .... (978) 934-3165
John_White@uml.edu

Associate Professors

Flood, H. William (Emeritus) .... (978) 934-3171
Lawton, Carl W. .... (978) 934-3158
Carl_Lawton@uml.edu

Accredited by: ABET

Degrees granted 2005-2006:

    B.S.: 4   M.S.: 7

Graduate advisor: Alfred Donatelli

Student organization: AIChE, ANS, TAPPI, OXE, Alpha Nu Sigma, ISPE
Advisor: John W. Walkinshaw(AIChE), Gilbert J. Brown (ANS, Alpha Nu Sigma), John W. Walkinshaw (TAPPI), Alfred A. Donatelli (Omega Chi Epsilon), Carl Lawton(ISPE)

Department reports to: John Ting (978) 934-2576, Dean of Engineering

Placement service: Patricia Yates (978) 934-2355, Director of Career Services

Massachusetts Institute of Technology

Department of Chemical Engineering
77 Massachusetts Avenue, 66-350
Cambridge, MA 02139-4307
Deliveries: 25 Ames Street, 66-350, Cambridge, MA 02142
University phone .... (617) 253-1000
University FAX .... (617) 253-3124
Department phone .... (617) 253-4561
Department FAX .... (617) 258-8992

Professors
Armstrong, Robert C. (Head) .... (617) 253-4581
  rca@mit.edu
Barton, Paul I. .... (617) 253-6526
  pib@mit.edu
Beér, János M. (Emeritus) .... (617) 253-6661
  jmbeer@mit.edu
Blankschtein, E. Daniel .... (617) 253-4594
  dblank@mit.edu
Brenner, Howard .... (617) 253-6687
  hbrenner@mit.edu
Chakraborty, Arup
Cohen, Robert E. .... (617) 253-3777
  recohen@mit.edu
Colton, Clark K. .... (617) 253-4585
  ckcolton@mit.edu
Cooney, Charles L. .... (617) 253-5066
  ccooney@mit.edu
Deen, William A. .... (617) 253-4535
  wmdeen@mit.edu
Gast, Alice P. .... (617) 253-1403
  gast@mit.edu
Gleason, Karen K. .... (617) 253-3108
  kkg@mit.edu
Hatton, T. Alan .... (617) 253-4588
  tahatton@mit.edu
Howard, Jack B. (Emeritus) .... (617) 253-4574
  jbhoward@mit.edu
Jensen, Klavs F. .... (617) 253-4589
  kfjensen@mit.edu
Karel, Marcus (Emeritus) .... (617) 253-6744
  karel@mit.edu
Langer, Robert S. .... (617) 253-3107
  rlanger@mit.edu
Lauffenburger, Douglas A. .... (617) 252-1629
  lauffen@mit.edu
McRae, Gregory J. .... (617) 253-6564
  mcrae@mit.edu
Merrill, Edward W. (Emeritus) .... (617) 253-4593
  emerrill@mit.edu
Reid, Robert C. (Emeritus) .... (617) 253-4571
  bobcreid@aol.com
Rutledge, Gregory C. (Associate Head) (617) 253-4561
  rutledge@mit.edu
Satterfield, Charles N. (Emeritus) ..... (617) 253-4584
  cnsatter@mit.edu
Sawin, Herbert H. .... (617) 253-4570
  hhsawin@mit.edu
Smith, Kenneth A. .... (617) 253-1973
  kas@mit.edu
Stephanopoulos, George .... (617) 253-3904
  geosteph@mit.edu
Stephanopoulos, Gregory .... (617) 253-4583
  gregstep@mit.edu
Tester, Jefferson W. .... (617) 253-3401
  testerel@mit.edu
Wang, Daniel I. C. .... (617) 253-2126
  dicwang@mit.edu
Wittrup, K. Dane .... (617) 253-4578
  wittrup@mit.edu

Associate Professors
Green, William H. .... (617) 253-4580
  whgreen@mit.edu
Hammond, Paula T. .... (617) 258-7577
  hammond@mit.edu
Trout, Bernhardt L. .... (617) 258-5021
  trout@mit.edu
Virk, Preetinder S. .... (617) 253-3177
  psvirk@mit.edu

Assistant Professors
Beers, Kenneth J. .... (617) 258-8986
  kbeers@mit.edu
Doyle, Patrick S. .... (617) 253-4534
  pdoyle@mit.edu
Prather, Kristala J. .... (617) 253-1950
  kljp@mit.edu

Other Faculty
Dalzell, William H. .... (617) 253-5273
  wdalzell@mit.edu
Johnston, Barry S. .... (617) 253-6600
  bsjohnst@mit.edu

Accredited by: ABET

Degrees granted 2005-2006:
  B.S.: 57   M.S.: 40   Ph.D.: 32

Graduate advisor: Daniel Blankschtein

Undergraduate advisor: Barry Johnston

Student organization: AIChE
Advisor: Barry Johnston

Department reports to: Thomas L. Magnanti, Dean of Engineering, (617) 253-6604

Placement service: Elizabeth Reed (617) 253-4733, Director, Career Planning and Placement

Northeastern University

Chemical Engineering Department - 342SN
360 Huntington Ave.
Boston, MA 02115
Deliveries: 342 Snell Engineering Center

University phone	(617) 373-2000
Department phone	(617) 373-2989
Department FAX	(617) 373-2209

Professors

Barabino, Gilda A. .... (617) 373-3900
g.barabino@neu.edu
BM,BT

Sacco Jr., Al (Chair) .... (617) 373-7912
asacco@coe.neu.edu
AM,NT,RE,SE,SU

Willey, Ronald J. .... (617) 373-3962
r.willey@neu.edu
AM,RE

Williams, John A. (Emeritus)
Wise, Donald L. (Emeritus)
BM

Associate Professors

Buonopane, Ralph A. (Emeritus) ...... (617) 373-2996
r.buonopane@neu.edu

Goodwin, Bernard M. (Emeritus) ..... (617) 373-2990
b.goodwin@neu.edu

Lee, Carolyn W. T. .... (617) 373-3634
clee@coe.neu.edu
BM,BT

Stewart, Richard R. (Emeritus) .... (617) 373-2990

Thorgerson, Eric (Adjunct) .... (617) 373-2991
e.thorgerson@neu.edu

Assistant Professors

Burkey, Daniel .... (617) 373-2989
d.burkey@neu.edu
AM,PO,SU

Carrier, Rebecca .... (617) 373-2989
r.carrier@neu.edu
BM,BT

Murthy, Shashi .... (617) 373-4017
s.murthy@neu.edu
AM,BM,PO,SU

Zeimer, Katherine S. .... (617) 373-2990
k.zeimer@neu.edu
AM,ME,NT,SU

Instructors

Satvat, Behrooz (Joint) .... (617) 373-3461
b.satvat@neu.edu

Other Faculty

Gutoff, Edgar B. (Adjunct) .... (617) 373-2989
ebgutof@coe.neu.edu

Manning, Michael (Adjunct) .... (617) 373-2991
mmanning@coe.neu.edu

Accredited by: ABET

Degrees granted 2005-2006:
    B.S.: 34   M.S.: 2   Ph.D.: 2

Graduate advisor: Al Sacco, Jr.

Undergraduate advisor: M. Bates

Student organization: AIChE
Advisor: K. S. Ziemer

Department reports to: Allen Soyster (617) 373-2685, Dean of Engineering

Placement service: Behrooz Satvat (617) 373-3461

Tufts University

Chemical & Biological Engineering Department
4 Colby Street, Room 148
Medford, MA 02155

University phone	(617) 628-5000
Department phone	(617) 627-3900
Department FAX	(617) 627-3991

Professors

Botsaris, Gregory B. (Emeritus) .... (617) 627-3445
gregory.botsaris@tufts.edu
SE,SU
Flytzani-Stephanopoulos, Maria .... (617) 627-3048
mflytzan@tufts.edu
AM,EN,NT
Georgakis, Christos (Chair) .... (617) 627-2573
Christos.Georgakis@tufts.edu
BT,MO,RE
Juda, Walter (Adjunct) .... (617) 627-2786
WJuda@hy9corp.com
RE
Kaplan, David .... (617) 627-3251
david.kaplan@tufts.edu
BM,BT
Saltsburg, Howard (Research) .... (617) 627-5658
howard.saltsburg@tufts.edu
EV,RE
Sung, Nakho .... (617) 627-3447
nsung@tufts.edu
PO
VanWormer, Kenneth A. .... (617) 627-2598
kenneth.vanwormer@tufts.edu

Associate Professors

Edwards, Aurelie (Research) .... (617) 627-3731
aurelie.edwards@tufts.edu
BT,MO
Kelley, Brian (Adjunct) .... (617) 627-3900
bkelley@wyeth.com
Meldon, Jerry .... (617) 627-3570
jerry.meldon@tufts.edu
PR,SE,TH
Ryder, Daniel .... (617) 627-3446
daniel.ryder@tufts.edu
SE,SU

Assistant Professors

Lee, Kyongbum .... (617) 627-4323
kyongbum.lee@tufts.edu
BT,RE
Pfeifer, Blaine .... (617) 627-2582
blaine.pfeifer@tufts.edu
BT
Yi, Hyunmin .... (617) 627-2195
hyunmin.yi@tufts.edu
BT,NT

Instructors

Anderson, Erik .... (617) 627-2784
Gyure, Dale C. .... (617) 627-2784
105365.2054@compuserve.com
Haghgooie, Ramin .... (617) 627-3900
Patel, Navin .... (617) 627-2784
navin.patel@tufts.edu

Accredited by: ABET

Degrees granted 2005-2006:
 B.S.: 24   M.S.: 16   Ph.D.: 6

Graduate advisor: Prof. Nak Ho Sung

Student organization: AIChE
Advisor: Blaine Pfeifer

Department reports to: Linda Abriola (617) 627-3237, Dean of Engineering

Placement service: Maureen Sakakeeny (617) 627-3299

United States

Worcester Polytechnic Institute

Chemical Engineering Department
100 Institute Road
Worcester, MA 01609-2280
Deliveries: Goddard Hall, Room 127

University phone	(508) 831-5000
Department phone	(508) 831-5250
Department FAX	(508) 831-5853

Professors

Datta, Ravindra .... (508) 831-6036
    rdatta@wpi.edu
Dixon, Anthony G. .... (508) 831-5350
    agdixon@wpi.edu
Ma, Yi Hua .... (508) 831-5398
    yhma@wpi.edu
Moser, William (Emeritus)
Thompson, Robert W. .... (508) 831-5525
    rwt@wpi.edu
Wagner, Robert (Emeritus) .... (508) 831-5250
Weiss, Alvin (Emeritus) .... (508) 831-5250

Associate Professors

Camesano, Terri A. .... (508) 831-5380
    terric@wpi.edu
Clark, William M. .... (508) 831-5259
    wmclark@wpi.edu
DiBiasio, David (Head) .... (508) 831-5372
    dibiasio@wpi.edu
Kazantzis, Nikolaos K. .... (508) 831-5666
    nikolas@wpi.edu

Assistant Professors

Nowick, Henry (Adjunct) .... (508) 831-5250
Starr, Thomas (Adjunct) .... (508) 831-5250
Wilcox, Jennifer L. .... (508) 831-5493
    jwilcox@wpi.edu
Zhou, H. Susan .... (508) 831-5275
    szhou@wpi.edu

Accredited by: ABET

Degrees granted 2005-2006:
    B.S.: 27   M.S.: 4   Ph.D.: 3

Student organization: AIChE
Advisor: Professor William Clark

Department reports to: Carol Simpson, Provost

Placement service: Yvonne V. Harrison (508) 831-5260, Director

Michigan

University of Michigan

The Herbert H. Dow Building
2300 Hayward St., Room 3074
Ann Arbor, MI 48109-2136

University phone .... (734) 764-2383
Department phone .... (734) 764-2383
Department FAX .... (734) 763-0459

Professors

Briggs, Dale E. (Emeritus) .... (734) 763-1331
briggs@umich.edu
Burns, Mark A. .... (734) 764-4315
maburns@umich.edu
BM,BT,ME,SE
Carnahan, Brice (Emeritus) .... (734) 764-3366
carnahan@umich.edu
MO,PR
Curl, Rane (Emeritus) .... (734) 764-3489
ranecurl@umich.edu
Donahue, Francis (Emeritus) .... (734) 764-4304
fdonahue@umich.edu
Fogler, H. Scott .... (734) 763-1361
sfogler@umich.edu
BM,EN,PR
Glotzer, Sharon C. .... (734) 615-6296
sglotzer@umich.edu
AM,EN,MO,NT,PO
Gulari, Erdogan .... (734) 763-5941
gulari@umich.edu
BT,EN
Kadlec, Robert H. (Emeritus) .... (734) 764-3362
rhkad@umich.edu
Larson, Ronald G. (Chair) .... (734) 936-0772
rlarson@umich.edu
BT,MO,PO,TP
Linderman, Jennifer J. .... (734) 763-5253
linderma@umich.edu
BM,MO
Powers, John E. (Emeritus) .... (801) 259-5060
Savage, Phillip E. .... (734) 764-3386
psavage@umich.edu
EN,EV,RE
Schwank, Johannes W. .... (734) 764-3374
schwank@umich.edu
AM,EN,ME,MO,NT,RE,SU
Thompson, Levi T. .... (734) 936-2015
ltt@umich.edu
Wang, Henry Y. .... (734) 763-5659
hywang@umich.edu
Weber, Walter J. Jr. .... (734) 763-1464
wjwjr@umich.edu
Wilkes, James O. (Emeritus) .... (734) 764-3378
wilkes@umich.edu
Yang, Ralph T. .... (734) 936-0771
yang@umich.edu
EN,EV,NT,SE,SU
Young, Edwin H. (Emeritus) .... (734) 764-4316
young@umich.edu
Ziff, Robert M. .... (734) 764-5498
rziff@umich.edu

Associate Professors

Kotov, Nicholas .... (734) 763-8768
kotov@umich.edu
Solomon, Michael J. .... (734) 764-3119
mjsolo@umich.edu

Assistant Professors

Eniola-Adefeso, Omolola .... (734) 764-2383
BM,BT,PO
Kim, Jingsang (Joint) .... (734) 936-4681
jingsang@umich.edu
Lahann, Joerg .... (734) 763-7543
lahann@umich.edu
Linic, Suljo .... (734) 647-7984
linic@umich.edu
Mayer, Michael (Joint) .... (734) 763-4609
mimayer@umich.edu
AM,BM,BT,EN,ME,NT,SE,SU
Violi, Angela (Joint) .... (734) 615-6448
avioli@engin.umich.edu
BM,EN,MO,NT
Woolf, Peter .... (734) 647-7985
pwoolf@umich.edu
BM,BT,MO

Instructors

Barkel, Barry (Adjunct) .... (734) 647-3093
bmbarkel@umich.edu
EN,EV,PR
Montgomery, Susan M. .... (734) 936-1890
smontgom@umich.edu

Accredited by: ABET

Degrees granted 2005-2006:

M.S.: 23   Ph.D.: 17

Graduate advisor: Robert Ziff

Undergraduate advisor: Susan Montgomery

Student organization: AIChE
Advisor: Sharon Glotzer

Department reports to: Dean College of Engineering

Placement service: Cynthia Redwine, Director (734) 647-7168, 230 Chrysler Center, 2121 Bonisteel Blvd. N. Campus

Michigan State University

Chemical Engineering & Materials Science Department
East Lansing, MI 48824-1226
Deliveries: 2527 Engineering Building
University phone .... (517) 355-1855
Department phone .... (517) 355-5135
Department FAX .... (517) 432-1105

Professors
Berglund, Kris A. .... (517) 353-4565
    berglund@cem.msu.edu
Case, Eldon .... (517) 353-6715
    casee@egr.msu.edu
Crimp, Martin .... (517) 355-0294
    crimp@egr.msu.edu
Dale, Bruce E. .... (517) 353-6777
    bdale@egr.msu.edu
Drzal, Lawrence T. (Director) .... (517) 353-7759
    drzal@egr.msu.edu
Grummon, David .... (517) 353-4688
    grummon@egr.msu.edu
Hawley, Martin C. (Chair) .... (517) 355-5135
    hawley@egr.msu.edu
Jayaraman, Krishnamurthy .... (517) 355-5138
    jayarama@egr.msu.edu
Mackay, Michael .... (517) 432-4495
    mackay@egr.msu.edu
Miller, Dennis J. .... (517) 353-3928
    millerd@egr.msu.edu
Narayan, Ramani .... (517) 432-0775
    narayan@msu.edu
Petty, Charles A. .... (517) 353-5486
    petty@egr.msu.edu
Subramanian, K.N. .... (517) 353-5397
    subraman@egr.msu.edu
Worden, R. Mark .... (517) 353-9015
    worden@egr.msu.edu

Associate Professors
Baumann, Melissa .... (517) 432-1243
    mbaumann@egr.msu.edu
Bieler, Thomas .... (517) 353-9767
    bieler@egr.msu.edu
Briedis, Daina M. .... (517) 353-3861
    briedis@egr.msu.edu
Chan, Christina .... (517) 432-4530
    krischan@egr.msu.edu
Lee, Andre .... (517) 355-5112
    leea@egr.msu.edu
Lira, Carl T. .... (517) 355-9731
    lira@egr.msu.edu
Lucas, James .... (517) 432-2883
    lucas@egr.msu.edu
Ofoli, Robert Y. .... (517) 432-1575
    ofoli@egr.msu.edu

Assistant Professors
Boehlert, Carl .... (517) 353-3703
    boehlert@egr.msu.edu
Calabrese Barton, Scott .... (517) 355-5135
    scb@egr.msu.edu
    BT,EN
Lee, Ilsoon .... (517) 355-9291
    leeil@egr.msu.edu
Walton, S. Patrick .... (517) 432-8733
    spwalton@egr.msu.edu

Other Faculty
Bender, Timothy .... (517) 355-3444
    bendert@egr.msu.edu

Accredited by: ABET

Degrees granted 2005-2006:
    B.S.: 45   M.S.: 3   Ph.D.: 8

Undergraduate advisor: Cynthia Sarver

Student organization: AIChE
Advisor: Robert Y. Ofoli

Department reports to: Satish Udpa, Acting Dean of Engineering

Placement service: Jim Novak, Director

Michigan Technological University

Department of Chemical Engineering
1400 Townsend Drive
Houghton, MI 49931-1295
Deliveries: Adminstration Building Room G31

University phone .... (906) 487-1885
Department phone .... (906) 487-3132
Department FAX .... (906) 487-3213

Professors

Barna, Bruce A. .... (906) 487-2569
bbarna@mtu.edu
EN,PR
Crowl, Daniel A. .... (906) 487-3221
crowl@mtu.edu
PR
Kawatra, S. Komar .... (906) 487-2064
skkawatr@mtu.edu
SE,SU
Mullins, Michael E. (Chair) .... (906) 487-3132
memullin@mtu.edu
AM,NT,RE
Shonnard, David R. .... (906) 487-3468
drshonna@mtu.edu
BT,EN,EV

Associate Professors

Caneba, Gerard T. .... (906) 487-2051
caneba@mtu.edu
AM,EN,ME,NT,PO
Co, Tomas B. .... (906) 487-2144
tbco@mtu.edu
MO,PR
Keith, Jason M. .... (906) 487-2106
jmkeith@mtu.edu
AM,EN,EV,MO,PO,RE
King, Julia A. .... (906) 487-3106
jaking@mtu.edu
PO
Morrison, Faith A. .... (906) 487-2050
fmorriso@mtu.edu
AM,PO
Nesbitt, Carl .... (906) 487-2796
cnesbitt@mtu.edu
AM,BT,EN,EV,SE,SU
Rogers, Tony N. .... (906) 487-2210
tnrogers@mtu.edu
EV,MO,PR,SE
Sandell, John .... (906) 487-2557
jfsandel@mtu.edu
EV

Assistant Professors

Holles, Joseph H. .... (906) 487-1956
jhholles@mtu.edu
RE

Instructors

Clancey, M. Sean .... (906) 487-3338
msclance@mtu.edu

Other Faculty

Fisher, Edward R. (Emeritus)
Kim, Nam K. (Emeritus)
Pintar, Anton J. (Emeritus)

Accredited by: ABET

Degrees granted 2005-2006:
    B.S.: 50   M.S.: 6   Ph.D.: 1

Undergraduate advisor: Katie Torrey

Student organization: AIChE
Advisor: Jason M. Keith

Department reports to: Dean Engr. (906) 487-2005

Placement service: Univ. Career Ctr. (906) 487-2313

United States

Wayne State University

Department of Chemical Engineering and Materials Science
5050 Anthony Wayne Drive
Detroit, MI 48202
Deliveries: Room 1103, 5050 Anthony Wayne Dr.

Department phone .... (313) 577-3800
Department FAX .... (313) 577-3810

Professors
Gulari, Esin .... (313) 577-3767
    egulari@che.eng.wayne.edu
Huang, Yinlun .... (313) 577-3771
    yhuang@che.eng.wayne.edu
Kummler, Ralph H. (Dean) .... (313) 577-3861
    rkummler@che.eng.wayne.edu
Louvar, Joeseph .... (313) 577-9358
    jlouvar@che.eng.wayne.edu
Manke, Charles (Chair) .... (313) 577-3849
    cmanke@che.eng.wayne.edu
Ng, K. Y. Simon .... (313) 577-3805
    sng@che.eng.wayne.edu
Putatunda, Susil K. .... (313) 577-3808
    sputa@che.eng.wayne.edu
Rothe, Erhard W. .... (313) 577-3865
    erothe@che.eng.wayne.edu

Associate Professors
Kannan, Rangaramanujam .... (313) 577-3879
    rkannan@che.eng.wayne.edu
Mao, Guangzhao .... (313) 577-3804
    gzmao@che.eng.wayne.edu
Matthew, Howard .... (313) 577-5238
    hmatthew@che.eng.wayne.edu
McMicking, James H. (Emeritus) ...... (313) 577-3802
    jmcmicking@che.eng.wayne.edu
Salley, Steven O. .... (313) 577-3755
    ssalley@che.eng.wayne.edu
Shreve, Gina S. .... (313) 577-3874
    gshreve@che.eng.wayne.edu

Assistant Professors
da Rocha, Sandro .... (313) 577-4669
    sdr@eng.wayne.edu
Potoff, Jeffrey J. .... (313) 577-9537
    jpotoff@che.eng.wayne.edu

Accredited by: ABET

Degrees granted 2004-2005:
    B.S.: 17   M.S.: 12   Ph.D.: 5

Graduate advisor: Yinlun Huang

Undergraduate advisor: Andrea Eisenberg

Student organization: AIChE
Advisor: Jeff Potoff

Placement service: John Crusoe (313) 577-3390

Western Michigan University

Department of Paper Engineering, Chemical Engineering, and Imaging
College of Engineering and Applied Sciences
Klamazoo, MI 49008

University phone .... (269) 2763500
Department phone .... (269) 276-3502
Department FAX .... (269) 2763501

Professors
Abubakr, Said (Chair) .... (269) 276-3502
    sabubakr@wmich.edu
Aravamuthan, Raja .... (269) 2763507
    raja.ravamuthan@wmich.edu
Cameron, John .... (269) 276-3508
    john.cameron@wmich.edu
Joyce, Tom .... (269) 276-3515
    thomas.joyce@wmich.edu

Associate Professors
Fleming, Dan .... (269) 276-3511
    da.fleming@wmich.edu
Joyce, Margaret .... (269) 276-3514
    margaret.joyce@wmich.edu
Parker, Peter .... (269) 276-3519
    peter.parker@wmich.edu
Pekarovicova, Alexandra .... (269) 276-3521
    a.pekarovicova@wmich.edu
Peterson, david .... (269) 276-3522
    david.peterson@wmich.edu
Qi, Dewie .... (269) 276-3523
    dewie.qi@wmich.edu

Assistant Professors
Kline, Andy .... (269) 276 3516
    andrew.kline@wmich.edu

Other Faculty
Ahleman, Larry .... (269) 276 3506
    larry.ahleman@wmich.edu
Hladky, Harold .... (269) 276 3513
    harold.hladky@wmich.edu
Lemon, Lois .... (269) 276 3517
    lois.lemon@wmich.edu

Accredited by: ABET

Degrees granted 2005-2006:
    B.S.: 14   M.S.: 4   Ph.D.: 2

Graduate advisor: Andew Kline

Undergraduate advisor: Peter Parker

Student organization: AIChE Chapter
Advisor: Peter Parker

Department reports to: Dean, College of Engineering and Applied Sciences

Placement service: Tracy Moon

Minnesota

University of Minnesota at Duluth

Chemical Engineering Department
176 Engineering Building
1303 Ordean Court
Duluth MN 55812-3025

University phone .... (218) 726-8000
Department phone .... (218) 726-7126
Department FAX .... (218) 726-6907

Professors

Davis, Richard .... (218) 726-6162
    rdavis@d.umn.edu
Hasan, A. Rashid (Head) .... (218) 726-7127
    rhasan@d.umn.edu

Associate Professors

Lodge, Keith B. .... (218) 726-6164
    klodge@d.umn.edu
Sternberg, Steven P.K. .... (218) 726-6165
    ssternbe@d.umn.edu

Assistant Professors

Rother, Michael .... (218) 726-6154
    mrother@d.umn.edu
Rutkowski, Gregory .... (218) 726-7828
    grutkows@d.umn.edu

Instructors

Horabik, Carol .... (218) 726-6578
    chorabik@d.umn.edu

Accredited by: ABET

Degrees granted 2005-2006:
    B.S.: 28

Graduate advisor: A. Rashid Hasan

Undergraduate advisor: A. Rashid Hasan, Richard Davis, Keith Lodge, Michael Rother, Gregory Rutkowski, Steven Sternberg

Student organization: AIChE
Advisor: Steven Sternberg

Department reports to: James Riehl (218) 726-6397, Dean, College of Science and Engineering

Placement service: Julie Westlund (218) 726-7985

University of Minnesota at Minneapolis St. Paul

Department of Chemical Engineering and Materials Science
151 Amundson Hall
421 Washington Avenue SE
Minneapolis, MN 55455

University phone .... (612) 625-5000
Department phone .... (612) 625-1313
Department FAX .... (612) 626-7246

Professors

Aydil, Eray .... (612) 625-8593
    aydil@umn.edu
    AM,EN,EV,ME,MO,NT,RE,SU
Bates, Frank S. (Head) .... (612) 625-4356
    bates@cems.umn.edu
    NT,PO
Caretta, Raul .... (612) 625-8066
    caretta@umn.edu
Carr, Robert W. Jr. (Emeritus) .... (612) 625-2551
    carrx002@umn.edu
    RE
Carter, C. Barry .... (612) 625-8805
    carter@cems.umn.edu
    AM,EN,NT,SU
Cussler, Edward L. .... (612) 625-1596
    cussler@cems.umn.edu
    EN,EV,SE
Dahler, John S. (Emeritus) .... (612) 625-1884
    dahler@chemsun.chem.umn.edu
    PO,RE,SU
Daoutidis, Prodromos .... (612) 625-8818
    daoutidi@cems.umn.edu
    BT,EN,MO,PR
Davis, H. Ted .... (612) 625-4088
    davis@cems.umn.edu
    MO,NT
Derby, Jeffrey J. .... (612) 625-8881
    derby@umn.edu
    AM,EN,ME,MO,SE
Francis, Lorraine F. .... (612) 625-0559
    lfrancis@umn.edu
    AM,NT,PO,SU
Fredrickson, Arnold G. (Emeritus) .... (612) 625-6588
    fredr001@umn.edu
Frisbie, C. Daniel .... (612) 625-0779
    frisbie@cems.umn.edu
    EN,ME,NT,PO,SU
Gerberich, William W. .... (612) 625-8548
    wgerb@umn.edu
Hu, Wei-Shou .... (612) 625-0546
    wshu@cems.umn.edu
    BM,BT
Keller, Kenneth H. .... (612) 626-9547
    khkeller@umn.edu

Lodge, Timothy P. .... (612) 625-0877
lodge@umn.edu
AM,NT,PO

Macosko, Christopher W. .... (612) 625-0092
macosko@umn.edu
PO,SU

McCormick, Alon V. .... (612) 625-1822
mccormic@cems.umn.edu
AM,MO,NT,PO,RE,SU

Norris, David .... (612) 625-2043
dnorris@cems.umn.edu
AM,EN,NT

Oriani, Richard A. (Emeritus) .... (612) 625-5862
peper001@umn.edu
EN,SU

Palmstrom, Christopher J. .... (612) 625-7558
palms001@umn.edu
AM,ME,NT,SU

Schmidt, Lanny D. .... (612) 625-9391
schmi001@umn.edu
EN,EV,RE,SU

Scriven, L. E. .... (612) 625-1058
pjensen@cems.umn.edu
AM,MO,PR,SU

Seidel, Robert W. .... (612) 624-8003
rws@umn.edu
BT,EN,EV,NT

Shores, David A. .... (612) 625-0014
dshores@umn.edu

Smyrl, William H. .... (612) 625-0717
smyrl001@umn.edu
AM,NT

Srienc, Friedrich .... (612) 624-9776
fried@biosci.cbs.umn.edu
AM,BT,EV,MO,PO,RE

Tranquillo, Robert T. .... (612) 625-6868
tranquillo@cems.umn.edu
BM,BT

Tsapatsis, Michael .... (612) 626-0920
tsapatsis@cems.umn.edu
AM,EN,EV,MO,NT,RE,SE,SU

Associate Professors

Morse, David C. .... (612) 625-0167
morse@cems.umn.edu
MO,PO

Schott, Jeffrey .... (612) 625-1420
Schott@cems.umn.edu
PR,SE

Sivertsen, John M. (Emeritus) .... (612) 625-1313

Wentzcovitch, Renata .... (612) 625-6345
wentzcov@cems.umn.edu

Assistant Professors

Cococcioni, Matteo .... (612) 624-9056
cococ002@umn.edu
AM,EN,MO,NT,RE

Dorfman, Kevin .... (612) 624-5560
dorfman@cems.umn.edu
BT,MO,SE

Holmes, Russell .... (612) 624-9058
rholmes@umn.edu
AM,EN,NT

Kaznessis, Yiannis .... (612) 624-4197
yiannis@cems.umn.edu

Kokkoli, Efrosini .... (612) 626-1185
kokkoli@cems.umn.edu
BM,BT,NT,PO,SU

Kumar, Satish .... (612) 625-2558
kumar@cems.umn.edu
MO,NT,PO

Leighton, Christopher .... (612) 624-4018
leighton@umn.edu
AM,NT,SU

Maynard, Jennifer .... (612) 626-0635
jmaynard@umn.edu

Accredited by: ABET

Degrees granted 2005-2006:
    B.S.: 70   M.S.: 4   Ph.D.: 24

Graduate advisor: David Norris

Undergraduate advisor: Alon McCormick

Student organization: AIChE
Advisor: David Norris

Department reports to: Steven L. Crouch, Dean, Institute of Technology

Placement service: Mark Sorenson-Wagner

Mississippi

University of Mississippi

Department of Chemical Engineering
134 Anderson Hall
University, MS 38677-1840
University phone .... (662) 915-7211
Department phone .... (662) 915-7023
Department FAX .... (662) 915-7023

Professors

Anderson, Frank (Emeritus) .... (662) 915-7023
Aven, Russell E. (Emeritus) .... (662) 915-7023
Chen, Wei-Yin .... (662) 915-5651
    cmchengs@olemiss.edu
Sadana, Ajit .... (662) 915-5349
    cmsadana@olemiss.edu
Stasaik, Raymond (Adjunct) .... (662) 915-7023
Sukanek, Peter C. (Chair) .... (662) 915-7023
    cmpcs@olemiss.edu

Associate Professors

O'Haver, John Howard .... (662) 915-5347
    johaver@olemiss.edu
Williford, Clint W. Jr. .... (662) 915-5348
    drwill@olemiss.edu

Assistant Professors

Scovazzo, Paul .... (662) 915-5354
    scovazzo@olemiss.edu

Accredited by: ABET

Degrees granted 2005-2006:
    B.S.: 8   M.S.: 1   Ph.D.: 1

Undergraduate advisor: Peter C. Sukanek

Student organization: AIChE
Advisor: Clint W. Williford, Jr.

Department reports to: Dean of Engineering

Placement service: Regina Starnes, Career Center
662-915-5984

Mississippi State University

Dave C. Swalm School of Chemical Engineering
Box 9595
Mississippi State, MS 39762-9595
Deliveries: 330 Swalm - President's Circle
University phone .... (662) 325-2131
Department phone .... (662) 325-2480
Department FAX .... (662) 325-2482

Professors

Cornell, David (Emeritus)
George, Clifford E. .... (662) 325-7205
    george@che.msstate.edu
Hall, William B. (Emeritus)
Hill, Donald O. (Emeritus)
Jefcoat, Irvin (Emeritus) .... (662) 325-7206
    jefcoat@che.msstate.edu
Koelling, Harold A. (Emeritus)
Kuo, Chiang-Hai (Emeritus)
Rogers, Rudy E. .... (662) 325-5106
    rogers@che.msstate.edu
Schulz, Kirk H. (Dean) .... (662) 325-2480
    schulz@che.msstate.edu
Sparrow, Charles A. (Emeritus) .... (662) 325-7323
    sparrow@che.msstate.edu
White, Mark G. (Chair) .... (662) 325-5797
    white@che.msstate.edu
    RE

Associate Professors

Bricka, R. Mark .... (662) 325-1615
    bricka@che.msstate.edu
Toghiani, Hossein .... (662) 325-8607
    hossein@che.msstate.edu
Toghiani, Rebecca K. .... (662) 325-8615
    rebecca@che.msstate.edu

Assistant Professors

Hernandez, Rafael .... (662) 325-0790
    rhernandez@che.msstate.edu
Hill, Priscilla J. .... (662) 325-8249
    phill@che.msstate.edu
Minerick, Adrienne R. .... (662) 325-7323
    minerick@che.msstate.edu
Walters, Keisha
    walters@che.msstate.edu

Other Faculty

French, W. Todd .... (662) 325-4308
    tfrench@che.msstate.edu

Accredited by: ABET

Degrees granted 2005-2006:
    B.S.: 25   M.S.: 3   Ph.D.: 1

Graduate advisor: Rudy Rogers

Undergraduate advisor: Clifford George

Student organization: AIChE
Advisor: Bill Elmore

Department reports to: Dean of the James Worth Bagley College of Engineering

Placement service: Dr. Luther Epting, Director; Career Center, P. O. Drawer P

Missouri

University of Missouri-Columbia

Department of Chemical Engineering
W2033 Lafferre Hall
Columbia, MO 65211-2200
Deliveries: W2030 Lafferre Hall

University phone	(573) 882-2121
Department phone	(573) 882-3563
Department FAX	(573) 884-4940

Professors

Bajpai, Rakesh K. .... (573) 882-3708
bajpair@missouri.edu
BM,BT,MO
Loyalka, Sudarshan .... (573) 882-3568
loyalkas@missouri.edu
Luecke, Richard (Emeritus) .... (573) 882-3691
luecker@missouri.edu
Marrero, Thomas R. .... (573) 882-3802
marrerot@missouri.edu
EN,EV
Storvick, Truman (Emeritus) .... (573) 882-3215
storvickt@missouri.edu
EN,EV,TH
Tan, Jinglu (Chair) .... (573) 882-7778
tanj@missouri.edu
BM,BT
Viswanath, Dabir S. (Emeritus) .... (573) 884-0707
viswanthd@missouri.edu
Yasuda, H. K. (Emeritus) .... (573) 882-9602
yasudah@missouri.edu
PO,SU

Associate Professors

Chan, Paul C. H. .... (573) 882-7684
chanp@missouri.edu
RE,TP
Jacoby, William A. .... (573) 882-3563
jacobyw@missouri.edu
RE,TH,TP
Lombardo, Stephen J. .... (573) 882-3563
lombardos@missouri.edu
AM,RE,TP
Retzloff, David G. .... (573) 882-4036
retzloffd@missouri.edu
AM,RE
Suppes, Galen J. .... (573) 884-0562
suppesg@missouri.edu
EN,EV,TH

Assistant Professors

Doskocil, Eric .... (573) 884-2273
doskocile@missouri.edu
RE
Yu, Qingsong .... (573) 882-8076
yuq@missouri.edu
SU

Accredited by: ABET

Degrees granted 2005-2006:
    B.S.: 33   M.S.: 3   Ph.D.: 3

Graduate advisor: William A. Jacoby

Undergraduate advisor: Paul C.H. Chan

Student organization: AIChE
Advisor: Paul C.H. Chan

Department reports to: James Thompson, Dean

Placement service: Matt Riske, Director

University of Missouri-Rolla

Chemical & Biological Engineering Department
143 Schrenk Hall
Rolla, MO 65409-1230
Deliveries: 143 Schrenk Hall, 11th and State Str.

University phone .... (573) 341-4111
University FAX .... (573) 341-4082
Department phone .... (573) 341-4416
Department FAX .... (573) 341-4377

Professors

Adams, Craig (Joint) .... (573) 341-4041
adams@umr.edu
EV,RE
Azbel, David S. (Emeritus)
Crosser, Orrin K. (Emeritus) .... (573) 341-4459
okc@umr.edu
MO,PR,RE,SE
Forciniti, Daniel .... (573) 341-4427
forcinit@umr.edu
BT,SE
Lee, Sunggyu .... (573) 882-4281
slee@missouri.edu
AM,BT,EV,PO,PR,RE
Liapis, Athanasios I. .... (573) 341-4414
ail@umr.edu
EN,PO,RE
Ludlow, Douglas K. .... (573) 341-6477
dludlow@umr.edu
EN,EV,NT,RE,SU
Manley, David B. (Emeritus) .... (573) 341-4428
dbm@umr.edu
Morosoff, Nicholas (Emeritus)
ncmoros@umr.edu
Neogi, Partho .... (573) 341-4460
neogi@umr.edu
SU
Patterson, Gary K. (Emeritus) .... (573) 341-6941
garyp@umr.edu
Raper, Judy A. (Chair) .... (573) 341-7518
raperj@umr.edu
EN,NT
Reed, X B Jr. (Emeritus) .... (573) 341-4423
xbreed@umr.edu
Rosen, Stephen L. (Emeritus) .... (573) 341-4443
slr@umr.edu
Strunk, Mailand R. (Emeritus)
Waggoner, Raymond C. (Emeritus) ... (573) 341-4425
wagg1r@umr.edu

Associate Professors

Book, Neil L. .... (573) 341-4422
nbook@umr.edu
PR,RE
Sitton, Oliver C. .... (573) 341-4426
ocs@umr.edu

Wang, Jee-Ching .... (573) 341-6705
jcwang@umr.edu
AM,MO
Westenberg, David (Joint) .... (573) 341-4798
djwesten@umr.edu

Assistant Professors

Henthorn, David .... (573) 341-7632
henthord@umr.edu
AM,BM,BT,PO
Henthorn, Kimberly .... (573) 341-7633
henthork@umr.edu
PR
Wan, Kai-Tak (Joint) .... (573) 341-4428
wankt@umr.edu
BM,ME,MO,NT,SU
Xing, Yangchuan .... (573) 341-6772
xingy@umr.edu
AM,EN,EV,NT,RE

Accredited by: ABET

Degrees granted 2005-2006:
    B.S.: 35   M.S.: 10   Ph.D.: 3

Graduate advisor: Daniel Forciniti

Undergraduate advisor: Douglas K. Ludlow

Student organization: AIChE
Advisor: Daniel Forciniti

Department reports to: Dean, School of Engineering

Placement service: Lea Ann Morton, Director

Washington University

Chemical Engineering Department
Campus Box 1198
St. Louis, MO 63130-4899
Deliveries: One Brookings Drive, Urbauer Hall 208

University phone .... (314) 935-5000
Department phone .... (314) 935-6082
Department FAX .... (314) 935-7211

Professors

Al-Dahhan, Muthanna .... (314) 935-7187
muthanna@wustl.edu
BT,EN,EV,RE
Biswas, Pratim (Chair) .... (314) 935-5482
pratim.biswas@seas.wustl.edu
EV,NT
Dudukovic, Milorad .... (314) 935-6021
dudu@wustl.edu
EN,EV,RE,TP
Khomami, Bamin .... (314) 935-6065
bam@che.wustl.edu
MO,NT,TP
Ramachandran, P. A. .... (314) 935-6531
rama@che.wustl.edu
EN,EV,RE,TP
Sureshkumar, R. .... (314) 935-4988
suresh@che.wustl.edu
MO,NT,PO,TP

Associate Professors

Gleaves, John T. .... (314) 935-4159
klatu_00@che.wustl.edu
EN,RE,SU
Turner, Jay R. .... (314) 935-5480
jrturner@seas.wustl.edu
EN,EV

Assistant Professors

Angenent, Lars .... (314) 935-5663
angenent@seas.wustl.edu
EN,EV

Other Faculty

Kardos, John L. (Emeritus) .... (314) 935-6062
kardos@che.wustl.edu
PO
McKelvey, James M. .... (314) 935-4836
mckelvey@che.wustl.edu
PO,TH
Thies, Curt (Emeritus) .... (314) 935-6082
thiesman@aol.com
Yablonsky, Gregory (Research) .... (314) 935-4367
gy@che.wustl.edu

Accredited by: ABET

Degrees granted 2005-2006:
    B.S.: 20   M.S.: 10   Ph.D.: 10

Student organization: AIChE
Advisor: Dr. Lars Angenent

Department reports to: Mary J. Sansalone, Dean, School of Engineering and Applied Science

Placement service: Mark W. Smith (314) 935-6489

Montana

Montana State University

Chemical and Biological Engineering Department
P.O. Box 173920
306 Cobleigh Hall
Bozeman, MT 59717-3920
Deliveries: 306 Cobleigh Hall

University phone .... (406) 994-0211
Department phone .... (406) 994-2221
Department FAX .... (406) 994-5308

Professors

Cokelet, Giles (Research) .... (406) 994-7048
    giles_c@coe.montana.edu
Larsen, Ronald W. (Head) .... (406) 994-2221
    ronl@coe.montana.edu
Mandell, John F. .... (406) 994-4543
    johnm@coe.montana.edu
McCandless, Frank P. (Emeritus) ...... (406) 994-5928
    philm@coe.montana.edu
Nickelson, Robert L. (Emeritus)
Scarrah, Warren P. (Emeritus)
Stewart, Phillip .... (406) 994-2890
    phil_s@erc.montana.edu

Associate Professors

Deibert, Max C. .... (406) 994-5990
    maxd@coe.montana.edu
Duffy, James E. .... (406) 994-5926
    james_d@coe.montana.edu
Peyton, Brent M. .... (406) 994-7419
    bpeyton@coe.montana.edu
Seymour, Joseph .... (406) 994-6853
    jseymour@coe.montana.edu
Shaffer, Daniel L. .... (406) 994-5922
    dans@coe.montana.edu

Assistant Professors

Carlson, Ross P. .... (406) 994-3631
    rossc@coe.montana.edu

Accredited by: ABET

Degrees granted 2005-2006:
    B.S.: 23   M.S.: 3

Graduate advisor: Ronald W. Larsen

Undergraduate advisor: Ronald W. Larsen

Student organization: AIChE
Advisor: Ross Carlson

Department reports to: Robert Marley, Dean of Engineering

Placement service: Ron Larsen, Shelley I. Thomas

Nebraska

University of Nebraska

Department of Chemical & Biomolecular Engineering
Lincoln, NE 68588-0643
Deliveries: Othmer 207

University phone .... (402) 472-7211
Department phone .... (402) 472-2750
Department FAX .... (402) 472-6989

Professors

Clements, L. Davis (Emeritus) .... (402) 472-2750
Gilbert, Richard E. (Emeritus) .... (402) 472-2750
Hendrix, James L. .... (402) 472-0697
    jhendrix1@unl.edu
Larsen, Gustavo .... (402) 472-9805
    glarsen1@unl.edu
Meagher, Michael M. .... (402) 472-2342
    mmeagher1@unl.edu
Saraf, Ravi .... (402) 472-8284
    rsaraf2@unl.edu
Tao, Luh C. (Emeritus) .... (402) 472-2600
Timm, Delmar C. .... (402) 472-3232
    dtimm1@unl.edu
Velander, William H (Chair) .... (402) 472-3697
    wvelander2@unl.edu
    BM,BT
Viljoen, Hendrik J. .... (402) 472-9318
    hviljoen1@unl.edu
Weber, James H. (Emeritus) .... (402) 472-2600

Associate Professors

Brand, Jennifer I. (Adjunct) .... (402) 472-9320
    jbrand1@unl.edu
Lauderback, Lee L. .... (402) 472-5725
    llauderback1@unl.edu
Noureddini, Hossein (Associate Chair) . (402) 472-2751
    hnoureddini@unl.edu
Subramanian, Anu .... (402) 472-3463
    asubramanian2@unl.edu
    BM,BT,SE,SU
Van Cott, Kevin .... (402) 472-1743
    kvancott2@unl.edu

Other Faculty

Demirel, Yasar .... (402) 472-2745
    ydemirel2@unl.edu
    EN,HT,MO,SE,TH,TP
Inan, Mehmet (Research) .... (402) 472-2616
    minan2@unl.edu
    BT
Padhye, Nisha (Research) .... (402) 472-2752
    npadhye2@unl.edu
Sinha, Jayanta (Research) .... (402) 472-9301
    jsinha2@unl.edu

Swanson, Todd (Research) .... (402) 472-0038
sswanson4@unl.edu

Accredited by: ABET

Degrees granted 2005-2006:
    B.S.: 11   M.S.: 3

Graduate advisor: Dr. Kevin Van Cott

Undergraduate advisor: Dr. J. L. Hendrix

Student organization: AIChE
Advisor: Dr. Hossein Noureddini

Department reports to: Dr. David Allen, Dean, College of Engineering

Placement service: Larry Routh, Nebraska Union 230

Nevada

University of Nevada, Reno

Chemical Engineering Program
Mailstop 388
Reno, NV 89557-0136
Deliveries: Room 474, LMR Building
University phone .... (775) 784-1110
Department phone .... (775) 784-6771
Department FAX .... (775) 327 5059

Professors

Baglin, Frank G. (Adjunct) .... (775) 784-6651
baglin@chem.unr.edu
Farmer, Richard (Adjunct) .... (775) 784-4307
rfarmer@unr.edu
Tsoulfanidis, Nicholas (Chair) .... (775) 784 8287
nikost@unr.edu
Whiting, Wallace B. (Emeritus) .... (775) 784-4307
wwhiting@unr.edu

Associate Professors

Coronella, Charles J. .... (775) 784-4253
coronell@unr.edu
Fuchs, Alan .... (775) 327-2227
afuchs@unr.edu
Gecol, Hatice .... (775) 784-1116
gecol@unr.edu

Assistant Professors

Subramanian, Ravi
BT,MO,NT,PO,SU
Vasquez, Victor R. .... (775) 784-6060
vvasquez@unr.edu

Accredited by: ABET

Degrees granted 2005-2006:
    B.S.: 20   M.S.: 3

Student organization: AIChE
Advisor: C. J. Coronella

Department reports to: Ted Batchman, Dean (775) 784-6925

Placement service: David A. Hansen (775) 784-4898

New Hampshire

Dartmouth College

Biotechnology and Biochemical Engineering Program
Thayer School of Engineering
8000 Cummings Hall
Hanover, NH 03755-8000

University phone	(603) 646-1110
University FAX	603 646-3856
Department phone	(603) 646-2230
Department FAX	(603) 646-2277

Professors
Baker, Ian .... (603) 646-2184
                Ian.Baker@Dartmouth.EDU
Collier, John .... (603) 646-2355
                John.P.Collier@Dartmouth.EDU
Converse, Alvin O. (Emeritus) .... (603) 646-3060
                Alvin.O.Converse@Dartmouth.EDU
Cushman-Roisin, Benoit .... (603) 646-3248
                Benoit.R.Roisin@Dartmouth.EDU
Kennedy, Francis .... (603) 646-2094
                Francis.E.Kennedy@Dartmouth.EDU
Lynd, Lee R. .... (603) 646-2231
                Lee.Lynd@Dartmouth.EDU
Petrenko, Victor .... (603) 646-3526
                Victor.F.Petrenko@Dartmouth.EDU
Queneau, Paul (Emeritus) .... (603) 646-2208
Richter, Horst .... (603) 646-2701
                Horst.J.Richter@Dartmouth.EDU
Schulson, Erland .... (603) 646-2888
                Erland.M.Schulson@Dartmouth.EDU
Sonnerup, Bengt (Emeritus) .... (603) 646-2883
Wallis, Graham B. (Emeritus) .... (603) 646-2789

Associate Professors
Frost, Harold .... (603) 646-3444
                Harold.J.Frost@Dartmouth.EDU
Gerngross, Tillman .... (603) 646-3161
                Tillman.Gerngross@Dartmouth.EDU
Gibson, Urusla .... (603) 646-3243
                Ursula.J.Gibson@Dartmouth.EDU

Assistant Professors
Proehl, Jeffrey A. .... (603) 646-3803
                Jeffrey.A.Proehl@Dartmouth.EDU

Accredited by: ABET

Graduate advisor: Lee Lynd

Department reports to: Joesph Helble (603) 646-2238, Dean of Engineering

Placement service: Chandlee Bryan (603) 646-2375, Director of Career Services

University of New Hampshire

Chemical Engineering Department
Kingsbury Hall, Rm W301
33 College Road
Durham, NH 03824

University phone	(603) 862-1234
Department phone	(603) 862-3654
Department FAX	(603) 862-3747

Professors
Barkey, Dale P. .... (603) 862-1918
                dpb@cisunix.unh.edu
Carr, Russell T. .... (603) 862-1429
                rtc@cisunix.unh.edu
Fan, Stephen (Chair) .... (603) 862-3656
                sstf@cisunix.unh.edu
Farag, Ihab H. .... (603) 862-2313
                Ihab.Farag@unh.edu
Mathur, Virendra K. .... (603) 862-1917
                vkm@cisunix.unh.edu
Ulrich, Gael D. (Emeritus)
                gdu@cisunix.unh.edu
Vasudevan, Palligarnai T. .... (603) 862-2298
                vasu@unh.edu

Assistant Professors
Gupta, Nivedita .... (603) 862-3655
                N.Gupta@unh.edu

Accredited by: ABET

Degrees granted 2005-2006:
    B.S.: 15   M.S.: 4   Ph.D.: 1

Graduate advisor: Stephen S.T. Fan

Undergraduate advisor: Stephen S.T. Fan

Student organization: AIChE
Advisor: P. T. Vasudevan

Department reports to: Joseph Klewicki, Dean, College of Engineering and Physical Sciences

Placement service: Bethany Cooper, Univ. Advis./Career Ctr, Hood House 102

New Jersey

New Jersey Institute of Technology

Otto H. York Department of Chemical Engineering
University Heights
Newark, NJ 07102
Deliveries: 150 Tiernan Hall, 161 Warren St.

University phone .... (973) 596-3000
Department phone .... (973) 596-3568
Department FAX .... (973) 596-8436

Professors

Armenante, Piero (Director) .... (973) 596-3548
armenante@adm.njit.edu
BM,EV,MO
Baltzis, Basil (Chair) .... (973) 596-3619
baltzis@adm.njit.edu
EV,MO,RE
Barat, Robert .... (973) 596-5605
barat@adm.njit.edu
EV,MO,RE
Dave, Rajesh .... (973) 596-5860
rajesh.n.dave@njit.edu
AM,MO,NT,PR
Greenstein, Teddy .... (973) 596-5706
greenstein@adm.njit.edu
BT
Hanesian, Deran .... (973) 596-3597
hanesian@adm.njit.edu
EV,RE
Huang, Ching-Rong (Emeritus)
ching-rong.huang@njit.edu
Kimmel, Howard .... (973) 596-3574
kimmel@adm.njit.edu
Knox, Dana .... (973) 596-3599
knoxd@adm.njit.edu
EV,MO
Lewandowski, Gordon .... (973) 596-3573
lewandows@adm.njit.edu
EV
Perna, Angelo .... (973) 596-3616
perna@adm.njit.edu
EV
Pfeffer, Robert (Emeritus) .... (973) 642-7496
pfeffer@adm.njit.edu
AM,NT,PR
Sirkar, Kamalesh .... (973) 596-8447
sirkar@adm.njit.edu
AM,BT,PR,SE
Tomkins, Reginald (Associate Chair) .. (973) 596-5656
tomkins@adm.njit.edu
PR
Xanthos, Marino .... (973) 642-4762
xanthos@admin.njit.edu
AM,EV,PO,PR
York, Otto (Honorary)

Associate Professors

Loney, Norman .... (973) 596-6598
loney@adm.njit.edu
BM,MO,SE

Assistant Professors

Bart, Ernest .... (973) 596-2998
bart@adm.njit.edu
MO
Huang, Chien-Yueh .... (973) 596-5613
chien-yueh.huang@njit.edu
AM,MO,NT,PO
Simon, Laurent .... (973) 596-5263
laurent.simon@njit.edu
BM,BT,MO,SE
Wu, Jing .... (973) 642-7064
wu@adm.njit.edu
AM,PO,PR

Instructors

Hendela, Art .... (973) 596-5388
ahh2@njit.edu

Other Faculty

Gogos, Costas (Research) .... (973) 642-7365
costas@polymers-ppi.org
AM,PO
Hyun, Kun Sup (Research) .... (973) 596-3267
kshyun@adm.njit.edu
PO,PR
Young, Ming-Wan (Research) .... (973) 596-5256
Ming-Wan.Young@njit.edu
AM,PO,PR

Accredited by: ABET

Degrees granted 2005-2006:
    B.S.: 14   M.S.: 12   Ph.D.: 2

Graduate advisor: Reginald Tomkins

Undergraduate advisor: Reginald Tomkins

Student organization: AIChE
Advisor: Angelo Perna

Department reports to: John Schuring (973) 596-5534, Dean of Engineering

Placement service: Greg Mass, Director of Placement (973) 596-5745

Princeton University

Chemical Engineering Department
Olden Street
Princeton, NJ 08544-5263
Deliveries: Prospect Street

University phone	(609) 258-3000
Department phone	(609) 258-4581
Department FAX	(609) 258-0211

Professors

Aksay, Ilhan A. .... (609) 258-4393
    iaksay@princeton.edu
Benziger, Jay B. .... (609) 258-5416
    benziger@princeton.edu
Debenedetti, Pablo G. .... (609) 258-5480
    pdebene@princeton.edu
Floudas, Christodoulos A. .... (609) 258-4595
    floudas@titan.princeton.edu
Gillham, John K. (Emeritus) .... (609) 258-4581
    gillham@princeton.edu
Graessley, William W. (Emeritus) ..... (609) 258-4581
    graessle@princeton.edu
Jackson, Roy (Emeritus) .... (609) 258-4581
    rjackson@always-online.com
Johnson, Ernest F. (Emeritus) .... (609) 258-4581
Kevrekidis, Yannis G. .... (609) 258-2818
    yannis@arnold.princeton.edu
Kostin, Morton D. .... (609) 258-4586
    kostin@princeton.edu
Panagiotopoulos, Athanassios Z. ...... (609) 258-4591
    azp@princeton.edu
Prud'homme, Robert K. .... (609) 258-4577
    prudhomm@princeton.edu
Register, Richard A. .... (609) 258-4691
    register@princeton.edu
Russel, William B. .... (609) 258-4590
    wbrussel@princeton.edu
Saville, Dudley A. .... (609) 258-4585
    dsaville@princeton.edu
Schowalter, William (Emeritus) .... (609) 258-3553
    schowalt@princeton.edu
Sundaresan, Sankaran .... (609) 258-4583
    sundar@princeton.edu
Troian, Sandra M. .... (609) 258-4574
    stroian@princeton.edu
Vanderlick, T. Kyle (Chair) .... (609) 258-4891
    vandertk@princeton.edu
Wei, James .... (609) 258-5618
    jameswei@princeton.edu

Assistant Professors

Shvartsman, Stanislav Y. .... (609) 258-4694
    stas@princeton.edu
Wood, David W. .... (609) 258-5721
    dwood@princeton.edu

Accredited by: ABET

Degrees granted 2005-2006:
    B.S.: 28   Ph.D.: 11

Graduate advisor: Sandra M. Troian

Undergraduate advisor: Jay B. Benziger

Student organization: AIChE
Advisor: David W. Wood

Department reports to: Maria Klawe, Dean, School of Engineering and Applied Science

Placement service: Minnie H. Reed, Director

Rowan University

Department of Chemical Engineering
201 Mullica Hill Road
Glassboro, NJ 08028-1701
Deliveries: 312 Henry M. Rowan Hall

University phone .... (856) 256-4000
Department phone .... (856) 256-5310
Department FAX .... (856) 256-5242

Professors

Dorland, Dianne (Dean) .... (856) 256-5300
    dorland@rowan.edu
Hesketh, Robert P. (Chair) .... (856) 256-5313
    hesketh@rowan.edu
Newell, James A. .... (856) 256-5316
    newell@rowan.edu
Slater, C. Stewart .... (856) 256-5312
    slater@rowan.edu

Associate Professors

Dahm, Kevin D. .... (856) 256-5318
    dahm@rowan.edu
Farrell, Stephanie .... (856) 256-5315
    farrell@rowan.edu
Gephardt, Zenaida Otero .... (856) 256-5314
    gephardtzo@rowan.edu
Savelski, Mariano J. .... (856) 256-5317
    savelski@rowan.edu

Assistant Professors

Lefebvre, Brian G. .... (856) 256-5338
    lefebvre@rowan.edu

Other Faculty

Gould, Ronald (Adjunct) .... (856) 256-5310
Grady, Michael (Adjunct) .... (856) 256-5310
Hoffman, Robert (Adjunct) .... (856) 256-5310
Natoli, John (Adjunct) .... (856) 256-5310
VanKirk, Jesse (Adjunct) .... (856) 256-5310

Accredited by: ABET

Degrees granted 2005-2006:
    B.S.: 12   M.S.: 3

Graduate advisor: Mariano Savelski

Undergraduate advisor: Robert P. Hesketh

Student organization: AIChE
Advisor: Brian Lefebvre

Department reports to: Dianne Dorland (856) 256-5300, Dean of Engineering

Placement service: Melanie Basantis, Director of Engineering Outreach (856) 256-5307

Rutgers, The State University of New Jersey

Department of Chemical and Biochemical Engineering
School of Engineering
98 Brett Road
Piscataway, NJ 08854-8058

University phone .... (732) 932-1766
Department phone .... (732) 445-2228
Department FAX .... (732) 445-2581

Professors

Chiew, Yee C. .... (732) 445-0315
    ychiew@soemail.rutgers.edu
Constantinides, Alkis .... (732) 445-3678
    constant@soemail.rutgers.edu
Couchman, Peter (Emeritus)
Davidson, Burton Z. .... (732) 445-2203
    burtond@soemail.rutgers.edu
Dittman, F. W. (Emeritus)
Hara, Masanori .... (732) 445-3817
    mhara@rci.rutgers.edu
Klein, Michael T. (Dean) .... (732) 445-2214
    mtklein@jove.rutgers.edu
Muzzio, Fernando J. .... (732) 445-3357
    muzzio@soemail.rutgers.edu
Pedersen, Henrik (Chair) .... (732) 445-2568
    hpederse@soemail.rutgers.edu
Scheinbeim, Jerry I. .... (732) 445-3669
    jis@rci.rutgers.edu
Wang, Shaw S. .... (732) 445-3360
    shaww@soemail.rutgers.edu
Yarmush, Martin L. (Joint) .... (732) 445-4346

Associate Professors

Buettner, Helen M. (Joint) .... (732) 445-2231
    buettner@soemail.rutgers.edu
Glasser, Benjamin J. .... (732) 445-4243
    bglasser@soemail.rutgers.edu
Ierapetritou, Marianthi G. .... (732) 445-2971
    marianth@soemail.rutgers.edu
Khinast, Johannes G. .... (732) 445-2970
    khinast@soemail.rutgers.edu
Moghe, Prabhas V. (Joint) .... (732) 445-4951
    moghe@rci.rutgers.edu

Assistant Professors

Androulakis, Ioannis (Yannis) P. (Joint) (732) 445-0099
    yannis@rci.rutgers.edu
Roth, Charles M. (Joint) .... (732) 445-4109
    cmroth@rci.rutgers.edu
Tomassone, M. Silvina .... (732) 445-2972
    silvina@soemail.rutgers.edu

Accredited by: ABET

Degrees granted 2004-2005:
    B.S.: 35   M.S.: 10   Ph.D.: 8

Graduate advisor: Yee Chiew

Undergraduate advisor: Alkis Constantinides

Student organization: AIChE
Advisor: M. Silvina Tomassone

Department reports to: Michael T. Klein, Dean, School of Engineering

Placement service: Janet Jones (732) 445-6127 x 0

Stevens Institute of Technology

Department of Chemical, Biomedical and Materials Engineering
McLean Hall
Castle Point on Hudson
Hoboken, NJ 07030
Deliveries: Wesley J. Howe Center (Receiving)
University phone .... (201) 216-5000
Department phone .... (201) 216-5546
Department FAX .... (201) 216-8306

Professors

Besser, Ronald .... (201) 216-5257
                rbesser@stevens.edu
DeLancey, George B. .... (201) 216-5060
                gdelance@stevens.edu
Du, Henry (Director) .... (201) 216-5262
                hdu@stevens.edu
Gallois, Bernard .... (201) 216-5041
                bgallois@stevens.edu
Kalyon, Dilhan .... (201) 216-8225
                dkalyon@stevens.edu
Kovenklioglu, Suphan .... (201) 216-5519
                skoven@stevens.edu
Lawal, Adeniyi .... (201) 216-8241
                alawal@stevens.edu
Lee, Woo Y. .... (201) 216-8307
                wlee@stevens.edu
Libera, Matthew R. .... (201) 216-5259
                mlibera@stevens.edu
Ritter, Arthur .... (201) 216-8290
                aritter@steven.edu
Rothberg, Gerald .... (201) 216-5269
                grothber@stevens.edu
Sheppard, Keith (Associate Dean) ..... (201) 216-5260
                ksheppar@stevens.edu

Assistant Professors

Wang, Hongjun .... (201) 216-5556
                hwang@stevens.edu
Yu, Xiaojun .... (201) 216-5256
                xyu@stevens.edu

Instructors

Hazelwood, Vikki .... (201) 216-5051
                vhazelwo@stevens.edu

Accredited by: ABET

Graduate advisor: Suphan Kovenklioglu

Undergraduate advisor: George DeLancey

Student organization: AIChE
Advisor: Ronald Besser

Department reports to: George Korfiatis (201) 216-5263

Placement service: Lynn Insley (201)216-8927

New Mexico

New Mexico Institute of Mining and Technology

Department of Chemical Engineering
801 Leroy Place
Socorro, NM 87801

University phone 800-428-8324
Department phone .... (505) 835-5412
Department FAX .... (505) 835-5210

Professors

Islam, Mohammad (Visiting) .... (505) 835-5293
    mislam@nmt.edu
Lee, Robert L. (Adjunct) .... (505) 835-5408
    lee@prrc.nmt.edu
McCoy, John (Adjunct) .... (505) 835-5379
    mccoy@nmt.edu
Sharma, Rajendra (Visiting) .... (505) 835-5761
    sharma@nmt.edu

Associate Professors

Bretz, Robert E. .... (505) 835-5436
    bretz@nmt.edu
Cal, Mark (Adjunct) .... (505) 835-5059
    mcal@nmt.edu
Dunston, Doug (Adjunct) .... (505) 835-5200
    ddunston@nmt.edu
Weinkauf, Donald H. (Chair) .... (505) 835-5689
    weinkauf@nmt.edu

Instructors

Marshall, John .... (505) 835-5926
    jmarshal@nmt.edu

Accredited by: ABET

Degrees granted 2005-2006:
    B.S.: 15

Undergraduate advisor: D. H. Weinkauf

Student organization: AIChE

Department reports to: VP of Academic Affairs

Placement service: D. H. Weinkauf

University of New Mexico

Chemical and Nuclear Engineering Department
MSC01 1120
1 University of New Mexico
Albuquerque, NM 87131-0001
Deliveries: 209 Farris Engineering Center
University phone .... (505) 277-0111
Department phone .... (505) 277-5431
Department FAX .... (505) 277-5433

Professors

Brinker, C. Jeffrey .... (505) 272-7627
jbrinker@unm.edu
AM,BM,BT,ME,NT,SE,SU
Cecchi, Joseph L. (Dean) .... (505) 277-5522
cecchi@unm.edu
Datye, Abhaya (Director) .... (505) 277-0477
datye@unm.edu
AM,EN,NT,RE,SU
El-Genk, Mohamed S. (Director) ...... (505) 277-5442
mgenk@unm.edu
Fulghum, Julia E. (Chair) .... (505) 277-8670
jfulghum@unm.edu
AM,BM,EN,NT,PO,SU
Ivnitski, Dmitri (Research) .... (505) 277-2563
ivnitski@unm.edu
BM,EN,NT
Kauffman, David (Emeritus) .... (505) 277-5431
kauffman@unm.edu
Lopez, Gabriel P. .... (505) 277-4939
gplopez@unm.edu
AM,BM,BT,EN,EV,NT,PO,SU
Nuttall, H. Eric .... (505) 277-6112
nuttall@unm.edu
Prinja, Anil K. (Associate Chair) ...... (505) 277-4600
prinja@unm.edu
Roderick, Norman F. (Emeritus) ...... (505) 277-2209
roderick@unm.edu
Sibbett, Scott S. (Research) .... (505) 277-2803
ssibbett@unm.edu
BM,BT,SE
Ward, Timothy L. .... (505) 277-2067
tlward@unm.edu
Weaver, Harry (Research) .... (505) 277-8446
hweaver@unm.edu
Whitten, David G. (Research) .... (505) 277-5736
whitten@unm.edu
AM,BM,BT,EN,NT,PO

Associate Professors

Anderson, Harold M. .... (505) 277-5661
anderson@unm.edu
Atanassov, Plamen .... (505) 277-2640
plamen@unm.edu
AM,BM,BT,EN,NT,RE
Cooper, Gary W. .... (505) 277-2557
garywc@unm.edu
Han, Sang .... (505) 277-3118
meister@unm.edu
AM,ME,NT,SE,SU
Mead, Richard W. (Emeritus) .... (505) 277-3221
rmead@unm.edu

Assistant Professors

Artyushkova, Kateryna (Research) .... (505) 277-0750
kartyush@unm.edu
AM,PO,RE,SU
Brevnov, Dimitri (Research) .... (505) 277-2563
dbrevnov@unm.edu
EN,ME,NT
Canavan, Heather .... (505) 277-8026
canavan@unm.edu
AM,BM,BT,TH
Challa, Sivakumar (Research) .... (505) 272-7218
challa@unm.edu
AM,MO,SU,TH
Dirk, Elizabeth LeBleu .... (505) 277-5906
edirk@unm.edu
BM,NT,PO,SU
Edwards, Jeremy S. .... (505) 272-5465
jsedwards@salud.unm.edu
Li, Qiming (Research) .... (505) 272-7932
liqiming@unm.edu
EN,ME
Petsev, Dimiter .... (505) 277-3221
dimiter@unm.edu
AM,BM,MO,NT,SE,TH,TP
Pham, Hien (Research) .... (505) 277-5717
hipham@unm.edu
NT,PO,RE
Tournier, Jean-Michel (Research) ..... (505) 277-7961
tournier@unm.edu
Ueki, Taro .... (505) 277-7964
tueki@unm.edu
Xomeritakis, George (Research) ...... (505) 272-7628
xomerita@unm.edu
AM,SE,TP

Instructors

Busch, Robert D. .... (505) 277-8027
busch@unm.edu
EN,MO

Other Faculty

Curro, John (Adjunct) .... (505) 277-5431
jgcurro@sandia.gov
AM,MO,PO,SU
Kroenke, William .... (505) 277-6824
yonder@unm.edu
Loehman, Ronald E. (Research) ...... (505) 272-7601
loehman@unm.edu
Morel, Jim (Adjunct) .... (505) 277-5431
morel@ne.tamu.edu
MO

Accredited by: ABET

Degrees granted 2005-2006:
    B.S.: 24   M.S.: 20   Ph.D.: 9
Graduate advisor: Jeff Brinker

Undergraduate advisor: Abhaya Datye

Student organization: AIChE, ANS
Advisor: Tim L. Ward, Robert Busch

Department reports to: Joseph L. Cecchi, Dean of Engineering

Placement service: Eleanor Sanchez

New Mexico State University

Chemical Engineering Department
Post Office Box 30001, MSC 3805
Las Cruces, NM 88003-8001
Deliveries: Jett Hall, Room 259

University phone .... (505) 646-0111
Department phone .... (505) 646-1214
Department FAX .... (505) 646-7706

Professors

Johnson, Charles L. .... (505) 646-8637
    cjohnson@nmsu.edu
    PO
Long, Richard L. Jr. (Associate Head) . (505) 646-2503
    rilong@nmsu.edu
    BM,SU
Munson-McGee, Stuart .... (505) 646-6439
    stumcgee@nmsu.edu
    AM,NT,PO,SU
Rockstraw, David A. .... (505) 646-7705
    drockstr@nmsu.edu
    AM,EV,PR,RE,SE

Associate Professors

Andersen, Paul .... (505) 646-8153
    pka@nmsu.edu
    EV
Deng, Shuguang .... (505) 646-4346
    sdeng@nmsu.edu
    AM,EN,EV,NT,RE,SE,SU
Mitchell, Martha C. (Head) .... (505) 646-2093
    martmitc@nmsu.edu
    AM,EN,EV,MO,NT,SE,SU

Other Faculty

Del Valle, Francisco .... (505) 646-4204
    fdelvall@nmsu.edu

Accredited by: ABET

Degrees granted 2005-2006:
    B.S.: 7   M.S.: 4   Ph.D.: 1
Graduate advisor: David Rockstraw

Undergraduate advisor: Paul Andersen

Student organization: AIChE
Advisor: Richard L. Long, Jr. (505) 646-2503

Department reports to: Steve Castillo, (505) 646-2914, Dean of Engineering

Placement service: Steve Salway (505) 646-1631, Box 3509

New York

The City College of The City University of New York

Chemical Engineering Department
Convent Avenue at 140th St.
Room 322
New York, NY 10031
Department phone .... (212) 650-7232
Department FAX .... (212) 650-6660

Professors

Acrivos, Andreas (Emeritus)
Couzis, Alexander .... (212) 650-6701
    acouzis@chemail.engr.ccny.cuny.edu
Denn, Morton M. (Director) .... (212) 650-7444
    denn@levdec.engr.ccny.cuny.edu
Graff, Robert A. (Emeritus)
Isaacs, Leslie L. .... (212) 650-7146
    chelli@chemail.engr.ccny.cuny.edu
Maldarelli, Charles .... (212) 650-8160
    charles@chemail.engr.ccny.cuny.edu
Rinard, Irven H. (Chair) .... (212) 650-7135
    rinard@chemail.engr.ccny.cuny.edu
Rumschitzki, David .... (212) 650-5430
    david@ees1s0.engr.ccny.cuny.edu
Shinnar, Reuel .... (212) 650-6679
    shinnar@chemail.engr.ccny.cuny.edu
Steiner, Carol .... (212) 650-7230
    steiner@chemail.engr.ccny.cuny.edu
Tardos, Gabriel I. .... (212) 650-6665
    tardos@chemail.engr.ccny.cuny.edu
Weinstein, Herbert (Emeritus)

Associate Professors

Lee, Jae W. .... (212) 650-6688
    lee@chemail.engr.ccny.cuny.edu
Morris, Jeffrey .... (212) 650-6844
    Jmorris@chemail.engr.ccny.cuny.edu

Assistant Professors

Gilchrist, M. Lane .... (212) 650-6664
    lane@chemail.engr.ccny.cuny.edu
Kretzschmar, Ilona .... (212) 650-6769
    Kretzschmar@ccny.cuny.edu
Tu, Raymond .... (212) 650-7031
    tu@chemail.engr.ccny.cuny.edu

Accredited by: ABET

Graduate advisor: David Rumschitzki (PhD) and Jae W. Lee (Master)

Student organization: AIChE
Advisor: Jae W Lee

Department reports to: Joseph Barba, Dean, Engr

Placement service: Sophia Demetriou (212) 650-6507

Clarkson University

Department of Chemical Engineering
Box 5705
Potsdam, NY 13699-5705
Deliveries: 8 Clarkson Ave.
University phone .... (315) 268-6400
Department phone .... (315) 268-6650
Department FAX .... (315) 268-6654

Professors

Babu, S. V. .... (315) 268-3999
    babu@clarkson.edu
Baltus, Ruth E. (Chair) .... (315) 268-2368
    baltus@clarkson.edu
Campbell, Gregory A. (Emeritus) ..... (315) 268-6577
    gac@clarkson.edu
Chin, Der-Tau (Emeritus) .... (315) 268-7930
    chin@clarkson.edu
Cole, Robert (Emeritus)
Hopke, Philip K. .... (315) 268-3861
    hopkepk@clarkson.edu
McLaughlin, John B. .... (315) 268-6663
    jmclau@clarkson.edu
Rasmussen, Don H. .... (315) 268-3820
    rasmu@clarkson.edu
Subramanian, R. Shankar .... (315) 268-6648
    shankar@clarkson.edu
Taylor, Ross .... (315) 268-6650
    taylor@clarkson.edu
Wilcox, William R. .... (315) 268-7672
    wilcox@clarkson.edu

Associate Professors

Harris, Sandra L. .... (315) 268-2284
    slharris@clarkson.edu
Jachuck, Roshan J.J. .... (315) 268 6325
    rjachuck@clarkson.edu
McCluskey, Richard J. (Associate Chair) (315) 268-2303
    bq02@clarkson.edu
Rengaswamy, Raghunathan .... (315) 268-4423
    raghu@clarkson.edu
Suni, Ian I. .... (315) 268-4471
    isuni@clarkson.edu

Accredited by: ABET

Degrees granted 2005-2006:
    B.S.: 26   M.S.: 8   Ph.D.: 5

Undergraduate advisor: Richard J. McCluskey

Student organization: AIChE
Advisor: I.I. Suni

Department reports to: Dr. Goodarz Ahmadi, Dean of Engineering

Placement service: Kathryn Briggs Johnson

Columbia University

Department of Chemical Engineering
801 S. W. Mudd, Mail Code 4721
500 W. 120th St.
New York, NY 10027
Deliveries: Same

University phone .... (212) 854-1754
Department phone 212-854-4453
Department FAX 212-854-3054

Professors

Durning, C. J. .... (212) 854-8161
    cjd2@columbia.edu
Flynn, George W. .... (212) 854-4162
    flynn@chem.columbia.edu
Gryte, Carl C. .... (212) 854-2470
    ccg2@columbia.edu
Ju, Jingyue .... (212) 854-2487
    dj222@columbia.edu
Koberstein, J. .... (212) 854-3120
    jk1191@columbia.edu
Kumar, Sanat (Visiting) 854-2193
    sk2794@columbia.edu
    BT,PO
Leonard, Edward F. .... (212) 854-4448
    leonard@columbia.edu
O'Shaughnessy, B. .... (212) 854-3203
    bo8@columbia.edu
Spencer, Jordan L. (Emeritus) .... (212) 854-4471
    spencer@columbia.edu
Turro, Nicholas J. .... (212) 854-2175
    turro@chem.columbia.edu
West, A. C. (Chair) .... (212) 854-4452
    acw17@columbia.edu

Assistant Professors

Banta, Scott .... (214) 854-7531
    sab2373@columbia.edu
Shapley, Nina .... (212) 854-1095
    ncs2101@columbia.edu

Accredited by: ABET

Degrees granted 2005-2006:
    B.S.: 32   M.S.: 9   Ph.D.: 2

Graduate advisor: Ben O'Shaughnessy

Undergraduate advisor: Christopher Durning

Student organization: AIChE

Cooper Union

Albert Nerken School of Engineering
Chemical Engineering Department
51 Astor Place
New York, NY 10003

University phone .... (212) 353-4100
Department phone .... (212) 353-4370
Department FAX .... (212) 353-4341

Professors

Ahmed, Zikri .... (212) 353-4379
    ahmed@cooper.edu
Brazinsky, Irv (Chair) .... (212) 353-4373
    brazin@cooper.edu
Cheng, Shang-I (Emeritus)

Associate Professors

Okorator, Charles .... (212) 353-4371
    okoraf@cooper.edu
Sidebotham, George .... (212) 353-4308
    sidebo@cooper.edu
Stock, Richard .... (212) 353-4317
    stock@cooper.edu

Assistant Professors

Lam-Anderson, Marca .... (212) 353-4393
    mjlam@cooper.edu

Accredited by: ABET

Student organization: AIChE
Advisor: Zikri Ahmed

Department reports to: E. Baum, Dean, School of Engineering

Placement service: Melissa Benca

Cornell University

School of Chemical and Biomolecular Engineering
Ithaca, NY 14853-5201
Deliveries: 120 Olin Hall

University phone .... (607) 255-2000
Department phone .... (607) 255-8656
Department FAX .... (607) 255-9166

Professors
Archer, Lynden A. .... (607) 254-8825
Laa25@cornell.edu
PO
Clancy, Paulette (Director) .... (607) 255-4430
pqc1@cornell.edu
EN,EV,ME
Cohen, Claude .... (607) 255-7292
cc112@cornell.edu
PO,TP
Engstrom, James R. .... (607) 255-9934
jre7@cornell.edu
ME
Finn, Robert K. (Emeritus)
Gubbins, Keith E. (Emeritus)
Harriott, Peter (Emeritus) .... (607) 255-3529
ph@cheme.cornell.edu
Koch, Donald L. .... (607) 255-3484
dLk15@cornell.edu
PO,TP
Olbricht, William L. .... (607) 255-4362
wLo1@cornell.edu
PO,TP
Rodriguez, Ferdinand (Emeritus)
fr@cheme.cornell.edu
Shuler, Michael L. .... (607) 255-7577
mLs50@cornell.edu
BM,BT
Smith, Julian C. (Emeritus)
jcs29@cornell.edu
Steen, Paul H. .... (607) 255-4749
phs7@cornell.edu
PO,TP
Streett, William B. (Emeritus)
VonBerg, Robert L. (Emeritus)
Wiegandt, H. F. (Emeritus)

Associate Professors
Anton, A. Brad .... (607) 255-3629
aba6@cornell.edu
EN,PO
Duncan, T. Michael (Associate Head) .. (607) 255-8715
tmd10@cornell.edu
PO,TP
Escobedo, Fernando .... (607) 255-8243
fe13@cornell.edu
PO,TP
Lee, Kelvin H. .... (607) 255-4215
khL9@cornell.edu
BM,BT

Assistant Professors
DeLisa, Matthew P. .... (607) 254-8560
md255@cornell.edu
BM,BT
Joo, Yong L. .... (607) 255-8591
yLj2@cornell.edu
PO,TP
Putnam, David A. .... (607) 255-4352
dap43@cornell.edu
BM,BT,PO,TP
Stroock, Abraham D. .... (607) 255-4276
ads10@cornell.edu
BT,ME,NT
Varner, Jeffrey D. .... (607) 255-4258
jdv27@cornell.edu
BT

Instructors
Center, Alfred M. .... (607) 255-3422
ac222@cornell.edu
EN

Other Faculty
Baxter, Gregory (Adjunct)
Hatzimanikatis, Vassily (Adjunct)
Vaeth, Kathleen (Adjunct)
Weinstein, Steven (Adjunct)

Accredited by: ABET

Degrees granted 2005-2006:
  B.S.: 58   M.S.: 19   Ph.D.: 8

Graduate advisor: F. Escobedo

Undergraduate advisor: T. M. Duncan

Student organization: AIChE
Advisor: J. R. Engstrom

Department reports to: W. Kent Fuchs, Dean of Engineering

Manhattan College

Chemical Engineering Department
Riverdale, NY 10471-4099
Deliveries: 3825 Corlear Ave., Riverdale, NY 10463

Department phone .... (718) 862-7185
Department FAX .... (718) 862-7819

Professors
Burris, Conrad T. (Emeritus) .... (718) 862-7185
    chmldept@manhattan.edu
Heist, Richard (Dean) .... (718) 862-7307
    richard.heist@manhattan.edu
Hollein, Helen C. (Emeritus) .... (718) 862-7281
    chmldept@manhattan.edu
Reynolds, Joseph .... (718) 862-7187
    jreynold@manhattan.edu
Theodore, Louis .... (718) 862-7188
    chmldept@manhattan.edu
Zenz, Frederick A. (Emeritus) .... (718) 862-7185
    chmldept@manhattan.edu

Associate Professors
Assaf-Anid, Nada (Chair) .... (718) 862-7420
    nada.assaf-anid@manhattan.edu
Famularo, Jack (Emeritus) .... (718) 862-7185
    jackf507@rcn.com
Marnell, Paul .... (718) 862-7194
    paul.marnell@manhattan.edu

Assistant Professors
Antonio, Vincitore (Adjunct) .... (718) 862-7185
    Antonio.vincitore@manhattan.edu
Castaldi, Marco (Adjunct) .... (718) 862-7185
    mc2352@columbia.edu
Feintuch, Howard (Adjunct) .... (718) 862-7185
    Howard_Feintuch@fwc.com
Flynn, Ann Marie .... (718) 862-7286
    annmarie.flynn@manhattan.edu

Instructors
Devine, Kevin (Adjunct) .... (718) 862-7185
    kdevine@kraft.com
Lucas, Robert (Adjunct) .... (718) 862-7185
    robert.lucas@manhattan.edu

Accredited by: ABET

Degrees granted 2005-2006:
    B.S.: 27   M.S.: 10

Graduate advisor: Dr. Nada Assaf-Anid

Undergraduate advisor: Richard Schneider

Student organization: AIChE
Advisor: Dr. Joseph Reynolds

Department reports to: Dr. Richard Heist, Dean of Engineering

Placement service: Marjorie Apel, Career Services

Polytechnic University

Dept. of Chemical and Biological Engineering
Six Metrotech Center
Brooklyn, NY 11201

University phone .... (718) 260-3600
Department phone .... (718) 260-3470
Department FAX .... (718) 260-3125

Professors
Mijovic, Jovan (Chair) .... (718) 260-3097
    jmijovic@duke.poly.edu
    AM,PO

Associate Professors
Pinto, Jose M. .... (718) 260-3470
    jpinto@poly.edu
    BM,MO,PR
Stiel, Leonard I. (Emeritus) .... (718) 260-3638
    lstiel@duke.poly.edu
    TH
Ziegler, Edward N. .... (718) 260-3276
    eziegler@duke.poly.edu
    EV,RE
Zurawsky, Walter P. .... (718) 260-3725
    zurawsky@duke.poly.edu
    AM,SU

Assistant Professors
Kim, Jin Ryoun .... (718) 260-3834
    BM,BT,MO
Levicky, Rastislav .... (718) 260-3682
    rlevicky@poly.edu
    BM,BT,NT

Accredited by: ABET

Degrees granted 2005-2006:
    B.S.: 10   M.S.: 7   Ph.D.: 4

Graduate advisor: Jose Pinto

Undergraduate advisor: Edward Ziegler

Student organization: AIChE
Advisor: Walter Zurawsky

Department reports to: Budd Griffis, Vice President & Dean of Engineering

Rensselaer Polytechnic Institute

Chemical & Biological Engineering Department
110-8th Street
Troy, NY 12180-3590
Deliveries: Ricketts Building

University phone	(518) 276-6000
Department phone	(518) 276-6377
Department FAX	(518) 276-4030

Professors

Altwicker, Elmar R. (Emeritus) .... (518) 276-6927
altwie@rpi.edu
EN,EV,RE
Belfort, Georges .... (518) 276-6948
belfog@rpi.edu
BT,PO,SE,SU,TP
Bequette, B. Wayne .... (518) 276-6683
bequette@rpi.edu
BM,BT,EN,EV
Bungay, Henry R. III (Emeritus) ...... (518) 276-6799
bungah@rpi.edu
AM,BT,EV
Cale, Timothy S. .... (518) 276-6059
calet@rpi.edu
AM,ME,MO,NT,RE
Coppens, Marc-Olivier
AM,BT,MO,NT,RE
Cramer, Steven M. .... (518) 276-6198
crames@rpi.edu
BT,NT,SE,SU
Dordick, Jonathan S. .... (518) 276-2899
dordick@rpi.edu
BT,NT
Fontijn, Arthur .... (518) 276-6508
fontia@rpi.edu
EV,RE,TH
Garde, Shekhar .... (518) 276-6048
gardes@rpi.edu
BT,MO,NT,TH
Gill, William N. .... (518) 276-2880
gillw@rpi.edu
AM,ME
Kumar, Sanat .... (528) 276-3032
kumar@rpi.edu
Linhardt, Robert (Joint) .... (518) 276-3404
linhar@rpi.edu
BT
Littman, Howard (Emeritus) .... (518) 276-6039
littmh@rpi.edu
AM,TP
Nauman, E. Bruce .... (518) 276-6726
nauman@rpi.edu
NT,PO,RE
Plawsky, Joel L. .... (518) 276-6049
plawsky@rpi.edu
AM,BT,EN,HT,ME,NT,SU,TP

Van Ness, Hendrick C. (Emeritus) ..... (518) 276-6727
vanneh@rpi.edu
TH
Wayner, Peter C. .... (518) 276-6199
wayner@rpi.edu
EN,HT,MO,NT,SU,TH,TP

Associate Professors

Kane, Ravi .... (518) 276-2356
kaner@rpi.edu
AM,BM,BT,NT,PO

Assistant Professors

Martin, Lealon .... (518) 276-3327
lealon@rpi.edu
BT,EN,NT,PR
Sharfstein, Susan .... (518) 276-2166
sharfs@rpi.edu
BT
Zhang, Fuming (Research) .... (518) 276-6839
zhangf2@rpi.edu
BT,NT,RE,SE

Accredited by: ABET

Degrees granted 2005-2006:
  B.S.: 33   M.S.: 5   Ph.D.: 9

Graduate advisor: Steven M. Cramer

Undergraduate advisor: Steven M. Cramer

Student organization: AIChE
Advisor: B. Wayne Bequette

Department reports to: Alan W. Cramb, Dean of Engineering

Placement service: Thomas Tarantelli (518) 276-6234

University of Rochester

Department of Chemical Engineering
CPU Box 270166
Rochester, NY 14627-0166
Deliveries: 206 Gavett Hall

University phone	(585) 275-2121
University FAX	(585) 273-1348
Department phone	(585) 275-4041
Department FAX	(585) 273-1348

Professors

Chen, Shaw H. (Chair) .... (585) 275-0909
    shch@lle.rochester.edu
Chimowitz, Eldred H. (Associate Chair) (585) 275-8497
    chim@che.rochester.edu
Eisenberg, Richard F. (Emeritus) ...... (585) 275-4041
Greener, Jehuda (Adjunct) .... (585) 275-4041
    jehuda.greener@kodak.com
Jacobs, Stephen J. (Joint) .... (585) 275-4837
    sjac@lle.rochester.edu
Jorne', Jacob .... (585) 275-4584
    jorne@che.rochester.edu
Rothberg, Lewis (Joint) .... (585) 275-7291
    rothberg@chem.rochester.edu
Shapir, Yonthan (Joint) .... (585) 275-7291
    ysha@pas.rochester.ed
Tang, Ching W. .... (585) 275-4041
    ching.tang@kodak.com
Wu, J. H. David .... (585) 275-8499
    davidwu@che.rochester.edu

Associate Professors

Foster, David (Adjunct) .... (585) 275-4041
    dafoster@che.rochester.edy
Harding, David R. (Joint) .... (585) 275-5850
    dhar@lle.rochester.edu

Assistant Professors

Anthamatten, Mitchell .... (585) 273-5526
    anthamatten@che.rochester.edu
King, Michael (Joint) .... (585) 275 3285
    mike_king@urmc.rochester.edu
Yang, Hong .... (585) 275-2110
    hongyang@che.rochester.edu
Yates, Matthew .... (585) 273-2335
    myates@che.rochester.edu

Instructors

Ebenhack, Ben .... (585) 275-9209
    bwe@che.rochester.edu
Hilborn, David (Adjunct) .... (585) 275-4041
    DnDHilborn@aol.com
Olsen, Thor .... (585) 275-7885
    thor@che.rochester.edu
Weinstein, Michael (Adjunct) .... (585) 275-4041
    mikejan@frontiernet.net
Weinstein, Steven (Adjunct) .... (585) 275-4041
    steven.weinstein@kodak.com

Accredited by: ABET

Graduate advisor: Matthew Z. Yates

Undergraduate advisor: Eldred H. Chimowitz

Student organization: AIChE
Advisor: Ben Ebenhack

Department reports to: Shaw H. Chen, Chairman, Department of Chemical Engineering

Placement service: Burton Nadler

State University of New York at Buffalo

Department of Chemical & Biological Engineering
Clifford C. Furnas Hall
Buffalo, NY 14260-4200
Deliveries: 303 Furnas Hall

University phone .... (716) 645-2000
Department phone .... (716) 645-2911
Department FAX .... (716) 645-3822

Professors

Alexandridis, Paschalis .... x2210 [†]
    palexand@eng.buffalo.edu
    AM,BM,NT,PO,SU
Good, Robert J. (Emeritus)
Hlavacek, Vladimir x2208
    hlavacek@acsu.buffalo.edu
    AM,RE
Kofke, David A. x2215
    kofke@eng.buffalo.edu
Lund, Carl R. F. (Associate Dean) x2211
    lund@eng.buffalo.edu
    RE,SU
Mountziaris, T. J. (Adjunct)
Nicholson, Bruce J. (Adjunct)
Nitsche, Johannes M. x2213
    nitsche@eng.buffalo.edu
    BM,MO
Petrou, Athos (Adjunct) .... (716) 645-2987
    petrou@acsu.buffalo.edu
Ruckenstein, Eli x2214
    feaeliru@acsu.buffalo.edu
    AM,BM,BT,NT,PO,SU
Ryan, Michael E. (Dean) .... (716) 645-6003
    meryan@eng.buffalo.edu
    EV,MO,PO
Sachs, Frederick (Adjunct) ...... (716) 829-3289 x105
    sachs@buffalo.edu
van Oss, Carel J. (Adjunct) .... (716) 831-2900
    cjvanoss@acsu.buffalo.edu
Weber, Thomas W. (Emeritus)
    twweber@eng.buffalo.edu
Weller, Sol W. (Emeritus)

Associate Professors

Andreadis, Stelios T. x2204
    sandread@eng.buffalo.edu
    BM,BT,NT,PO
Neelamegham, Sriram x2220
    neel@acsu.buffalo.edu
    BM,BT,EN,MO
Swihart, Mark T. x2205
    swihart@eng.buffalo.edu
    AM,EN,ME,MO,NT,RE
Zhou, Yaoqi (Adjunct) .... (716) 829-2985
    yqzhou@acsu.buffalo.edu

Assistant Professors

Errington, Jeffrey R. x2222
    jerring@eng.buffalo.edu
    AM,BT,EN,MO,NT,SU
Koffas, Mattheos x2221
    mkoffas@eng.buffalo.edu
    BT,EN
Park, Sheldon x2212
    BM,BT,MO,NT
Tzanakakis, Emmanouhl x2206
    emtzan@buffalo.edu
    BM,BT

Instructors

Kofke, Tamara G. x2168
    tgkofke@eng.buffalo.edu

[†] Unless noted faculty numbers are extensions of (716) 645-2911.

Accredited by: ABET

Degrees granted 2005-2006:
    B.S.: 43   M.S.: 10   Ph.D.: 10

Graduate advisor: M. T. Swihart

Undergraduate advisor: J. R. Errington

Student organization: AIChE
Advisor: Manolis Tzanakakis

Department reports to: Harvey Stenger, Dean of Engineering

Placement service: Daniel J. Ryan, Director

Syracuse University

Department of Biomedical and Chemical Engineering
121 Link Hall
Syracuse University
Syracuse, NY 13244-1240

University phone .... (315) 443-1870
Department phone .... (315) 443-1931
Department FAX .... (315) 443-9175

Professors

Carney, Laurel .... (315) 443-9749
lacarney@syr.edu
BM,MO
Engbretson, Gustav A (Chair) .... (315) 443-1931
gengbret@syr.edu
BM
Gilbert, Jeremy (Associate Dean) ..... (315) 443-2105
gilbert@syr.edu
BM,PO,SU
Hiiemae, Karen .... (315) 443-9709
khiiemae@syr.edu
BM
Martin, George C. .... (315) 443-4467
gcmartin@syr.edu
PO
Rice, Philip A. (Emeritus) .... (315) 443-9444
parice@syr.edu
RE
Sangani, Ashok .... (315) 443-4502
asangani@syr.edu
MO,TP
Santanam, Suresh (Adjunct) .... (315) 443-4445
ssantana@syr.edu
EV
Schroder, Klaus (Emeritus) .... (315) 443-2846
kschrode@syr.edu
AM,SU
Smith, Robert L .... (315) 443-9702
bob_smith@isr.syr.edu
BM,MO
Tavlarides, Lawrence L. .... (315) 443-1883
lltavlar@syr.edu
EN,SE
Tien, Chi (Emeritus) .... (315) 443-4050
ctien@syr.edu

Associate Professors

Heydweiller, John C. .... (315) 443-4468
jcheydwe@syr.edu
MO

Assistant Professors

Hasenwinkel, Julie .... (315) 443-3064
jmhasen@syr.edu
BM,PO

Luk, Yan Yeung (Adjunct) .... (315) 443-7440
yluk@syr.edu
AM,NT,SU
Ren, Dacheng .... (315) 443-1931
dren@syr.edu
BT

Accredited by: ABET

Degrees granted 2005-2006:
    B.S.: 28   M.S.: 20   Ph.D.: 2

Graduate advisor: George C. Martin

Undergraduate advisor: George C. Martin

Student organization: AIChE
Advisor: George C. Martin

Department reports to: Shiu-Kai Chin (315) 443-4341, Interim Dean

Placement service: Karen Davis, ECS Opportunities Center, (315) 443-2239

North Carolina

North Carolina Agricultural and Technical State University

Department of Mechanical and Chemical Engineering
618 McNair Hall
Greensboro, NC 27411

University phone	(336) 334-7500
Department phone	(336) 334-7564
Department FAX	(336) 334-7417

Professors

Adewuyi, Yusuf G.	x107 [†]
	adewuyi@ncat.edu
Ilias, Shamsuddin	x317
	ilias@ncat.edu
Kabadi, Vinayak N. (Director)	x327
	kabadi@ncat.edu
King, Franklin G.	x311
	king@ncat.edu
Tatterson, Gary B.	x320
	gbt@ncat.edu
Uitenham, Leonard (Chair)	x310
	u10ham@ncat.edu

Associate Professors

Lou, Jianzhong	x323
	lou@ncat.edu
Roberts, Kenneth L.	x330
	kroberts@ncat.edu
Schimmel, Keith A.	(336)256-2341 ext.2493
	schimmel@ncat.edu

[†] Faculty numbers are extensions of department phone

Accredited by: ABET

Degrees granted 2005-2006:
  B.S.: 22   M.S.: 6

Graduate advisor: Vinayak N. Kabadi

Undergraduate advisor: Vinayak N. Kabadi

Student organization: AIChE
Advisor: Kenneth Roberts

Department reports to: Joseph Monroe (336) 334-7589, Dean, College of Engineering

Placement service: Joyce Edwards (336) 334-7755, Director

North Carolina State University

Department of Chemical & Biomolecular Engineering
Campus Box 7905
Raleigh, NC 27695-7905
Deliveries: Room 2001, EB1, 911 Partners Way

University phone .... (919) 515-2011
Department phone .... (919) 515-2324
Department FAX .... (919) 515-3465

Professors

Beatty, Kenneth O. (Emeritus) .... (919) 515-6398
kobeatty@eos.ncsu.edu
Carbonell, Ruben G. .... (919) 515-5118
ruben@ncsu.edu
AM,BM,BT,PO
DeSimone, Joseph M. .... (919) 962-2166
desimone@unc.edu
AM,BM,EN,EV,NT,PO
Fedkiw, Peter S. (Associate Head) ..... (919) 515-3572
fedkiw@eos.ncsu.edu
AM,EN,NT,RE
Felder, Richard M. (Emeritus) .... (919) 515-2327
felder@eos.ncsu.edu
Genzer, Jan .... (919) 515-2069
jgenzer@eos.ncsu.edu
AM,BT,NT,PO
Grant, Christine S. .... (919) 515-2317
grant@eos.ncsu.edu
EV
Gubbins, Keith E. .... (919) 513-2262
keg@ncsu.edu
MO,NT,TH,TP
Hall, Carol K. .... (919) 515-3571
hall@turbo.che.ncsu.edu
MO,NT,TH
Hopfenberg, Harold B. (Emeritus) ..... (919) 515-2318
hbh@ncsu.edu
Kelly, Robert M. .... (919) 515-6396
rmkelly@eos.ncsu.edu
BT
Khan, Saad A. .... (919) 515-4519
khan@eos.ncsu.edu
AM,EN,PO
Kilpatrick, Peter K. (Head) .... (919) 515-7121
peter-k@eos.ncsu.edu
BT,NT,SU
Lim, Phooi K. .... (919) 515-2328
lim@eos.ncsu.edu
RE
Ollis, David F. .... (919) 515-2329
ollis@eos.ncsu.edu
EN,RE
Overcash, Michael R. .... (919) 515-2325
overcash@eos.ncsu.edu
EV
Parsons, Gregory N. .... (919) 515-7553
gnp@ncsu.edu
ME
Roberts, George W. .... (919) 515-7328
groberts@eos.ncsu.edu
PO,RE
Setzer, C. John (Emeritus) .... (919) 515-2520
setzer@eos.ncsu.edu
Spontak, Richard J. .... (919) 515-4200
rich_spontak@ncsu.edu
AM,NT,PO

Associate Professors

Haugh, Jason M. .... (919) 513-3851
jmhaugh@eos.ncsu.edu
BT
Lamb, H. Henry .... (919) 515-6395
lamb@eos.ncsu.edu
RE
Peretti, Steven W. .... (919) 515-6397
peretti@eos.ncsu.edu
BT
Velev, Orlin D. .... (919) 513-4318
odvelev@unity.ncsu.edu
BT,NT

Assistant Professors

Rao, Balaji
bmrao@ncsu.edu
BT

Instructors

Bullard, Lisa G. .... (919) 515-7455
lgbullar@eos.ncsu.edu

Accredited by: ABET

Degrees granted 2005-2006:
    B.S.: 61   M.S.: 20   Ph.D.: 18

Graduate advisor: Saad Khan

Undergraduate advisor: Lisa Bullard

Student organization: AIChE
Advisor: Peter Kilpatrick

Department reports to: . Louis Martin-Vega (919) 515-2311, Dean of Engineering

Placement service: Beverly J. Marchi (919) 515-2396

North Dakota

University of North Dakota

Department of Chemical Engineering
Harrington Hall Room 323
241 Centennial Drive, Stop 7101
Grand Forks, ND 58202-7101

University phone .... (701) 777-2011
Department phone .... (701) 777-4244
Department FAX .... (701) 777-3773

Professors

Owens, Thomas C. (Emeritus) .... (701) 777-4245
        tom.owens@mail.und.nodak.edu
Watson, John (Dean) .... (701) 777-3411
        john.watson@mail.und.nodak.edu

Associate Professors

Mann, Michael D. (Chair) .... (701) 777-3852
        mike.mann@mail.und.nodak.edu
        EN,EV,MO
Muggli, Darrin S. .... (701) 777-2337
        darrin.muggli@und.nodak.edu
        EN,EV,NT,RE,SE
Seames, Wayne S. .... (701) 777-2958
        wayne.seames@mail.und.nodak.edu
        BT,EN,EV,PR

Assistant Professors

Bowman, Frank .... (701) 777-4244
        frank.bowman@mail.und.nodak.edu
        EN,EV,MO,RE,SE
Kolodka, Edward B. .... (701) 777-3798
        edward.kolodka@mail.und.nodak.edu
        AM,PO
Tande, Brian .... (701) 777-3797
        briantande@mail.und.edu
        AM,EN,EV,NT,PO

Accredited by: ABET

Degrees granted 2005-2006:
    B.S.: 21   M.S.: 8   Ph.D.: 1

Graduate advisor: Mann, Michael

Undergraduate advisor: Mann, Michael

Student organization: AIChE
Advisor: Muggli, Darrin S.

Department reports to: John L. Watson, Dean, School of Engineering and Mines

Placement service: Mark Thompson, Director, Career Services

Ohio

University of Akron

Department of Chemical and Biomolecular Engineering
Whitby Hall 211
Akron, OH 44325-3906
Deliveries: 200 E. Buchtel Commons

University phone	(330) 972-7111
Department phone	(330) 972-7250
Department FAX	(330) 972-5856

Professors

Atwood, Glenn A. (Emeritus) .... (330) 972-7250
Chase, George G. .... (330) 972-7943
    gchase@uakron.edu
    PO,SE,TP
Cheung, H. Michael .... (330) 972-7282
    cheung@uakron.edu
    AM,PO,SU
Chuang, Steven .... (330) 972-6993
    schuang@uakron.edu
    EN,NT,RE
Elliott, J. Richard .... (330) 972-7253
    jelliott@uakron.edu
    MO,TH
Greene, Howard (Emeritus) .... (330) 972-7250
Ju, Lu-Kwang (Chair) .... (330) 972-7252
    lukeju@uakron.edu
    BT,EV
Roberts, Robert W. (Emeritus) .... (330) 972-7250

Associate Professors

Evans, Edward A. .... (330) 972-8292
    evanse@uakron.edu
    AM,RE,SU
Lopina, Stephanie T. .... (330) 972-7255
    lopina@uakron.edu
    BM,NT,PO
Qammar, Helen C. .... (330) 972-5917
    hqammar@uakron.edu
    MO,PR
Wang, Ping (Adjunct) .... (330) 972-2096
    wangp@uakron.edu
    BT,NT

Assistant Professors

Newby, Bi-min Zhang .... (330) 972-2510
    bmznewby@uakron.edu
    AM,NT,SU

Accredited by: ABET

Degrees granted 2005-2006:
    B.S.: 26   M.S.: 8   Ph.D.: 4

Graduate advisor: J Richard Elliott

Undergraduate advisor: Helen Qammar

Student organization: AIChE
Advisor: J Richard Elliott

Department reports to: Dr. George K. Haritos, Dean of Engineering

Placement service: Dr. Kim Beyer, Interim Co-Director of Center for Career Management

Case Western Reserve University

Department of Chemical Engineering
10900 Euclid Avenue
Cleveland, OH 44106-7217
Deliveries: A.W. Smith Building, Room 116

University phone	(216) 368-2000
Department phone	(216) 368-4182
Department FAX	(216) 368-3016

Professors

Anderson, John L. .... (216) 368-4346
jla24@case.edu
SU,TP

Angus, John C. (Emeritus) .... (216) 368-4133
jca3@case.edu
AM,NT,TH

Edwards, Robert V. .... (216) 368-4151
rve2@case.edu
MO

Feke, Donald L. .... (216) 368-2750
dlf4@case.edu
SE,SU,TP

Harris, Robert E. (Adjunct) .... (216) 368-1145
reh@case.edu
MO

Landau, Uziel .... (216) 368-4132
uxl@case.edu
AM,EN,MO,TP

Liu, Chung-Chiun .... (216) 368-2935
cxl9@case.edu
AM,BM,EN,ME

Mann, J. Adin Jr. .... (216) 368-4122
jam12@case.edu
SU

Pintauro, Peter N. (Chair) .... (216) 368-4150
pnp3@case.edu
EN,NT,PO,SE,TP

Qutubuddin, Syed .... (216) 368-2764
sxq@case.edu
PO,SE,SU

Savinell, Robert F. (Dean) .... (216) 368-4436
rfs2@case.edu
EN,ME

Wnek, Gary .... (216) 368-2782
gew5@case.edu
EN,NT,PO

Zawodzinski, Thomas .... (216) 368-5547
taz5@case.edu
AM,EN,NT,PO,RE,TP

Assistant Professors

Baskaran, Harihara .... (216) 368-1029
hari@case.edu
BM,BT,TP

Martin, Heidi B. .... (216) 368-3810
hbm@case.edu
AM,BM,EN,NT

Sankaran, R. Mohan .... (216) 368-4589
rxs192@case.edu
AM,BT,EN,ME,NT,RE

Accredited by: ABET

Degrees granted 2005-2006:
  B.S.: 19   M.S.: 12   Ph.D.: 8

Graduate advisor: Heidi Martin

Student organization: AIChE
Advisor: Peter N. Pintauro

Department reports to: Robert F. Savinell, Dean of Engineering

Placement service: Patrick Keebler

University of Cincinnati

Department of Chemical and Materials Engineering
Mail Location 12
Cincinnati, Ohio 45221-0012
Deliveries: 497 Rhodes Hall

University phone .... (513) 556-6000
Department phone .... (513) 556-3096
Department FAX .... (513) 556-2569

Professors

Boerio, James F. .... (513) 556-3111
    F.james.boerio@uc.ed
Buchanan, Relva C. .... (513) 556-3190
    relva.buchanan@uc.edu
Clarson, Stephen J. .... (513) 556-5430
    stephen.carlson@uc.edu
Delcamp, Robert M. (Emeritus) .... (513) 556-2761
Donglu, Shi .... (513) 556-3100
    donglu.shi@uc.edu
Fried, Joel R. .... (513) 556-2767
    Joel.Fried@uc.edu
Govind, Rakesh .... (513) 556-2666
    Rakesh.Govind@UC.Edu
Greenberg, David B. (Emeritus) .... (513) 556-2741
    dgreenbe@uceng.uc.edu
Hershey, Daniel .... (513) 556-2740
    Daniel.Hershey@UC.edu
Hwang, Sun-Tak .... (513) 556-2791
    Sun-Tak.Hwang@UC.Edu
Kao, Y. K. .... (513) 556-2762
    ykao@alpha.che.uc.edu
Khang, Soon-Jai .... (513) 556-2789
    skhang@alpha.che.uc.edu
Lemlich, Robert (Emeritus) .... (513) 556-2761
Licht, William (Emeritus) .... (513) 556-2761
Lin, Jerry Y.S. (Adjunct) .... (513) 556-2769
    jlin@alpha.che.uc.edu
Phillips, Paul .... (513) 556-1178
    paul.phillips@uc.edu
Pinto, Neville G. (Associate Dean) .... (513) 556-2770
    neville.pinto@uc.edu
Schaefer, Dale W. .... (513) 556-5431
    dale.schaefer@uc.edu
Sekhar, Jainagesh A. .... (513) 556-3105
    jainagesh.sekhar@uc.edu
Smirniotis, Panagiotis (Head) .... (513) 556-1474
    Panagiotis.Smirniotis@uc.edu
Van Ooij, Wim J. .... (513) 556-3194
    wim.vanooij@uc.edu
Vasudevan, Vijay K. .... (513) 556-3103
    vijay.vasudevan@uc.edu

Associate Professors

Beaucage, Gregory .... (513) 556-3063
    gregory.beaucage@uc.edu
Cosgrove, Stanley (Emeritus) .... (513) 556-2761
Guliants, Vadim (Director) .... (513) 556-0203
    vguliant@alpha.che.uc.edu
Iroh, Jude O. .... (513) 556-3115
    jude.iroh@uc.edu
Roseman, Rodney D. (Associate Head) . (513) 556-3187
    rodney.roseman@uc.edu

Assistant Professors

Co, Carlos .... (513) 556-2731
    cco@alpha.che.uc.edu
Ho, Chia-Chi .... (513) 556-2438
    cho@alpha.che.uc.edu

Other Faculty

Krantz, William (Emeritus) .... (513) 556-4021

Accredited by: ABET

Degrees granted 2005-2006:
    B.S.: 37   M.S.: 11   Ph.D.: 4

Graduate advisor: V. Guliants

Undergraduate advisor: Carlos Co

Student organization: AIChE
Advisor: Carlos Co

Department reports to: Carlo Montemagno, Dean

Placement service: Stephen Gilby

Cleveland State University

Chemical and Biomedical Engineering Department
2121 Euclid Avenue, Stilwell Hall 455
Cleveland, OH 44115
Deliveries: Stilwell Hall 455

University phone . . . . (216) 687-2000
Department phone . . . . (216) 687-2569
Department FAX . . . . (216) 687-9220

Professors

Coulman, George A. (Emeritus) . . . . (216) 687-3526
Gatica, Jorge E. . . . . (216) 523-7274
j.gatica@csuohio.edu
BM,MO,RE,SU
Ghorashi, Bahman (Associate Dean) . . . (216) 687-2562
b.ghorashi@csuohio.edu
EN,MO,PR
Godleski, Edward S. (Emeritus) . . . . (216) 687-2555
Shah, Dhananjai B. (Head) . . . . (216) 687-3569
d.shah@csuohio.edu
AM,PR,SE
Talu, Orhan . . . . (216) 687-3539
o.talu@csuohio.edu
AM,EN,EV,SE,SU
Tewari, Surendra N. . . . . (216) 687-7342
s.tewari@csuohio.edu
AM,EN,PO

Associate Professors

Belovich, Joanne M. . . . . (216) 687-3502
j.belovich@csuohio.edu
BM,BT,MO
Chatzimavroudis, George P. . . . . (216) 687-5396
g.chatzimavroudis@csuohio.edu
BM,MO
Lustig, Rolf . . . . (216) 687-3526
r.lustig@csuohio.edu
MO
Ungarala, Sridhar . . . . (216) 687-9368
s.ungarala@csuohio.edu
MO,PR,SE

Assistant Professors

Holland, Nolan B. . . . . (216) 687-2569
N.Holland@csuohio.edu
BM,PO

Other Faculty

Rekhson, Simon . . . . (216) 687-5283
s.rekhson@csuohio.edu
AM,EN

Accredited by: ABET

Degrees granted 2005-2006:
 B.S.: 19   M.S.: 18   Ph.D.: 1
Graduate advisor: Dr. Jorge E. Gatica

Undergraduate advisor: Jorge E. Gatica

Student organization: AIChE
Advisor: Sridhar Ungarala

Department reports to: Dr. Charles Alexander, Dean of Engineering

Placement service: Paul B. Klein (216) 687-2246, Director

University of Dayton

Chemical & Materials Engineering Department
Kettering Lab - Room 445
300 College Park
Dayton, OH 45469-0246

University phone .... (937) 229-4411
Department phone .... (937) 229-2627
Department FAX .... (937) 229-3433

Professors

Browning, Charles E. .... (937) 229-2679
    Charles.Browning@notes.udayton.edu
    AM,NT
Dai, Liming .... (937) 229-2679
    Liming.Dai@notes.udayton.edu
    NT,PO
Eylon, Daniel (Director) .... (937) 229-2679
    Daniel.Eylon@notes.udayton.edu
    AM,SU
Flach, Lawrance .... (937) 229-2627
    Larry.Flach@notes.udayton.edu
    MO,PR
Lee, C. William .... (937) 229-2627
    Chih-Kuo.Lee@notes.udayton.edu
    MO,PR
Myers, Kevin J. .... (937) 229-2627
    Kevin.Myers@notes.udayton.edu
    RE
Saliba, Tony E. (Chair) .... (937) 229-2627
    TSaliba@udayton.edu
    MO,SE
Sandhu, Sarwan S. .... (937) 229-2627
    Sarwan.Sandhu@notes.udayton.edu
    EN,TP
Snide, James A. (Emeritus) .... (937) 229-2679
    James.Snide@notes.udayton.edu

Associate Professors

Lu, Christopher C. (Emeritus) .... (937) 229-2627
    Chris.Lu@notes.udayton.edu
Wilkens, Robert .... (937) 229-2627
    Wilkens@notes.udayton.edu
    TP

Assistant Professors

Dewitt, Matthew (Joint) .... (937) 229-2627
    Matthew.Dewitt@wpafb.af.mil
Elsass, Michael (Adjunct) .... (937) 229-2627
    Michael.Elsass@notes.udayton.edu
Johnson, Jay (Joint) .... (937) 229-2569
    Jay.Johnson@udri.udayton.edu
    BT
Klosterman, Donald (Joint) .... (937) 229-4365
    Donald.Klosterman@udri.udayton.edu
    NT,PO

Instructors

Ciric, Amy .... (937) 229-2627
    Amy.Ciric@notes.udayton.edu
    MO,PR

Other Faculty

Hart, Elizabeth .... (937) 229-2627
    Elizabeth.Hart@notes.udayton.edu

Accredited by: ABET

Degrees granted 2005-2006:
    B.S.: 33   M.S.: 5

Graduate advisor: Dr. Kevin J. Myers

Student organization: AIChE
Advisor: Robert J. Wilkens

Department reports to: Dr. Joseph Saliba, Dean of Engineering

Placement service: Gregory Hayes, Director

Ohio State University

Chemical & Biomolecular Engineering Department
125 Koffolt Labs
140 West 19th Avenue
Columbus, OH 43210-1180

University phone .... (614) 292-3980
Department phone .... (614) 292-7907
Department FAX .... (614) 292-3769

Professors
Bakshi, Bhavik .... (614) 292-4904
    bakshi@chbmeng.ohio-state.edu
Brodkey, Robert (Emeritus) .... (614) 292-2609
    brodkey@chbmeng.ohio-state.edu
Chalmers, Jeffrey .... (614) 292-2727
    chalmers@chbmeng.ohio-state.edu
Cooper, Stuart (Chair) .... (614) 292-7907
    coopers@chbmeng.ohio-state.edu
Fan, L. S. .... (614) 688-3262
    fan@chbmeng.ohio-state.edu
Hershey, Harry C. (Emeritus) .... (614) 292-6591
    hershey@che.eng.ohio-state.edu
Ho, W.S. Winston .... (614) 292-9970
    ho@chbmeng.ohio-state.edu
Koelling, Kurt W. .... (614) 292-2256
    koelling@chbmeng.ohio-state.edu
Lee, James Ly .... (614) 292-2408
    leelj@chbmeng.ohio-state.edu
Ozkan, Umit .... (614) 292-6623
    ozkan@chbmeng.ohio-state.edu
Paulaitis, Michael .... (614) 247-8847
    paulaitis@chbmeng.ohio-state.edu
Rathman, James F. .... (614) 292-3760
    rathman@chbmeng.ohio-state.edu
Slider, H. C. (Emeritus) .... (614) 292-2698
    slider@che.eng.ohio-state.edu
Smith, Edwin E. (Emeritus) .... (614) 292-6033
    smith@che.eng.ohio-state.edu
Sweeney, Thomas L. (Emeritus) .... (614) 292-7907
    sweeney.3@osu.edu
Tomasko, David .... (614) 292-4249
    tomasko@chbmeng.ohio-state.edu
Wyslouzil, Barbara .... (614) 688-3583
    wyslouzil@chbmeng.ohio-state.edu
Yang, Shang-Tian .... (614) 292-6611
    yangst@chbmeng.ohio-state.edu
Zakin, Jacques L. (Emeritus) .... (614) 688-4113
    zakin@chbmeng.ohio-state.edu

Associate Professors
Lee, Stephen .... (614) 688-5447
    lee@bme.ohio-state.edu

Assistant Professors
Kusaka, Isamu .... (614) 292-7907
    kusaka@chbmeng.ohio-state.edu
Winter, Jessica .... (614) 247-7668
    winter.63@osu.edu

Instructors
Corn, John (Adjunct) .... (614) 688-8254
    corn@chbmeng.ohio-state.edu

Accredited by: ABET

Degrees granted 2005-2006:
    B.S.: 49   M.S.: 3   Ph.D.: 21

Graduate advisor: Kurt Koelling

Undergraduate advisor: Mary Hoy

Student organization: AIChE
Advisor: Barbara Wyslouzil

Department reports to: Bud Baeslack (614) 292-2836, Dean of Engineering

Placement service: Rosemary Hill (614) 292-6651

Ohio University

Department of Chemical and Biomolecular Engineering
172 Stocker Center
Athens, OH 45701-2979

University phone .... (740) 593-1000
Department phone .... (740) 593-1492
Department FAX .... (740) 593-0873

Professors

Clement, Gilles (Visiting) .... (740) 593-0033
    clement@ohio.edu
Crist, Kevin C. .... (740) 593-4751
    cristk@ohio.edu
Goetz, Douglas J. .... (740) 593-1494
    goetzd@ohio.edu
    BM,BT
Nesic, Srdjan .... (740) 593-1498
    nesic@ohio.edu
Prudich, Michael E. .... (740) 593-1501
    prudich@ohio.edu
    EN,EV,SE

Associate Professors

Botte, Gerardine G. .... (740) 593-9670
    botte@ohio.edu
Chen, Wen-Jia .... (740) 593-1497
    chenw@ohio.edu
Gu, Tingyue .... (740) 593-1499
    gu@ohio.edu
    BT,SE,SU
Gulino, Daniel A. .... (740) 593-1495
    gulino@ohio.edu
    AM,NT
Ridgway, Darin (Associate Chair) ..... (740) 593-1504
    ridgway@ohio.edu
    BT
Sampson, Kendree J. (Associate Dean) . (740) 593-1503
    sampson@ohio.edu
Young, Valerie L. (Chair) .... (740) 593-1496
    youngv@ohio.edu
    EN,EV

Accredited by: ABET

Degrees granted 2005-2006:
    B.S.: 7   M.S.: 3   Ph.D.: 1

Graduate advisor: Dan Gulino

Undergraduate advisor: Darin Ridgway

Student organization: AIChE
Advisor: Darin Ridgway

Department reports to: Dennis Irwin, Dean, Russ College of Engineering and Technology

Placement service: Thomas F. Korvas, Career Services

University of Toledo

Chemical and Environmental Engineering Department
NI 3048, Mail Stop 305
2801 West Bancroft Street
Toledo, OH 43606-3390
Deliveries: 3048 Nitschke Hall

Department phone .... (419) 530-8080
Department FAX .... (419) 530-8086

Professors

Abraham, Martin A. (Dean) .... (419) 530-8092
    martin.abraham@utoledo.edu
Bennett, Gary F. (Emeritus)
    gbennett@eng.utoledo.edu
Brace, John (Research) .... (419) 530-5003
    jbrace@utnet.utoledo.edu
Coleman, Maria R. .... (419) 530-8091
    mcolema@utnet.utoledo.edu
DeWitt, Kenneth J. (Emeritus) .... (419) 530-8094
    kdewitt@utnet.utoledo.edu
Dismukes, John P. .... (419) 530-8065
    john.dismukes@utoledo.edu
Jabarin, Saleh A. (Director) .... (419) 530-5005
    sjabari@utnet.utoledo.edu
LeBlanc, Steven E. (Director) .... (419) 530-8264
    steven.leblanc@utoledo.edu
Lipscomb, G. Glenn (Chair) .... (419) 530-8088
    glenn.lipscomb@utoledo.edu
Nadarajah, Arunan .... (419) 530-8060
    anadara@utnet.utoledo.edu
Poling, Bruce E. (Associate Dean) ..... (419) 530-8255
    bpoling@utnet.utoledo.edu
Varanasi, Sasidhar .... (419) 530-8093
    svarana@utnet.utoledo.edu

Associate Professors

Azad, Abdul-Majeed .... (419) 530-8103
    aazad@utnet.utoledo.edu
Cameron, Michael R. (Research) ...... (419) 530-5009
    mcamero@utnet.utoledo.edu
Escobar, Isabel C. .... (419) 530-8267
    isabel.escobar@utoledo.edu
Schall, Constance A. .... (419) 530-8097
    constance.schall@utoledo.edu

Assistant Professors

Kim, Dong-Shik .... (419) 530-8084
    dong.kim@utoledo.edu

Accredited by: ABET

Degrees granted 2005-2006:
    B.S.: 34   M.S.: 4   Ph.D.: 4

Student organization: AIChE
Advisor: Dong-Shik Kim

Placement service: Andrea Joldrichsen

Youngstown State University

Chemical Engineering Program
One University Plaza
Youngstown, OH 44555

University phone .... (330) 941-3000
Department phone .... (330) 941-3019
Department FAX .... (330) 941-3265

Professors

Garr, Jeanette .... (330) 941-1737
                                         jmgarr@ysu.edu
Hirtzel, Cynthia (Dean) .... (330) 941-3009
                                         cshirtzel@ysu.edu
Lim, Soon-Sik .... (330) 941-3022
                                         sslim@ysu.edu

Associate Professors

Price, Douglas (Director) .... (330) 941-3019
                                         dmprice@ysu.edu
                                         AM,EN

Accredited by: ABET

Degrees granted 2005-2006:
      B.S.: 6   M.S.: 1

Graduate advisor: Dr. Douglas M. Price

Undergraduate advisor: Dr. Douglas M. Price

Student organization: AIChE
Advisor: Dr. Soon-Sik Lim

Department reports to: Dr. Cynthia Hirtzel, Dean, College of Engineering & Technology (330) 941-3009

Placement service: Diane Hritz, Career Services (330) 742-3515

Oklahoma

University of Oklahoma

School of Chemical, Biological and Materials Engineering
100 E. Boyd, T-335
Sarkeys Energy Center
Norman, OK 73019-1004

University phone .... (405) 325-0311
Department phone .... (405) 325-5811
Department FAX .... (405) 325-5813

Professors

Bagajewicz, Miguel J. .... (405) 325-5458
bagajewicz@ou.edu
EN,EV,MO,PR,SE
Block, Robert J. (Emeritus) .... (405) 325-5811
Crynes, Billy L. (Emeritus) .... (405) 325-5811
crynes@ou.edu
Daniels, Raymond D. (Emeritus) ...... (405) 325-5811
Grady, Brian P. .... (405) 325-4369
bpgrady@ou.edu
AM,NT
Harwell, Jeffrey H. .... (405) 325-4375
jharwell@ou.edu
AM,EN,EV,NT,SE,TH
Lee, Lloyd L. (Emeritus) .... (405) 325-5811
lle@ou.edu
Lobban, Lance (Director) .... (405) 325-5814
llobban@ou.edu
EN,RE
Mallinson, Richard G. .... (405) 325-4378
mallinson@ou.edu
EN,RE,TH
O'Rear, Edgar A. III .... (405) 325-4379
eorear@ou.edu
BM,SU,TP
Resasco, Daniel E. .... (405) 325-4370
resasco@ou.edu
AM,EN,NT,SU
Scamehorn, John F. .... (405) 325-4382
scamehor@ou.edu
EV,SE,SU,TH
Shambaugh, Robert L. .... (405) 325-6070
shambaugh@ou.edu
AM,HT,PO,TP
Sliepcevich, C. M. (Emeritus) .... (405) 325-5811
Starling, Kenneth E. (Emeritus) .... (405) 325-5811

Associate Professors

Harrison, Roger G. .... (405) 325-4367
rharrison@ou.edu
BM,BT,NT,SE
Nollert, Matthias U. .... (405) 325-4366
nollert@ou.edu
AM,BM,BT,TP
Papavassiliou, Dimitrios .... (405) 325-0574
dvpapava@ou.edu
EN,MO,NT,TP
Schmidtke, David W. .... (405) 325-7944
dschmidtke@ou.edu
BM,BT,NT

Assistant Professors

McFetridge, Peter S. .... (405) 325-7193
mcfetridge@ou.edu
AM,BM,BT,NT,SU
Sikavitsas, Vassilios I. .... (405) 325-1511
vis@ou.edu
BM,BT,RE,TP
Striolo, Alberto .... (405) 325-5716
astriolo@ou.edu
AM,BM,EN,EV,MO,NT,PO,SU

Accredited by: ABET

Degrees granted 2005-2006:
  B.S.: 41   M.S.: 11   Ph.D.: 2

Graduate advisor: M. U. Nollert

Undergraduate advisor: V. I. Sikavitsas

Student organization: AIChE
Advisor: V. I. Sikavitsas

Department reports to: Dr. Tom Landers (405) 325-2621, Dean of Engineering

Placement service: Bette Scott (405) 325-1974

Oklahoma State University

School of Chemical Engineering - 423EN
Stillwater, OK 74078-5021
Deliveries: 423 Engineering North

University phone .... (405) 744-5000
Department phone .... (405) 744-5280
Department FAX .... (405) 744-6338

Professors

Bell, Kenneth J. (Emeritus) .... (405) 744-5280
kbell@okstate.edu
HT
Foutch, Gary L. .... (405) 744-5280
foutch@okstate.edu
MO,SE,TH
Gasem, Khaled A. M. .... (405) 744-5280
gasem@okstate.edu
EV,SE,TH
Halligan, James E. (Emeritus) .... (405) 744-5000
halligan@okstate.edu
Johannes, Arland H. .... (405) 744-9118
aj@okstate.edu
EV,MO,SE
Maddox, Robert N. (Emeritus) .... (405) 744-5280
rmaddox@okstate.edu
EN,PR,TH
Rhinehart, R. Russell (Head) .... (405) 744-5280
rrr@okstate.edu
MO,PR
Robinson, Robert L. Jr. (Emeritus) ... (405) 744-5280
rrobins@okstate.edu
SU,TH
Tree, D. Alan (Associate Dean) .... (405) 744-5280
tree@okstate.edu
PO,TH
Wagner, Jan .... (405) 744-5280
jwagner@okstate.edu
PR,SE,TH

Associate Professors

High, Karen A. .... (405) 744-9112
high@okstate.edu
EV,MO
High, Martin S. .... (405) 744-9125
mhigh@okstate.edu
PO,TH
Whiteley, James R. .... (405) 744-9117
whitele@okstate.edu
MO,PR

Assistant Professors

Gappa-Fahlenkamp, Heather .... (405) 744-5280
BM,BT
Madihally, Sundararajan V. .... (405) 744-5280
svm@okstate.edu
BM,BT,RE

Smay, James E. .... (405) 744-5280
smay@okstate.edu
AM,NT,SU

Other Faculty

Moshfeghian, Mahmood (Adjunct)
EN,PR,SE
Shariat, Ahmad (Adjunct)
EN,SE

Accredited by: ABET

Degrees granted 2005-2006:
  B.S.: 23   M.S.: 13   Ph.D.: 3

Graduate advisor: Khaled A. M. Gasem

Undergraduate advisor: R. Russell Rhinehart

Student organization: AIChE
Advisor: Sundar V. Madihally

Department reports to: Karl N. Reid, Dean, College of Engineering, Architecture, and Technology

Placement service: Cathy Southwick, Coordinator, Career Services

University of Tulsa

Chemical Engineering Department
600 South College
Tulsa, OK 74104-3189
Deliveries: 600 S. College

University phone .... (918) 631-2000
Department phone .... (918) 631-2226
Department FAX .... (918) 631-3268

Professors

Buthod, Paul (Emeritus)
Luks, Kraemer D. .... (918) 631-2974
                kraemer-luks@utulsa.edu
Manning, Francis S. .... (918) 631-2977
                frank-manning@utulsa.edu
Price, Geoffrey L. (Chair) .... (918) 631-2575
                price@utulsa.edu
Sublette, Kerry L. .... (918) 631-3085
                kerry-sublette@utulsa.edu
Thompson, Richard E. (Emeritus)
                richardthompson@webzone.net

Associate Professors

Ford, Laura P. .... (918) 631-2227
                laura-ford@utulsa.edu
Wisecarver, Keith D. .... (918) 631-2975
                keith-wisecarver@utulsa.edu

Assistant Professors

Crunkleton, Daniel .... (918) 631-2644
                daniel-crunkleton@utulsa.edu

Other Faculty

Patton, Christi L. .... (918) 631-2978
                christi-patton@utulsa.edu

Accredited by: ABET

Degrees granted 2005-2006:
    B.S.: 12   M.S.: 6   Ph.D.: 2

Graduate advisor: Laura P. Ford

Undergraduate advisor: Christi Patton-Luks

Student organization: AIChE
Advisor: Daniel Crunkleton

Department reports to: Steve Bellovich (918) 631-2478, Dean of Engineering and Natural Science

Placement service: Mike Mills, Director of Placement

Oregon

Oregon State University

Chemical Engineering Department
103 Gleeson Hall
Corvallis, OR 97331-2702

Department phone .... (541) 737-2491
Department FAX .... (541) 737-4600

Department reports to: Ron Adams, Dean

Placement service: Tom Munnerlyn (541) 737-4085

Professors
Kimura, Shoichi .... (541) 737-4831
    kimuras@che.orst.edu
Knudsen, James G. (Emeritus) .... (541) 737-4791
    knudsejg@che.orst.edu
Levenspiel, Octave (Emeritus) .... (541) 737-3618
    levenspo@peak.org
McGuire, Joseph .... (541) 737-6306
    mcguirej@engr.oregonstate.edu
Wicks, Charles E. (Emeritus) .... (541) 737-4797
    wicksc@ucs.orst.edu
Williamson, Kenneth (Head) .... (541) 737-6836
    kenneth.williamson@oregonstate.edu

Associate Professors
Bothwell, Michelle K. .... (541) 737-6313
    bothwell@engr.orst.edu
Jovanovic, Goran .... (541) 737-3614
    goran@che.orst.edu
Kelly, Christine .... (541) 737-6755
    christine.kelly@oregonstate.edu
Koretsky, Milo D. .... (541) 737-4591
    koretsm@engr.orst.edu
Levien, Keith L. .... (541) 737-3155
    levienk@engr.orst.edu
Rochefort, Willie E. .... (541) 737-2408
    rochefsk@engr.orst.edu
Rorrer, Gregory L. .... (541) 737-3370
    rorrergl@che.orst.edu

Assistant Professors
Chang, Chih-hung (Alex) .... (541) 737-8548
    changch@che.orst.edu
Yokochi, Alexandre .... (541) 737-9357
    alex.yokochi@orst.edu

Other Faculty
Hackleman, David .... (541) 737-8988
    david.hackleman@oregonstate.edu

Accredited by: ABET

Degrees granted 2005-2006:
    B.S.: 60   M.S.: 6   Ph.D.: 3

Graduate advisor: Goran Jovanovic

Undergraduate advisor: W. E. Rochefort

Student organization: AIChE
Advisor: W. E. Rochefort

Pennsylvania

Bucknell University

Chemical Engineering Department
Lewisburg, PA 17837

University phone .... (570) 577-1271
Department phone .... (570) 577-1114
Department FAX .... (570) 577-1141

Professors

Csernica, Jeffrey (Chair) .... (570) 577-1114
    csernica@bucknell.edu
Hanyak, Michael E. Jr. .... (570) 577-3547
    hanyak@bucknell.edu
King, William E. Jr. .... (570) 577-3403
    wking@bucknell.edu
Pommersheim, James M. (Emeritus)
    pommrshm@bucknell.edu
Prince, Michael J. .... (570) 577-1781
    prince@bucknell.edu
Slonaker, Robert E. Jr. (Emeritus)
    slonaker@bucknell.edu
Snyder, William J. .... (570) 577-1293
    bsnyder@bucknell.edu

Associate Professors

Cavanagh, Daniel P. .... (570) 577-3402
    dcavanag@bucknell.edu
Maneval, James E. .... (570) 577-1669
    maneval@bucknell.edu

Assistant Professors

Jablonski, Erin L. .... (570) 577-1644
    ejablons@bucknell.edu
Raymond, Timothy M. .... (570) 577-3192
    traymond@bucknell.edu
Vigeant, Margot A.S. .... (570) 577-1646
    mvigeant@bucknell.edu

Accredited by: ABET

Degrees granted 2005-2006:
    B.S.: 30   M.S.: 3

Student organization: AIChE
Advisor: Timothy Raymond

Department reports to: James G. Orbison (570) 577-3711, Dean of Engineering

Placement service: Career Development (570) 577-1238, Director

Carnegie Mellon University

Chemical Engineering Department
Pittsburgh, PA 15213
Deliveries: Doherty Hall 1105, attn: Toni McIltrot

University phone	(412) 268-2000
Department phone	(412) 268-2230
Department FAX	(412) 268-7139

Professors

Anderson, John L. (Adjunct) .... (412) 268-2537
johna@cmu.edu
PO

Biegler, Lorenz T. .... (412) 268-2232
lb01@andrew.cmu.edu
PR

Domach, Michael M. .... (412) 268-2246
md0q@andrew.cmu.edu
BM,BT

Gellman, Andrew J. (Head) .... (412) 268-3848
ag4b@andrew.cmu.edu
RE,SU

Grossmann, Ignacio E. .... (412) 268-2230
grossmann@cmu.edu
PR

Jhon, Myung S. .... (412) 268-2233
mj3a@andrew.cmu.edu
SU,TP

Li, Kun (Emeritus) .... (412) 268-2226
kli@andrew.cmu.edu

Pandis, Spyros .... (412) 268-3531
spyros@andrew.cmu.edu
EV

Powers, Gary J. .... (412) 268-3569
gp0c@andrew.cmu.edu
PR

Prieve, Dennis C. .... (412) 268-2247
dcprieve@andrew.cmu.edu
PO

Przybycien, Todd .... (412) 268-3857
todd@andrew.cmu.edu
BM,BT

Sholl, David .... (412) 268-4207
sholl@andrew.cmu.edu
EN,RE,SU

Sides, Paul .... (412) 268-3846
ps7r@andrew.cmu.edu
AM,HT,RE,SU

Tilton, Robert .... (412) 268-1159
tilton@andrew.cmu.edu
BM,BT,TP

Toor, Herbert L. (Emeritus) .... (412) 268-2225
ht16@andrew.cmu.edu

Walker, Lynn .... (412) 268-3020
lwalker@andrew.cmu.edu
PO

Westerberg, Arthur W. (Emeritus) ... (412) 268-2230
a.westerberg@cmu.edu
PR

White, Lee .... (412) 268-2227
white@andrew.cmu.edu
PO

Ydstie, B. Erik .... (412) 268-2235
ydstie@andrew.cmu.edu
PR

Associate Professors

Donahue, Neil M. .... (412) 268-4415
nmd@andrew.cmu.edu
EV

Hauan, Steinar .... (412) 268-4393
steinhau@andrew.cmu.edu
PR

Schneider, James .... (412) 268-4314
jamess3@andrew.cmu.edu
BM,BT,PO

Assistant Professors

Dahl, Kris Noel .... (412) 268-9609
kndahl@andrew.cmu.edu
BM,BT

Islam, Mohammad .... (412) 268-2230
islam@andrew.cmu.edu
PO

Instructors

Jacobson, Annette .... (412) 268-2244
jacobson@andrew.cmu.edu
PO

Other Faculty

Miller, James B (Research) .... (412) 268-9517
jbmiller@andrew.cmu.edu
RE,SU

Accredited by: ABET

Degrees granted 2005-2006:
   B.S.: 43   M.S.: 11

Graduate advisor: Paul Sides

Undergraduate advisor: Lynn Walker

Student organization: AIChE
Advisor: James Schneider

Department reports to: Andrew Gellman, Department Head, Chemical Engineering Department

Placement service: Lisa Dickter, Career Services

Drexel University

Department of Chemical and Biological Engineering
3141 Chestnut Street
Philadelphia, PA 19104
Deliveries: 34th and Ludlow Streets

University phone .... (215) 895-2000
Department phone .... (215) 895-2227
Department FAX .... (215) 895-5837

Professors

Grossmann, Elihu D. .... (215) 895-2229
    mailto:grossman@coe.drexel.edu
Mutharasan, Raj .... (215) 895-2236
    Raj.Mutharasan@coe.drexel.edu
Palmese, Giuseppe R. (Head) .... (215) 895-5814
    palmese@cbis.ece.drexel.edu
Soroush, Masoud .... (215) 895-1710
    Masoud.Soroush@coe.drexel.edu
Weinberger, Charles B. .... (215) 895-2226
    weinbecb@cbis.ece.drexel.edu

Associate Professors

Cairncross, Richard A. .... (215) 895-2230
    cairncross@drexel.edu
Dan, Nily R. .... (215) 895-6624
    dan@coe.drexel.edu
Lowman, Anthony M. (Associate Dean) (215) 895-2228
    alowman@cbis.ece.drexel
Wrenn, Steven .... (215) 895-6694
    wrenn@coe.drexel.edu

Assistant Professors

Abrams, Cameron F. .... (215) 895-2231
    cfabrams@drexel.edu
Elabd, Yossef .... (215) 895-0986
    elabd@drexel.edu

Instructors

Meyer, Stephen P. .... (215) 895-1855
    smeyer@cbis.ece.drexel
Rowell, George .... (215) 895-0987
    rowell@cbis.ece.drexel.edu

Accredited by: ABET

Degrees granted 2004-2005:
    B.S.: 45   M.S.: 8   Ph.D.: 4

Graduate advisor: Richard A. Cairncross

Undergraduate advisor: Stephen P. Meyer

Student organization: AIChE
Advisor: Cameron F. Abrams

Department reports to: S. Guceri, Dean, College of Engineering

Placement service: Anna Bruno (215) 895-1630

Lafayette College

Acopian Engineering Center
Chemical Engineering Department
Easton, PA 18042-1775
Deliveries: Room 122, Acopian Engineering Center

University phone .... (610) 330-5000
Department phone .... (610) 330-5435
Department FAX .... (610) 330-5059

Professors

Martin, J. Ronald .... (610) 330-5430
    martinj@lafayette.edu
    PO
Schaffer, James P. (Director) .... (610) 330-5406
    schaffej@lafayette.edu
    AM
Uz, Mehmet (Head) .... (610) 330-5408
    metmuz@lafayette.edu
    AM

Associate Professors

Piergiovanni, Polly .... (610) 330-5431
    piergiop@lafayette.edu
    BT
Tavakoli, Javad .... (610) 330-5433
    tavakoli@lafayette.edu
    RE

Assistant Professors

Darcy, Patricia .... (610) 330-5432
    darcyp@lafayette.edu
    BT
Ferri, James .... (610) 330-5820
    ferrij@lafayette.edu
    NT
Morton, Samuel .... (610) 330-5417
    mortons@lafayette.edu
    EV

Accredited by: ABET

Degrees granted 2005-2006:
    B.S.: 19

Student organization: AIChE
Advisor: Samuel Morton

Department reports to: James P. Schaffer (610) 330-5403, Director of Engineering

Placement service: Linda Arra (610) 330-5116, Director, Career Services

Lehigh University

Chemical Engineering Department
Iacocca Hall, 111 Research Dr.
Bethlehem, PA 18015-4791

University phone .... (610) 758-3000
Department phone .... (610) 758-4260
Department FAX .... (610) 758-5057

Professors

Blythe, Philip A. (Joint) .... (610) 758-3782
pab0@lehigh.edu
Caram, Hugo S. .... (610) 758-4259
hsc0@lehigh.edu
AM,EN,PR,RE,TP
Charles, Marvin (Emeritus)
mc02@lehigh.edu
BM,BT
Chaudhury, Manoj K. .... (610) 758-4471
mkc4@lehigh.edu
AM,NT,PO,SU,TP
Chen, John C. (Emeritus) .... (610) 758-4091
jcc0@lehigh.edu
EN,HT,MO,PR,TP
El-Aasser, Mohamed S. .... (610) 758-3605
mse0@lehigh.edu
AM,NT,PO,SU
Hsu, James T. .... (610) 758-4257
jth0@lehigh.edu
AM,BM,BT,SE,TP
Jagota, Anand .... (610) 758-4396
anj6@lehigh.edu
AM,BM,BT,NT,SU
Klein, Andrew .... (610) 758-4219
ak04@lehigh.edu
AM,PO
Luyben, William L. .... (610) 758-4256
wll0@lehigh.edu
MO,PR,RE,SE
McHugh, Anthony J. (Chair) .... (610) 758-4470
ajm8@lehigh.edu
AM,BT,MO,PO,TP
Schiesser, William (Emeritus) .... (610) 758-4264
wes1@lehigh.edu
MO
Sengupta, Arup K. (Joint) .... (610) 758-3534
aks0@lehigh.edu
AM,EV
Silebi, Cesar A. .... (610) 758-4267
cas5@lehigh.edu
AM,PO,TP
Sircar, Shivaji .... (610) 758-4469
shs3@lehigh.edu
AM,EN,SE
Sperling, Leslie H. (Emeritus) .... (610) 758-3845
lhs0@lehigh.edu
AM,PO
Stein, Fred P. (Emeritus)
fps0@lehigh.edu
TH
Tuzla, Kemal (Associate Chair) .... (610) 758-4729
kt01@lehigh.edu
EN,HT,TP
Wachs, Israel E. .... (610) 758-4274
iew0@lehigh.edu
AM,EN,EV,NT,RE,SU

Associate Professors

Kothare, Mayuresh V. .... (610) 758-6654
mvk2@lehigh.edu
ME,MO,PR

Assistant Professors

Gilchrist, James F. .... (610) 758-4781
jfg204@lehigh.edu
AM,NT,PR,RE,TP
Laurenzi, Ian J. .... (610) 758-6835
ijl205@lehigh.edu
BM,BT,MO
Rajagopalan, Padma .... (610) 758-6834
par205@lehigh.edu
AM,BM,BT,PO

Accredited by: ABET

Degrees granted 2005-2006:
 B.S.: 24

Graduate advisor: James T. Hsu

Undergraduate advisor: Kemal Tuzla

Student organization: AIChE
Advisor: Harvey G. Stenger

Department reports to: David Wu (610) 758-5308, Dean

Placement service: Donna Goldfeder (610) 758-3712

University of Pennsylvania

Department of Chemical and Biomolecular Engineering
311A Towne Building
220 South 33rd Street
Philadelphia, PA 19104-6393

University phone .... (215) 898-5000
Department phone .... (215) 898-8351
Department FAX .... (215) 573-2093

Professors

Churchill, Stuart W. (Emeritus) .... (215) 898-5579
churchil@seas.upenn.edu
EN,EV,MO,RE,TP
Composto, Russell J. .... (215) 898-4451
composto@seas.upenn.edu
AM,BM,NT,PO,SU,TH,TP
Diamond, Scott L. .... (215) 573-5702
sld@seas.upenn.edu
BT
Discher, Dennis .... (215) 898-9950
discher@seas.upenn.edu
AM,BM,BT,MO,NT,PO,SU
Glandt, Eduardo D. (Dean) .... (215) 898-7244
eglandt@seas.upenn.edu
EN,SU,TH
Gorte, Raymond J. .... (215) 898-4439
gorte@seas.upenn.edu
AM,EN,RE
Hammer, Daniel A. .... (215) 573-6761
hammer@seas.upenn.edu
AM,BM,BT,MO,NT,PO
Myers, Alan L. (Emeritus) .... (215) 898-7078
amyers@seas.upenn.edu
SE,SU
Perlmutter, Daniel D. (Emeritus) ...... (215) 898-8350
perlmutt@seas.upenn.edu
EN,EV,PR,RE
Quinn, John A. (Emeritus) .... (215) 898-8503
quinn@seas.upenn.edu
BT,SU
Seider, Warren D. .... (215) 898-7953
seider@seas.upenn.edu
MO,PR,SE
Shieh, Wen K. .... (215) 898-4634
shieh@seas.upenn.edu
EN,EV,PR
Vohs, John M. (Chair) .... (215) 898-6318
vohs@seas.upenn.edu
AM,EN,RE,SU
Winey, Karen I. .... (215) 898-0593
winey@seas.upenn.edu
AM,NT,PO

Associate Professors

Graves, David J. .... (215) 898-7951
graves@seas.upenn.edu
BM,BT,NT,PO,SE,SU

Sinno, Talid R. .... (215) 898-2511
talid@seas.upenn.edu
AM,ME,MO,NT,TP

Assistant Professors

Baumgart, Tobias .... (215) 573-7539
baumgart@sas.upenn.edu
AM,BM,BT,MO,PO,SU
Boder, Eric T. .... (215) 898-5658
boder@seas.upenn.edu
BM,BT
Crocker, John C. .... (215) 898-9188
jcrocker@seas.upenn.edu
BM,MO,NT,SU,TP
Radhakrishnan, Ravi .... (215) 898-0487
rradhak@seas.upenn.edu
BM
Sarkar, Casim A. .... (215) 573-4072
casarkar@seas.upenn.edu
BM,BT
Yang, Shu .... (215) 898-9645
shuyang@seas.upenn.edu
AM,NT,PO,SU

Other Faculty

Fabiano, Leonard A. (Adjunct) .... (215) 898-7077
lfabiano@seas.upenn.edu
MO,PR,SE
Huff, Marylin C. (Adjunct) .... (215) 898-9879
huff@seas.upenn.edu
EV,RE
Olson, David H. (Adjunct) .... (215) 573-4201
dholson@seas.upenn.edu
RE
Ramanarayanan, Trikur A. (Adjunct) .. (215) 573-0294
trikur@seas.upenn.edu
AM,EN,SU

Accredited by: ABET

Degrees granted 2005-2006:
    B.S.: 20   M.S.: 18   Ph.D.: 5

Graduate advisor: Dr. Dennis E. Discher

Undergraduate advisor: Dr. David J. Graves

Student organization: AIChE
Advisor: Dr. David J. Graves

Department reports to: Eduardo D. Glandt (215) 898-7244, Dean

Placement service: Patricia Rose (215) 898-3012

Pennsylvania State University

Chemical Engineering Department
158 Fenske Laboratory
University Park, PA 16802
University phone .... (814) 865-4700
Department phone .... (814) 865-2574
Department FAX .... (814) 865-7846

Professors
Ben-Jebria, Aziz .... (814) 863-8049
  axb23@psu.edu
  BM,MO,RE
Borhan, Ali .... (814) 865-7847
  borhan@psu.edu
  BM,MO,SU
Braun, Walter G. (Emeritus) .... (814) 234-9888
  wgb1@psu.edu
Curtis, Wayne R. .... (814) 863-4805
  wrc2@psu.edu
  BM,BT,RE
Danner, Ronald P. (Emeritus) .... (814) 863-4814
  rpd@psu.edu
  AM,PO,SE
Daubert, Thomas E. (Emeritus) .... (814) 863-4816
  ted@psu.edu
Duda, J. Larry .... (814) 865-1640
  jld@psu.edu
  AM,PO,SE
Engel, Alfred J. (Emeritus) .... (814) 865-2574
  aje@psu.edu
Fichthorn, Kristen A. .... (814) 863-4807
  fichthorn@psu.edu
  MO,NT,RE,SU
Foley, Henry C. .... (814) 863-9580
  hcf2@psu.edu
  NT,PO,RE,SE
Kabel, Robert L. (Emeritus) .... (814) 863-4815
  r8k@psu.edu
Maranas, Costas .... (814) 863-9958
  costas@psu.edu
  BM,MO,PO,SU
Nagarajan, R. (Emeritus) .... (814) 863-1375
  rxn@psu.edu
Pishko, Michael V. .... (814) 863-4810
  mpishko@engr.psu.edu
  BM,NT
Ultman, James S. .... (814) 863-4802
  jsu@psu.edu
  BM,MO,RE
Vannice, M. Albert (Emeritus) .... (814) 863-1375
  mavche@engr.psu.edu
  RE
Vrentas, James S. .... (814) 863-4808
  jsv1@psu.edu
  MO,PO,SE
Zydney, Andrew L. (Head) .... (814) 863-7113
  zydney@engr.psu.edu
  AM,BM,BT,NT,SE

Associate Professors
Maranas, Janna K. .... (814) 863-6228
  jmaranas@psu.edu
  AM,EN,MO,NT,PO
Matsoukas, Themistoklis .... (814) 863-2002
  matsoukas@psu.edu
  NT
Velegol, Darrell .... (814) 865-8739
  velegol@psu.edu
  AM,NT,SU

Assistant Professors
Armaou, Antonios .... (814) 863-5316
  armaou@engr.psu.edu
  ME,MO,PR,RE,SU
Cirino, Patrick C. .... (814) 865-5790
  cirino@engr.psu.edu
  BM,BT,RE
Hahm, Jong-in .... (814) 863-4801
  jhahm@engr.psu.edu
  AM,BM,BT,ME,NT,PO,RE,SU
Kim, Seong Han .... (814) 863-4809
  shkim@engr.psu.edu
  BT,NT,PO,RE,SU
Nedwick, Robert .... (814) 863-6229
  nedwick@engr.psu.edu

Other Faculty
Perez, Joseph M. .... (814) 865-0340
  jmp13@psu.edu

Accredited by: ABET

Degrees granted 2005-2006:
  B.S.: 103 M.S.: 5 Ph.D.: 15

Graduate advisor: Ali Borhan

Undergraduate advisor: Themis Matsoukas

Student organization: AIChE
Advisor: Wayne Curtis

Department reports to: D. Wormley, Harold and Inge Marcus Dean of Engineering

Placement service: Career Development, Jack Rayman

University of Pittsburgh

Chemical and Petroleum Engineering Department
1249 Benedum Hall
Pittsburgh, PA 15261
Deliveries: 3700 O'Hara Street

University phone .... (412) 624-4141
Department phone .... (412) 624-9631
Department FAX .... (412) 624-9639

Professors

Ataai, Mohammad M. .... (412) 624-9648
    ataai@pitt.edu
Balazs, Anna C. .... (412) 648-9250
    balazs1@engr.pitt.edu
Beckman, Eric J. .... (412) 624-4828
    beckman@engr.pitt.edu
Borovetz, Harvey S. (Joint) .... (412) 624-4725
    borovetzhs@upmc.edu
Chiang, Shiao-Hung (Emeritus) .... (412) 624-9636
    chiang@engr.pitt.edu
Enick, Robert M. (Chair) .... (412) 624-9649
    enick@engr.pitt.edu
Federspiel, William .... (412) 383-9499
    federspielwj@upmc.edu
Holder, Gerald D. (Dean) .... (412) 624-9809
    holder@engr.pitt.edu
Johnson, J. Karl .... (412) 624-5644
    karlj@engr.pitt.edu
Klinzing, George E. .... (412) 624-0784
    klinzing@engr.pitt.edu
Kovalchuk, Vladimir (Research) .... (412) 624-4754
    vkovalch@pitt.edu
Lindt, Jan T. (Emeritus) .... (412) 624-9729
    jtlindt@pitt.edu
Morsi, Badie I. .... (412) 624-9650
    morsi@engr.pitt.edu
Russell, Alan J. (Joint) .... (412) 235-5109
    russellaj@upmc.edu
Tierney, John W. (Emeritus) .... (412) 624-9645
    tierney@engr.pitt.edu
Wender, Irving (Research) .... (412) 624-9644
    wender@engr.pitt.edu

Associate Professors

Beroes, Charles (Emeritus) .... (412) 624-9630
    beroes@engr.pitt.edu
Cobb, James T. Jr. (Emeritus) .... (412) 624-7443
    cobb@engr.pitt.edu
d'Itri, Julie L. .... (412) 624-9634
    jditri@engr.pitt.edu
McCarthy, Joseph .... (412) 624-7362
    mccarthy@engr.pitt.edu
Parker, Robert S. .... (412) 624-7364
    rparker@engr.pitt.edu
Wagner, William R. (Joint) .... (412) 647-8311
    wagnerwr@upmc.edu

Assistant Professors

Gao, Di .... (412) 624-8488
    gaod@engr.pitt.edu
Little, Steven .... (412) 624-9614
    slittle@engr.pitt.edu
Patzer, John F. II (Joint) .... (412) 624-9819
    patzer@engr.pitt.edu
Velankar, Sachin .... (412) 624-9984
    velankar@pitt.edu
Veser, Goetz .... (412) 624-1042
    gveser@pitt.edu

Accredited by: ABET

Degrees granted 2005-2006:
    B.S.: 34   M.S.: 9   Ph.D.: 8

Graduate advisor: Robert S. Parker

Undergraduate advisor: Joseph J. McCarthy

Student organization: AIChE
Advisor: Joseph McCarthy

Department reports to: Gerald D. Holder, US Steel Dean of Engineering

Placement service: Chris Frankovic

Villanova University

Department of Chemical Engineering
800 Lancaster Avenue
Villanova, PA 19085
University phone . . . . (610) 519-4500
Department phone . . . . (610) 519-4950
Department FAX . . . . (610) 519-7354

Professors

Joye, Donald D. . . . . (610) 519-4966
donald.joye@villanova.edu
PO,TP
Kelly, C. Michael (Chair) . . . . (610) 519-4953
c.michael.kelly@villanova.edu
MO
Muske, Kenneth R. . . . . (610) 519-6195
kenneth.muske@villanova.edu
MO,PR
Myers, John A. (Emeritus)
Punzi, Vito L. . . . . (610) 519-4946
vito.punzi@villanova.edu
EV
Rice, William J. (Emeritus)
Sweeny, Robert F. (Emeritus)

Associate Professors

Kelly, William J. . . . . (610) 519-4947
william.j.kelly@villanova.edu
BT
Ritter, Edward R. . . . . (610) 519-4948
edward.ritter@villanova.edu
RE
Skaf, Dorothy W. . . . . (610) 519-4952
dorothy.skaf@villanova.edu
AM
Weinstein, Randy D. . . . . (610) 519-4954
randy.weinstein@villanova.edu
HT,NT,SU,TH

Accredited by: ABET

Degrees granted 2005-2006:
B.S.: 21   M.S.: 7

Graduate advisor: Vito L. Punzi

Student organization: AIChE
Advisor: Randy D. Weinstein

Department reports to: Dr. Gary Gabrielle, Dean of Engineering

Placement service: N. Dudak

Widener University

Chemical Engineering Department
Chester, PA 19013-5792
Deliveries: Kirkbride Hall, 17th and Walnut Sts.
University phone . . . . (610) 499-4000
Department phone . . . . (610) 499-4051
Department FAX . . . . (610) 499-4059

Professors

Maffia, Gennaro J. . . . . (610) 499-4089
GJMaffia@widener.edu
BM,BT,HT,MO,NT,PR,RE,SE,TP
McWilliams, Thomas G. Jr. (Emeritus) (610) 499-4608
tgm0001@widener.edu

Associate Professors

Chen, David H. T. . . . . (610) 499-4049
David.H.Chen@widener.edu
EN,HT,RE,SE,TH,TP
McNeil, Kenneth M. . . . . (610) 499-4056
Kenneth.M.McNeil@widener.edu
EN,EV,HT,RE,TP
Nippert, Charles R. (Chair) . . . . (610) 499-4050
Charles.R.Nippert@widener.edu
MO,PO,PR,SE,TH

Other Faculty

Chitra, Surya (Adjunct) . . . . (610) 499-4051
Henderson, Louis (Adjunct) . . . . (610) 499-4051
Hoopes, John W. Jr. (Adjunct) . . . . (610) 499-4051
EN,MO,PR,SE,TP
Kopatsis, Alexander (Adjunct) . . . . (610) 499-4051
Pourki, Forouza (Adjunct) . . . . (610) 499-4051
Rowell, George (Adjunct) . . . . (610) 499-4051

Accredited by: ABET

Degrees granted 2005-2006:
B.S.: 13   M.S.: 10

Graduate advisor: Gennaro J. Maffia

Undergraduate advisor: Gennaro J. Maffia

Student organization: AIChE
Advisor: Kenneth M. McNeil

Department reports to: Dr. Fred Akl (610) 499-4036, Dean of Engineering

Placement service: Barbara Buckley (610) 499-4509

Puerto Rico

University of Puerto Rico

Department of Chemical Engineering
PO Box 9046
Mayaguez PR 00681-9046

University phone .... (787) 832-4040
Department phone .... (787) 832-4040 x2593
Department FAX .... (787) 834-3655

Professors
Aguayo, Guillermo .... x3605 [†]
    gaguayo@uprm.edu
Benítez, Jaime x3209
    jbenrod@caribe.net
Briano, Julio G. (Associate Chair) x2545
    jbriano@uprm.edu
Cardona-Martínez, Nelson (Chair) x2592
    ncardona@uprm.edu
Colucci-Rios, José A. x3749
    jcolucci@uprm.edu
Colón, Guillermo x3727
    gcolon@uprm.edu
Estévez, L. Antonio x2573
    estevez@uprm.edu
Mandavilli, Satya N. x3589
    s_mandavilli@rumac.uprm.edu
Ramírez, Carlos A. x2561
    cramirez@uprm.edu
Rodríguez-Ramírez, Abraham x2581
    arodrig@uprm.edu
Saliceti-Piazza, Lorenzo x2549
    saliceti@uprm.edu
Sridhar, Lakshmi N. x3173
    l_sridhar@rumac.uprm.edu
Villafañe-Ruiz, Gilberto x2569
    gvilla@coqui.net

Associate Professors
Bogere, Moses N. x3607
    mbogere@caribe.net
Suleiman-Rosado, David x2685
    dsuleiman@uprm.edu
Torres-Lugo, Madeline x2585
    madeline@ece.uprm.edu
Velázquez-Figueroa, Carlos x2576
    cvelazqu@uprm.edu

Assistant Professors
Acevedo Nazario, Aldo x2473
    aldacevedo@uprm.edu
    AM,TP
Hernández-Maldonado, Arturo J. x3748
    arturojh@uprm.edu
    AM
Martínez-Iñesta, María M. x3606
    mariam@uprm.edu
    NT
Rinaldi-Ramos, Carlos Manuel 3585
    crinaldi@uprm.edu

Instructors
Avilés-Molina, Misael O.
Córdova-Figueroa, Ubaldo M.
Siberio-Pérez, Diana Y.

Other Faculty
Mehta, Narinder K. x3157
    nmehta@uprm.edu

[†] Faculty numbers are extensions of (787) 832-4040.

Accredited by: ABET

Degrees granted 2005-2006:
    B.S.: 119   M.S.: 6   Ph.D.: 3

Graduate advisor: Dr. Julio Briano

Undergraduate advisor: Lourdes Fabregas

Student organization: AIChE and IIQPR
Advisor: Prof. Luis A. Estevez

Department reports to: Ramón Vásquez, Dean, College of Engineering

Placement service: Carmen Sol Ramírez, Director

110 United States

Rhode Island

Brown University

Chemical and Biochemical Engineering Program
Division of Engineering, Box D
Providence, RI 02912
Deliveries: 182 Hope Street

University phone	(401) 863-2677
University FAX	(401) 863-1157
Department phone	(401) 863-2677
Department FAX	(401) 863-1157

Professors

Breuer, Kenneth .... (401) 863-2870
    kbreuer@brown.edu
    AM,BT,NT,SU
Calo, J.M. .... (401) 863-1421
    JMCalo@brown.edu
    AM,EN,EV,RE,SE,SU
Caswell, Bruce (Emeritus) .... (401) 863-1448
    Bruce_Caswell@brown.edu
    AM,PO
Dobbins, Richard A. (Emeritus) .... (401) 863-2867
    Richard_Dobbins@brown.edu
    EN,SU
Hurt, Robert H. .... (401) 863-2685
    Robert_Hurt@brown.edu
    AM,BT,EN,EV,NT,SU
Karlsson, Sture K. F. (Emeritus) ...... (401) 863-3200
    Sture_Karlsson@brown.edu
    EN
Liu, Joseph T. C. .... (401) 863-2654
    Joseph_Liu@brown.edu
    EN,MO
Richardson, Peter D. .... (401) 863-2687
    Peter_Richardson@brown.edu
    AM,BM,BT
Sheldon, Brian W. .... (401) 863-2866
    Brian_Sheldon@brown.edu
    AM,BT,NT,SU
Sibulkin, Merwin (Emeritus) .... (401) 863-2867
    Merwin_Sibulkin@brown.edu
    EN
Suuberg, Eric M. .... (401) 863-1420
    Eric_Suuberg@brown.edu
    EN,EV,RE,SU

Assistant Professors

Tripathi, Anubhav .... (401) 863-3063
    Anubhav_Tripathi@brown.edu
    AM,BM,BT,NT,PO,SE

Accredited by: ABET

Degrees granted 2005-2006:
    B.S.: 3   Ph.D.: 3

Graduate advisor: Robert Hurt

Undergraduate advisor: J. M. Calo

Student organization: AIChE
Advisor: J. M. Calo

Department reports to: Gregory Crawford (401) 863-2276, Dean of Engineering

Placement service: Sheila Curran (401) 863-3326, Director

University of Rhode Island

Department of Chemical Engineering
205 Crawford Hall
Kingston, RI 02881-0805

University phone .... (401) 874-1000
Department phone .... (401) 874-2655
Department FAX .... (401) 874-4689

Professors

Barnett, Stanley M. .... (401) 874-2443
                barnett@egr.uri.edu
Bose, Arijit (Chair) .... (401) 874-2804
                bose@egr.uri.edu
Brown, Richard .... (401) 874-2707
                brown@egr.uri.edu
Gregory, Otto J. (Associate Dean) ..... (401) 874-2085
                gregory@egr.uri.edu
Knickle, Harold N. .... (401) 874-2678
                knickle@egr.uri.edu
Lucia, Angelo .... (401) 874-2814
                lucia@egr.uri.edu
Rose, Vincent C. (Emeritus) .... (401) 874-5924
                rose@egr.uri.edu

Associate Professors

Gray, Donald J. .... (401) 874-2651
                gray@egr.uri.edu
Greenfield, Michael L. .... (401) 874-9289
                greenfield@egr.uri.edu
Rivero-Hudec, Mercedes (Associate Dean) (401) 874-2271
                rivero@egr.uri.edu

Other Faculty

Crisman, Everett .... (401) 874-5076
                crisman@egr.uri.edu
Park, Eugene .... (401) 874-4303
                epark@earthlink.net

Accredited by: ABET

Degrees granted 2004-2005:
    B.S.: 14   M.S.: 1   Ph.D.: 3

Graduate advisor: M. L. Greenfield

Undergraduate advisor: S. M. Barnett

Student organization: AIChE
Advisor: D. J. Gray

Department reports to: Bahram Nassersharif (401) 874-2186, Dean, College of Engineering

Placement service: Roberta K. Koppel (401) 874-2311, Director

South Carolina

Clemson University

Department of Chemical & Biomolecular Engineering
Box 340909
Clemson, SC 29634-0909
Deliveries: 127 Earle Hall

University phone .... (864) 656-3311
Department phone .... (864) 656-3055
Department FAX .... (864) 656-0784

Professors

Barron, Charles H. (Emeritus)
    charles.barron@ces.clemson.edu
Beard, John N. (Emeritus)
Beckwith, William F. (Emeritus)
Edie, Danny D. (Emeritus) .... (864) 656-4535
    dan.edie@ces.clemson.edu
    PO
Gooding, Charles H. .... (864) 656-2621
    charles.gooding@ces.clemson.edu
    PR
Goodwin, James G. (Chair) .... (864) 656-6614
    james.goodwin@ces.clemson.edu
    EN,RE
Guiseppi-Elie, Anthony 864-656-1712
    guiseppi@clemson.edu
    BM,ME,PO
Melsheimer, Stephen S. (Emeritus) .... (864) 656-5706
    stephen.melsheimer@ces.clemson.edu
Mullins, Joseph C. (Emeritus) .... (864) 656-5426
Ogale, Amod A. .... (864) 656-5483
    amod.ogale@ces.clemson.edu
    AM,PO
Thies, Mark C. .... (864) 656-5424
    mark.thies@ces.clemson.edu
    EN,SE
Torres, Walter (Research) 864-656-5419
    wtorres@clemson.edu
    BT

Associate Professors

Bruce, David A. .... (864) 656-5425
    david.bruce@ces.clemson.edu
    EN,MO,RE
Harrison, Graham M. .... (864) 656-6399
    graham.harrison@ces.clemson.edu
    PO
Hirt, Douglas E. .... (864) 656-0822
    doug.hirt@ces.clemson.edu
    PO
Husson, Scott M. .... (864) 656-4502
    scott.husson@ces.clemson.edu
    SE
Kilbey, S. Michael .... (864) 656-5423
    mike.kilbey@ces.clemson.edu
    PO
Rice, Richard W. .... (864) 656-5428
    richard.rice@ces.clemson.edu
    RE

Assistant Professors

Kitchens, Christopher
    NT
Lotero, Edgar (Research) 864-656-5419
    alegria@CLEMSON.EDU
    RE
Metters, Andrew .... (864) 656-3055
    metters@clemson.edu
    BM,PO

Accredited by: ABET

Degrees granted 2005-2006:
    B.S.: 33   M.S.: 1   Ph.D.: 7

Graduate advisor: G.M. Harrison

Undergraduate advisor: S.M. Husson

Student organization: AIChE
Advisor: S.M. Husson

Department reports to: Esin Gulari (864) 656-3202, Dean of Engineering and Science

Placement service: Flora Riley (864) 656-2152

University of South Carolina

Department of Chemical Engineering
Swearingen Engineering Center - Room 2C02
Columbia, SC 29208

University phone .... (803) 777-7000
Department phone .... (803) 777-4181
Department FAX .... (803) 777-0973

Professors

Amiridis, Michael D. (Dean) .... (803) 777-7294
    amiridis@engr.sc.edu
    EN,NT,RE
Gibbons, Joseph H. (Emeritus) .... (803) 777-8978
    gibbons@engr.sc.edu
    EN
Matthews, Michael A. (Chair) .... (803) 777-0556
    matthews@engr.sc.edu
    BM,EN,EV,TH
Ploehn, Harry J. .... (803) 777-7307
    ploehn@engr.sc.edu
    AM,NT,PO,TP
Popov, Branko N. .... (803) 777-7314
    popov@engr.sc.edu
    AM,EN,MO,NT,TH
Ritter, James A. .... (803) 777-3590
    ritter@engr.sc.edu
    AM,BM,EN,MO,NT,SE,SU
Van Brunt, Vincent .... (803) 777-3115
    vanbrunt@engr.sc.edu
    MO,PR,SE
Van Zee, John W. .... (803) 777-2285
    vanzee@engr.sc.edu
    EN,MO,PO
Weidner, John W. (Associate Chair) ... (803) 777-3207
    weidner@engr.sc.edu
    AM,EN,EV,MO,NT,RE,TP
White, Ralph E. .... (803) 777-3270
    white@engr.sc.edu
    EN,EV,MO,PR,RE,SU

Associate Professors

Gadala-Maria, Francis A. .... (803) 777-3182
    gadala-m@engr.sc.edu
    EN,PO,TP
Jabbari, Esmaiel .... (803) 777-8022
    jabbari@engr.sc.edu
    AM,BM,MO,NT,PO,SU
Papathanasiou, Thanasis D. .... (803) 777-7219
    papathan@engr.sc.edu
    AM,MO,PO,TP
Williams, Christopher T. .... (803) 777-0143
    willia84@engr.sc.edu
    NT,RE

Assistant Professors

Bender, Jonathan W. .... (803) 777-5025
    benderjw@engr.sc.edu
    BM,NT,TP
Delhommelle, Jerome P. .... (803) 777-7316
    delhomm@engr.sc.edu
    MO,TH
Gatzke, Edward P. .... (803) 777-1159
    gatzke@engr.sc.edu
    MO,PR
Moss, Melissa A. .... (803) 777-5604
    mossme@engr.sc.edu
    BM
Stanford, Thomas G. .... (803) 777-4101
    stanford@engr.sc.edu
    EN,MO,PR,RE,TH

Accredited by: ABET

Degrees granted 2005-2006:
    B.S.: 31   M.S.: 3   Ph.D.: 9

Student organization: AIChE
Advisor: Ed Gatzke, Harry Ploehn

Department reports to: Michael D. Amiridis, Dean - College of Engineering and Information Technology

Placement service: Robert Rinehart, Engineering Career Services (803) 777-3972

South Dakota

South Dakota School of Mines and Technology

Chemical and Biological Engineering Department
501 E. Saint Joseph Street
Rapid City, SD 57701-3995

University phone .... (605) 394-2511
Department phone .... (605) 394-2421
Department FAX .... (605) 394-1232

Professors

Bang, Sookie .... (605) 394-2426
                sookie.bang@sdsmt.edu
                BT

Bauer, Larry G. (Emeritus)

Dixon, David J. (Chair) .... (605) 394-1235
                David.Dixon@sdsmt.edu
                AM,BT,EN,EV,MO,NT,PO,RE,SE

Munro, James M. (Emeritus) .... (605) 394-2421
                James.Munro@sdsmt.edu

Puszynski, Jan A. .... (605) 394-1230
                Jan.Puszynski@sdsmt.edu
                AM,EN,MO,NT,RE

Sandvig, Robert L. (Emeritus)

Winter, Robb M. (Chair) .... (605) 394-1237
                Robb.Winter@sdsmt.edu
                AM,BT,NT,PO,SU

Assistant Professors

Gilcrease, Patrick .... (605) 394-1239
                Patrick.Gilcrease@sdsmt.edu
                BT,EV,RE

Menkhaus, Todd .... (605) 394-2422
                Todd.Menkhaus@sdsmt.edu
                BT,EN,SE

Sani, Rajesh .... (605) 394-1240
                rajesh.sani@sdsmt.edu
                BM,BT,EN,EV,RE,SE

Accredited by: ABET

Degrees granted 2005-2006:
    B.S.: 20   M.S.: 11

Graduate advisor: J.A. Puszynski

Student organization: AIChE
Advisor: Patrick C. Gilcrease

Department reports to: Dr. David Dixon

Placement service: Darrell Sawyer (605) 394-2667, Director

Tennessee

Christian Brothers University

Chemical and Biochemical Engineering Department
650 East Parkway South
Memphis, TN 38104-5581
Deliveries: Nolan Engineering Building
University phone .... (901) 321-3000
Department phone .... (901) 321-3401
Department FAX .... (901) 321-3402

Professors
Ray, Asit K. .... (901) 321-3418
    aray@cbu.edu

Associate Professors
Pourhashemi, Ali (Chair) .... (901) 321-3401
    apourhas@cbu.edu
Price, Randel M. .... (901) 321-3412
    rprice@cbu.edu

Instructors
Zhou, Youngquan (Adjunct) .... (901) 321-3405

Accredited by: ABET, SACS

Degrees granted 2005-2006:
    B.S.: 6

Undergraduate advisor: Dr. Ali Pourhashemi

Student organization: AIChE
Advisor: Dr. Asit Ray

Department reports to: Dr. Eric Welch, Dean of Engineering

Placement service: Ms. Betty McWillie, Director of Creer Center

The University of Tennessee at Chattanooga

College of Engineering and Computer Science
Engineering: Chemical
615 McCallie Avenue
Chattanooga, TN 37403-2598
Deliveries: Dept # 2502, EMCS Building Room 430
University phone .... (423) 425-4411
Department phone .... (423) 425-5217
Department FAX .... (423) 425-5229

Professors
Cunningham, James .... (423) 425-4361
    jim-cunningham@utc.edu
Henry, Jim .... (423) 425-4398
    jim-henry@utc.edu

Associate Professors
Jones, Frank (Director) .... (423) 425-4366
    frank-jones@utc.edu

Accredited by: ABET

Degrees granted 2005-2006:
    B.S.: 4   M.S.: 2

Graduate advisor: Dr. Neslihan Alp

Undergraduate advisor: Dr. Gary H. McDonald

Student organization: AIChE
Advisor: James Cunningham

Department reports to: Dr. Ronald Bailey, Dean of Engineering

Placement service: Jean Dake (423) 425-4184, Director

University of Tennessee at Knoxville

1512 Middle Drive
419 Dougherty Engineering Building
Knoxville, TN 37996-2200
Deliveries: 419 Dougherty Engineering Building

University phone	(865) 974-1000
Department phone	(865) 974-2421
Department FAX	(865) 974-7076

Professors

Bienkowski, Paul R. .... (865) 974-1618
prb@utk.edu
BT,EV
Counce, Robert M. .... (865) 974-5318
counce@utk.edu
EV,PR,SE
Daw, C. Stuart (Adjunct)
Khomami, Bamin (Head) .... (865) 974-9596
bkhomami@utk.edu
AM,MO,PO,TP
McFarlane, Joanna (Adjunct) .... (865) 574-4941
mcfarlanej@ornl.gov
EN,EV,RE,SE
Prados, John W. (Emeritus) .... (865) 974-6053
jprados@utk.edu
Rials, Timothy G. (Adjunct) .... (865) 984-3850
trials@chartertn.net
Spencer, Barry B. (Adjunct) .... (865) 574-7143
spencerbb@ornl.gov
EN,MO,SE
Steele, William V. (Adjunct) .... (865) 974-4820
steekwv@ornl.gov

Associate Professors

Bruns, Duane D. .... (865) 974-5317
dbruns@utk.edu
BT,EN,MO,PR,RE
Cox, Chris D. (Adjunct) .... (865) 974-7729
ccox9@utk.edu
Davison, Brian (Adjunct)
davisonbh@ornl.gov
DePaoli, David (Adjunct) .... (865) 574-6817
Edwards, Brian J. (Associate Head) ... (865) 974-9596
bjedwards@chem.engr.utk.edu
AM,EN,MO,PO,TH,TP
Frymier, Paul .... (865) 974-4961
pdf@ukt.edu
BT,EN
Handegama, Naresh (Adjunct)
nbhandagama@tva.gov
EN,EV,MO,PR,RE
Hayes, Douglas G. (Adjunct) .... (865) 974-7991
dhayes1@utk.edu
BT,SE,SU
Keffer, David .... (865) 974-5322
dkeffer@utk.edu
AM,EN,MO,NT,SE

Petrovan, Simioan .... (865) 974-5137
petrovan@chem.engr.utk.edu
AM,BT,MO,NT,PO,RE,SE,SU
Siirola, Jeffrey (Adjunct)
Visco, Donald P. (Adjunct) .... (931) 372-3606
dvisco@tntech.edu
Wang, Tse-Wei .... (865) 974-6769
twang@utk.edu
PR
Weber, Frederick E. .... (865) 974-6362
fweber@utk.edu

Other Faculty

Cui, Shengting .... (865) 974-4820
scui@utk.edu
Sheth, Atul .... (931) 393-7427
asheth@utsi.edu
AM,EN,EV,MO,RE,SE,SU

Accredited by: ABET

Degrees granted 2005-2006:
    B.S.: 16   M.S.: 5   Ph.D.: 6

Graduate advisor: David J. Keffer

Undergraduate advisor: Brian J. Edwards

Student organization: AIChE
Advisor: David J. Keffer

Department reports to: Way Kuo (865) 974-5321, Dean

Placement service: Russ Coughenour 974-5435

Tennessee Technological University

Department of Chemical Engineering
P.O. Box 5013
Cookeville, TN 38505
Deliveries: 1020 Stadium Drive, Prescott Hall, Room 214

University phone	(931) 372-3101
Department phone	(931) 372-3297
Department FAX	(931) 372-6352

Professors

Arce, Pedro E. (Chair) .... (931) 372-3189
parce@tntech.edu
AM,BM,BT,EV,MO,NT,RE,SE,TP
Biernacki, Joseph J. .... (931) 372-3667
jbiernacki@tntech.edu
AM,ME,SE,SU,TP
Holland, William D. (Emeritus) .... (931) 372-3297
Kerr, Clayton P. (Emeritus) .... (931) 372-3297
McGee, John C. (Emeritus) .... (931) 372-3297
Yarbrough, David W. (Emeritus) ...... (931) 372-3297
HT

Associate Professors

Eliassen, John D. (Adjunct) .... (931) 372-3267
jeliassen@tntech.edu
Visco, Donald P. .... (931) 372-3606
dvisco@tntech.edu
BM,BT,MO,TH

Assistant Professors

Carpen, Ileana .... (931) 372-3474
icarpen@tntech.edu
AM,BM,TP
Stretz, Holly A. .... (931) 372-3495
hstretz@tntech.edu
AM,EN,NT,PO,SU,TP
Subramanian, Venkat R. .... (931) 372-3494
vsubramanian@tntech.edu
EN,MO,PR,TP
Wang, Chunsheng .... (931) 372-3678
cswang@tntech.edu
AM,EN,NT,SU

Instructors

Clark, Yvette .... (931) 372-3004
yrclark@tntech.edu

Accredited by: ABET

Degrees granted 2005-2006:
  B.S.: 13   M.S.: 5

Graduate advisor: Venkat Subramanian

Undergraduate advisor: Donald P. Visco, Jr.

Student organization: AIChE
Advisor: Dr. Holly Stretz

Department reports to: Glen Johnson (931) 372-3172, Dean of Engineering

Placement service: Alice Camuti (931) 372-3232, Director

Vanderbilt University

Chemical Engineering Department
VU Station B Box 351604
Nashville, TN 37235-1604
Deliveries: Olin Hall-107, 24th and Garland Avenues
Department phone .... (615) 322-2441
Department FAX .... (615) 343-7951

Professors

Bayuzick, Robert J. (Emeritus) .... (615) 322-7047
    robert.j.bayuzick@vanderbilt.edu
Cummings, Peter T. .... (615) 322-8129
    peter.cummings@vanderbilt.edu
Fort, Tomlinson (Emeritus) .... (615) 343-6992
    tomlinson.fort@vanderbilt.edu
    PO,SU
Godbold, Thomas M. (Emeritus)
Laibinis, Paul E. .... (615) 936-8431
    paul.e.laibinis@vanderbilt.edu
    NT,SU
LeVan, M. Douglas (Chair) .... (615) 322-2441
    m.douglas.levan@vanderbilt.edu
    AM,EV,MO,NT,PR,SE,SU
Roth, John A. (Emeritus) .... (615) 322-3517
    john.roth@vanderbilt.edu
Schnelle, Karl B. Jr. .... (615) 322-3370
    karl.b.schnelle@vanderbilt.edu
    EV,SE
Tanner, Robert D. (Emeritus) .... (615) 322-2061
    robert.d.tanner@vanderbilt.edu
    BT,SE

Associate Professors

Debelak, Kenneth A. .... (615) 322-2088
    kenneth.a.debelak@vanderbilt.edu
Jennings, G. Kane .... (615) 322-2707
    kane.g.jennings@vanderbilt.edu
    AM,NT,PO,SE,SU
Rogers, Bridget R. .... (615) 343-3269
    bridget.rogers@vanderbilt.edu
    AM,ME,NT,SU
Sharp, Julie E. .... (615) 322-3700
    julie.e.sharp@vanderbilt.edu

Assistant Professors

Balcarcel, R. Robert .... (615) 322-2095
    r.robert.balcarcel@vanderbilt.edu
Guelcher, Scott A. .... (615) 322-2097
    scott.guelcher@vanderbilt.edu
    AM,BM,BT,NT,PO,SU
McCabe, Clare .... (615) 322-6853
    c.mccabe@vanderbilt.edu
    BT,MO,NT,SE

Other Faculty

Prokop, Ales (Research) .... (615) 343-3515
    ales.prokop@vanderbilt.edu

Accredited by: ABET

Degrees granted 2005-2006:
    B.S.: 28   Ph.D.: 10

Graduate advisor: Bridget R. Rogers

Undergraduate advisor: Kenneth A. Debelak

Student organization: AIChE
Advisor: Scott A. Guelcher

Department reports to: Kenneth Galloway, Dean

Placement service: Francene Gilmer, Director

Texas

University of Houston

Chemical and Biomolecular Engineering Department
S-222, Engineering Bldg. 1
4800 Calhoun Road
Houston, TX 77204-4004

University phone .... (713) 743-1000
Department phone .... (713) 743-4300
Department FAX .... (713) 743-4323

Professors

Amundson, Neal R. .... (713) 743-3492
   NAmundson@aol.com
Balakotaiah, Vemuri .... (713) 743-4318
   Bala@uh.edu
Bidani, Akhil (Joint) .... (713) 743-4066
   abidani@central.uh.edu
Daneshy, Ali (Adjunct) .... (713) 743-4328
   AliD@uh.edu
Donnelly, Vincent M. .... (713) 743-4313
   VMDonnelly@uh.edu
Economides, Michael J. (Adjunct) ..... (713) 743-4330
   MJE@uh.edu
Economou, Demetre J. (Associate Chair) (713) 743-4320
   Economou@uh.edu
Fleischer, Miguel T. "Micky" (Adjunct) (713) 743-4388
   MFleischer@uh.edu
Flumerfelt, Raymond W. (Dean) ...... (713) 743-4200
   RWF@uh.edu
Fox, George E. (Joint) .... (713) 743-8363
   Fox@uh.edu
Harold, Michael P. (Chair) .... (713) 743-4307
   MHarold@uh.edu
Henley, Ernest J. (Emeritus) .... (713) 743-4326
   EHenley@bayou.uh.edu
Jacobson, Allan J. (Joint) .... (713) 743-2785
   AJJacob@uh.edu
Khoury, Fouad M. (Adjunct) .... (713) 466-4686
   FKhoury@sbcglobal.net
Krishnamoorti, Ramanan (Assoc. Dean) (713) 743-4312
   Ramanan@uh.edu
Lee, T. Randall "Randy" (Joint) ...... (713) 743-2724
   TRLee@uh.edu
Luss, Dan .... (713) 743-4305
   DLuss@uh.edu
Mohanty, Kishore K. .... (713) 743-4331
   Mohanty@uh.edu
Richardson, James T. .... (713) 743-4324
   JTR@uh.edu
Rooks, Charles W. "Mickey" (Adjunct) (713) 743-4316
   CWRooks@uh.edu
Willson, Richard C. .... (713) 743-4308
   Willson@uh.edu

Associate Professors

Advincula, Rigoberto C. (Joint) .... (713) 743-1760
   RAdvincula@uh.edu
Annapragada, Ananth (Adjunct) ...... (713) 500-3982
   ananth.annapragada@uth.tmc.edu
Briggs, James M. (Joint) .... (713) 743-8366
   JBriggs@uh.edu
Chellam, Shankar (Joint) .... (713) 743-4265
   Chellam@uh.edu
Litvinov, Dmitri (Joint) .... (713) 743-4168
   dlitvinov@uh.edu
Marple, Stanley Jr. (Adjunct) .... (713) 743-4326
Nikolaou, Michael .... (713) 743-4309
   Nikolaou@uh.edu
Vekilov, Peter G. .... (713) 743-4315
   Vekilov@uh.edu

Assistant Professors

Galkin, Oleg N. (Research) .... (713) 743-4344
   OGalkin@mail.uh.edu
Martirosyan, Karen S. (Research) ..... (713) 743-4366
   KMartiro@mail.uh.edu
Strasser, Peter .... (713) 743-4310
   PStrasser@uh.edu

Accredited by: ABET

Degrees granted 2005-2006:
  B.S.: 47   M.S.: 14   Ph.D.: 15

Undergraduate advisor: Demetre Economou

Student organization: AIChE
Advisor: Richard Willson

Department reports to: Raymond Flumerfelt (713) 743-4200, Dean of Engineering

Placement service: David B. Small, (713) 743-5100

Lamar University

Chemical Engineering Department
P.O. Box 10053
Beaumont, TX 77710
Deliveries: 4400 Martin Luther King Dr., Beaumont, TX 77705

University phone	(409) 880-7011
Department phone	(409) 880-8784
Department FAX	(409) 880-2197

Professors
Chen, Daniel H. .... (409) 880-8786
    chendh@hal.lamar.edu
Cocke, David L. .... (409) 880-8372
    cockdl@hal.lamar.edu
Ho, Thomas C. .... (409) 880-8790
    hotc@hal.lamar.edu
Hopper, Jack R. (Dean) .... (409) 880-8785
    hopperjr@hal.lamar.edu
Li, Ku-Yen (Chair) .... (409) 880-8789
    liku@hal.lamar.edu
Yaws, Carl L. .... (409) 880-8787
    yawscl@hal.lamar.edu

Assistant Professors
Gossage, John L. .... (409) 880-8788
    gossagejl@hal.lamar.edu
Lin, Sidney .... (409) 880-2314
    sidney.lin@lamar.edu
Lou, Helen H. .... (409) 880-8207
    louhh@hal.lamar.edu
Tadmor, Rafael .... (409) 880-7791
    tadmorrx@hal.lamar.edu
Xu, Qiang .... (409) 880-7818
    qiang.xu@lamar.edu

Accredited by: ABET

Degrees granted 2005-2006:
    B.S.: 31   M.S.: 38   Ph.D.: 1

Graduate advisor: Dr. Helen H. Lou

Undergraduate advisor: Mrs. Rebecca Caddy

Student organization: AIChE
Advisor: Dr. John L. Gossage

Department reports to: Jack R. Hopper, Dean of Engineering

Placement service: Dr. Jim Thomas

Prairie View A&M University

Chemical Engineering Department
P.O. Box 519, Mailstop 2505
Prairie View, TX 77446-0519
Deliveries: Preston and Obanion Streets

University phone	(936) 857-3311
Department phone	(936) 857-2427
Department FAX	(936) 857-4540

Professors
Fotouh, Kamel H. .... (936) 857-2427
    khfotouh@pvamu.edu
    BT,EN,EV,PR

Associate Professors
Gabitto, Jorge F. .... (936) 857-2427
    jgabitto@aol.com
    AM,BT,EN,ME,MO,RE
Osborne-Lee, Irvin W. (Head) .... (936) 857-2427
    oslee@pvamu.edu
    EV,MO,NT,PR,SU

Assistant Professors
Aghara, Sukesh .... (936) 857-4606
    skaghara@pvamu.edu
    AM,EN,EV,MO,SU
Gyamerah, Michael .... (936) 857-2427
    michael_gyamerah@pvamu.edu
    BT,EN,EV,NT,PR,RE
Nave, Felecia M. .... (936) 857-2427
    fmnave@pvamu.edu
    BM,BT,PO,RE,SE

Accredited by: ABET

Graduate advisor: Sukesh Aghara

Undergraduate advisor: Kamel H. Fotouh

Student organization: AIChE
Advisor: Irvin W. Osborne-Lee

Department reports to: Milton R. Bryant, Dean of College of Engineering

Placement service: Glenda Jones (936) 857-2120, Assistant Director

Rice University

Chemical and Biomolecular Engineering
Department
MS-362
P.O. Box 1892
Houston, TX 77251-1892
Deliveries: 6100 Main Street; Houston, TX 77005

University phone .... (713) 348-0000
Department phone .... (713) 348-4902
Department FAX .... (713) 348-5478

Professors

Akers, William W. (Emeritus)
  wwakers@rice.edu
Armeniades, Constantine (Emeritus) ... (713) 348-3495
  cda@rice.edu
Chapman, Walter G. .... (713) 348-4900
  wgchap@rice.edu
  EN,MO,TH
Colvin, Vicki 713-348-5471
  colvin@rice.edu
  AM,NT
Davis, Sam H. (Emeritus) .... (713) 348-3536
  shdavis@rice.edu
Dyson, Derek C. (Emeritus)
  dyson@rice.edu
Hellums, J. David (Emeritus) .... (713) 348-5116
  jhellums@rice.edu
Hightower, Joe W. (Emeritus) .... (713) 348-5906
  jhigh@rice.edu
Hirasaki, George J. .... (713) 348-5416
  gjh@rice.edu
  EN,TP
Kobayashi, Riki (Emeritus)
  lkobay@aol.com
Mikos, Antonios G. .... (713) 348-5355
  mikos@rice.edu
  BM
Miller, Clarence A. .... (713) 348-4904
  camill@pop.rice.edu
  SU,TP
Robert, Marc A. .... (713) 348-3515
  mrobert@rice.edu
  SU,TH
San, Ka-Yiu .... (713) 348-5361
  ksan@rice.edu
  BT
West, Jennifer .... (713) 348-5955
  jwest@rice.edu
  BM
Zygourakis, Kyriacos (Chair) .... (713) 348-5208
  kyzy@rice.edu
  BM,MO,RE

Associate Professors

Pasquali, Matteo .... (713) 348-5830
  mp@rice.edu
  MO,NT,TP

Assistant Professors

Biswal, Sibani Lisa .... (713) 348-6055
  biswal@rice.edu
  AM,NT
Gonzalez, Ramon .... (713) 348-4893
  ramon.gonzalez@rice.edu
  BT
Mantzaris, Nikolaos .... (713) 348-2955
  nman@rice.edu
  BM,BT,MO
Segatori, Laura
  BM,BT
Wong, Michael .... (713) 348-3511
  mswong@rice.edu
  NT,RE

Instructors

Cox, Kenneth .... (713) 348-3529
  krcox@rice.edu
  EN,TH

Accredited by: ABET

Degrees granted 2005-2006:
  B.S.: 23   M.S.: 2   Ph.D.: 12

Graduate advisor: Clarence A. Miller

Undergraduate advisor: Kenneth R. Cox

Student organization: AIChE
Advisor: Kenneth R. Cox

Department reports to: Sallie Keller-McNulty, Dean (713) 348-4009

Placement service: Cheryl Matherly, Director (713) 348-4055, Career Services Center

Texas A&M University

Artie McFerrin Department of Chemical Engineering
3122 TAMU
College Station, TX 77843-3122
Deliveries: 200 Jack E. Brown Engineering Building

University phone .... (979) 845-3211
Department phone .... (979) 845-3361
Department FAX .... (979) 845-6446

Professors

Anthony, Rayford G. .... (979) 845-3370
    rg-anthony@tamu.edu
Appleby, Anthony J. .... (979) 845-5571
    ajappleby@tamu.edu
Balbuena, Perla (Associate Head) ..... (979) 845-3375
    balbuena@tamu.edu
Bukur, Dragomir B. .... (979) 845-3401
    d-bukur@tamu.edu
Bullin, Jerry A. (Emeritus)
    J.Bullin@bre.com
Cagin, Tahir .... (979) 862-1449
    cagin@chemail.tamu.edu
Darby, Ronald (Emeritus) .... (979) 845-3301
    R-Darby@tamu.edu
Davison, Richard R. (Emeritus) ...... (979) 845-3391
    rdavison@tamu.edu
Durbin, Leonel D. (Emeritus)
    madurbin99@aol.com
El-Halwagi, Mahmoud M. .... (979) 845-3484
    el-halwagi@tamu.edu
Eubank, Philip T. (Emeritus) .... (979) 845-3339
    p-eubank@tamu.edu
Froment, Gilbert (Visiting) .... (979) 845-0406
    g.froment@che.tamu.edu.edu
Glover, Charles J. (Associate Head) ... (979) 845-3389
    c-glover@tamu.edu
Hall, Kenneth R. .... (979) 845-3357
    krhall@tamu.edu
Holland, Charles D. (Emeritus) ...... (979) 845-3371
    cdholland@tamu.edu
Holste, James C. .... (979) 845-3384
    j-holste@tamu.edu
Holtzapple, Mark T. .... (979) 845-9708
    m-holtzapple@tamu.edu
Kuo, Yue .... (979) 845-9807
    yuekuo@tamu.edu
Mannan, Sam .... (979) 862-3985
    mannan@tamu.edu
Seminario, Jorge .... (979) 845-3301
    seminario@tamu.edu
Wood, Thomas .... (979) 862-1588
    thomas.wood@chemail.tamu.edu

Assistant Professors

Bevan, Michael .... (979) 847-8766
    mabevan@tamu.edu
Cheng, Zhengdong .... (979) 845-3413
    cheng@chemail.tamu.edu
Hahn, Juergen .... (979) 845-3568
    hahn@che.tamu.edu
Hahn, Mariah .... (979) 862-1454
    mariah.hahn@chemail.tamu.edu
Jayaraman, Arul .... (979) 845-3306
    arulj@tamu.edu
Shantz, Daniel F. .... (979) 845-3492
    shantz@che.tamu.edu
Silas, James .... (979) 862-1615
    james.silas@chemail.tamu.edu
Ugaz, Victor .... (979) 458-1002
    ugaz@che.tamu.ed

Other Faculty

Baldwin, John T. .... (979) 845-9803
    jt-baldwin@tamu.edu
Bradshaw, Jerry L. .... (979) 845-0610
    j-bradshaw@tamu.edu
Yurttas, Lale .... (979) 847-9316
    l-yurttas@tamu.edu

Accredited by: ABET

Degrees granted 2005-2006:
    B.S.: 86   M.S.: 17   Ph.D.: 25

Graduate advisor: Perla Balbuena

Undergraduate advisor: Lale Yurttas

Student organization: AIChE
Advisor: Lale Yurttas

Department reports to: G. K. Bennett, Vice Chancellor for Engineering, and Dean of Engineering, Wisenbaker Engineering Research Center

Placement service: Janice Leigh Turner, Executive Director of Placement, Rudder Tower

Texas A&M University-Kingsville

Chemical and Natural Gas Engineering Department
700 University Boulevard
MSC 193
Kingsville, TX 78363
Deliveries: Engineering Complex 917 W. Ave. B Suite 303

University phone .... (361) 593-2111
Department phone .... (361) 593-2002
Department FAX .... (361) 593-4026

Professors

Al-Saadoon, Faleh T. .... (361) 593-2380
KFFTA00@tamuk.edu
EN,TP
Heenan, William A. (Dean) .... (361) 593-2002
W_HEENAN@tamuk.edu
MO,PR
Mills, Patrick .... (361) 593-4827
Patrick.Mills@tamuk.edu
ME,PR,RE
Pilehvari, Ali (Chair) .... (361) 593-2089
A-PILEHVARI@tamuk.edu
PO,PR,TP
Serth, Robert W. .... (361) 593-2093
R-SERTH@tamuk.edu
HT,MO,SE

Associate Professors

Chisholm, John .... (361) 593-2092
CHISHOLM@tamuk.edu
EN,TP
Duarte, Horacio .... (361) 593-2008
Horacio.Duarte@tamuk.edu
PO,PR,TH
Lee, Sang-Yong .... (361) 593-2629
S-Lee@tamuk.edu
MO,RE,TH
Murphy, David (Visiting) .... (361) 593-4004
kfcdm00@tamuk.edu
PR,SE

Accredited by: ABET

Graduate advisor: R. W. Serth

Student organization: AIChE
Advisor: Ali A. Pilehvari

Department reports to: William A. Heenan (361) 593-2001, Dean of Engineering

Placement service: Leticia Silquero (361) 593-2217

University of Texas at Austin

Chemical Engineering Department
1 University Station, C0400
Austin, TX 78712-1062
Deliveries: Speedway at E. Dean Keeton

University phone .... (512) 471-3434
Department phone .... (512) 471-5238
Department FAX .... (512) 471-7060

Professors

Allen, David T. .... (512) 471-0049
allen@che.utexas.edu
EV
Bonnecaze, Roger T. (Chair) .... (512) 471-1497
bonnecaze@che.utexas.edu
TP
Brock, James R. (Emeritus) .... (512) 471-3348
chju344@uts.cc.utexas.edu
Chelikowsky, James R. .... (512) 232-9083
jrc@ices.utexas.edu
MO
Edgar, Thomas F. .... (512) 471-3080
edgar@che.utexas.edu
MO
Ekerdt, John G. .... (512) 471-4689
ekerdt@che.utexas.edu
RE,SU
Fair, James R. (Emeritus) .... (512) 471-3689
fair@che.utexas.edu
Freeman, Benny D. .... (512) 232-2803
freeman@che.utexas.edu
PO
Georgiou, George .... (512) 471-6975
gg@che.utexas.edu
Heller, Adam .... (512) 471-8874
heller@che.utexas.edu
Himmelblau, David M. (Emeritus) ..... (512) 471-7445
himmelblau@che.utexas.edu
Johnston, Keith P. .... (512) 471-4617
johnston@che.utexas.edu
BM,EV,NT
Jose-Yacaman, Miguel .... (512) 232-9111
yacaman@che.utexas.edu
AM,NT
Lloyd, Douglas R. .... (512) 471-4985
lloyd@che.utexas.edu
PO,SE
McKetta, John J. (Emeritus) .... (512) 471-5227
mcketta@che.utexas.edu
Mullins, C. Buddie .... (512) 471-5817
mullins@che.utexas.edu
NT,SU
Paul, Donald R. .... (512) 471-5392
paul@che.utexas.edu
PO

Peppas, Nicholas A. .... (512) 471-6644
    peppas@che.utexas.edu
    BM,BT,PO
Qin, S. Joe (Associate Chair) .... (512) 471-4417
    qin@che.utexas.edu
    MO,PR
Rase, Howard F. (Emeritus) .... (512) 471-3251
    rase@che.utexas.edu
Rochelle, Gary T. .... (512) 471-7230
    rochelle@che.utexas.edu
    EV
Rossky, Peter J. .... (512) 471-3555
    rossky@mail.utexas.edu
Sanchez, Isaac C. .... (512) 471-1020
    sanchez@che.utexas.edu
    PO,TH
Schechter, Robert S. (Emeritus) .... (512) 471-3245
    robert_schechter@pe.utexas.edu
Steinfink, Hugo (Emeritus) .... (512) 471-5233
    steinfink@che.utexas.edu
Stice, James E. (Emeritus) .... (512) 471-5527
    stice@che.utexas.edu
Willson, C. Grant .... (512) 471-4342
    willson@che.utexas.edu
    NT,PO
Wissler, Eugene H. (Emeritus) .... (512) 471-5353
    wissler@che.utexas.edu

Associate Professors
Korgel, Brian A. .... (512) 471-5633
    korgel@che.utexas.edu
    NT,TP
Schmidt, Christine .... (512) 471-1690
    schmidt@che.utexas.edu
    BM

Assistant Professors
Ganesan, Venkat .... (512) 471-4856
    ganesan@che.utexas.edu
    AM,MO,NT,TH
Hwang, Gyeong S. .... (512) 471-4847
    hwang@che.utexas.edu
    MO,NT,SU
Loo, Yueh Lin .... (512) 471-6300
    lloo@che.utexas.edu
    ME,PO
Truskett, Thomas M. .... (512) 471-6308
    truskett@che.utexas.edu
    MO,NT

Instructors
Eldridge, R. Bruce .... (512) 232-1407
    rbeldr@che.utexas.edu
    SE
Poehl, Michael .... (512) 471-4438
    poehlme@che.utexas.edu
Randall, D'Arcy .... (512) 471-8280
    darcyr@mail.utexas.edu
Swinnea, J. Steven .... (512) 471-3173
    swinnea@che.utexas.edu

Other Faculty
Sciance, Carroll T.
    scscorp@earthlink.net

Accredited by: ABET

Degrees granted 2005-2006:
    B.S.: 94   M.S.: 15   Ph.D.: 35

Graduate advisor: Isaac C. Sanchez

Undergraduate advisor: Douglas R. Lloyd

Student organization: AIChE
Advisor: Thomas M. Truskett

Department reports to: Ben G. Streetman (512) 471-1166, Dean of Engineering

Placement service: Nancy A. Evans (512) 471-7461, Director, Engineering Career Assistance Center, ECJ 2.400

Texas Tech University

Chemical Engineering Department
P. O. Box 43121
Lubbock, TX 79409-3121
Deliveries: 6th and Canton

University phone .... (806) 742-2011
Department phone .... (806) 742-3553
Department FAX .... (806) 742-3552

Professors

Bethea, Robert (Emeritus) .... (806) 742-3553
Robert.Bethea@ttu.edu
Heichelheim, H. R. (Emeritus) .... (806) 742-3553
Hoo, Karlene (Associate Dean) .... (806) 742-4079
Karlene.Hoo@ttu.edu
Karim, M. Nazmul (Chair) .... (806) 742-3553
naz.karim@ttu.edu
Mann, Uzi .... (806) 742-4086
uzi.mann@ttu.edu
McKenna, Greg .... (806) 742-4136
Greg.McKenna@ttu.edu
Riggs, James B. .... (806) 742-1765
jim.riggs@ttu.edu
Simon, Sindee .... (806) 742-1763
Sindee.Simon@ttu.edu
Tock, R. W. (Adjunct) .... (806) 742-3998
richard.tock@ttu.edu

Associate Professors

Abbott, James (Adjunct)
James.abbott@ttu.edu
Wiesner, Theodore F. .... (806) 742-1448
ted.wiesner@ttu.edu

Assistant Professors

Dai, Lenore .... (806) 742-1757
lenore.dai@ttu.edu
Khare, Rajesh .... (806) 742-3553
Rajesh.Khare@ttu.edu
Leggoe, Jeremy .... (806) 742-1767
Jeremy.Leggoe@ttu.edu
Shin, Jong-shik .... (806) 742-3553
Siviniah, Easan .... (806) 742-5158
easan.siviniah@ttu.edu
Vaughn, Mark .... (806) 742-5158
Mark.Vaughn@ttu.edu
Weeks, Brandon .... (806) 742-3998
Brandon.Weeks@ttu.edu

Accredited by: ABET

Degrees granted 2004-2005:

   B.S.: 20   M.S.: 4   Ph.D.: 5

Graduate advisor: M. Nazmul Karim

Undergraduate advisor: Sindee L. Simon

Student organization: AIChE
Advisor: Uzi Mann

Department reports to: Pamela Eibeck, Dean of Engineering

Placement service: Shelli Crockett

Utah

Brigham Young University

Department of Chemical Engineering
Brigham Young University
350 Clyde Building
Provo, UT 84602

University phone	(801) 422-1211
Department phone	(801) 422-2586
Department FAX	(801) 422-0151

Professors

Barker, Dee H. (Emeritus) .... (801) 422-4620
barkerd@et.byu.edu
Bartholomew, Calvin H. .... (801) 422-4162
bartc@et.byu.edu
Baxter, Larry L. .... (801) 422-8616
larry_baxter@byu.edu
Fletcher, Thomas H. (Associate Chair) . (801) 422-6236
tom_fletcher@byu.edu
Harb, John N. .... (801) 422-4393
jharb@et.byu.edu
Lewis, Randy S. .... (801) 422-7863
randy.lewis@byu.edu
Oscarson, John L. .... (801) 422-6243
oscarj@et.byu.edu
Pitt, William G. .... (801) 422-2589
pitt@byu.edu
Rowley, Richard L. .... (801) 422-2590
rowley@byu.edu
Smoot, L. Douglas (Emeritus) .... (801) 422-8930
lds@byu.edu
Solen, Kenneth A. .... (801) 422-6237
kenneth_solen@byu.edu
Terry, Ronald E. .... (801) 422-4297
ron_terry@byu.edu
Wilding, W. Vincent (Chair) .... (801) 422-2393
wildingv@et.byu.edu

Associate Professors

Hecker, William C. .... (801) 422-6235
hecker@byu.edu

Assistant Professors

Knotts, Thomas A. .... (801) 422-9158
tommy_knotts@qmail.com
Wheeler, Dean .... (801) 422-4126
dean_wheeler@byu.edu

Other Faculty

Hales, Hugh B. .... (801) 422-3749
hugh_hales@byu.edu

Accredited by: ABET

Degrees granted 2005-2006:
    B.S.: 41   M.S.: 5   Ph.D.: 7

Graduate advisor: Larry L. Baxter

Undergraduate advisor: Richard L. Rowley

Student organization: AIChE
Advisor: William C. Hecker

Department reports to: Alan Parkinson, Dean, Engineering and Technology

Placement service: Lloyd Hawkins (801) 422-6932

University of Utah

Department of Chemical Engineering
50 S. Central Campus Drive Rm 3290
Salt Lake City, UT 84112

University phone	(801) 581-7200
Department phone	(801) 581-6915
Department FAX	(801) 585-9291

Professors

Deo, Milind D. (Associate Dean) ...... (801) 581-7629
mddeo@eng.utah.edu
EN,EV,MO,RE
Hanson, Francis V. .... (801) 581-6591
Francis.Hanson@utah.edu
EN,PR,RE,SU
Lighty, JoAnn S. .... (801) 581-5763
jlighty@coe.utah.edu
EN,EV
Pershing, David W. .... (801) 581-5057
david.pershing@utah.edu
EN,EV
Pugmire, Ronald J. .... (801) 581-7236
pug@vpres.adm.utah.edu
EN,PO,SU
Ring, Terry A. .... (801) 585-5705
ring@eng.utah.edu
ME,MO,NT,RE,SE,SU
Sarofim, Adel .... (801) 585-9258
sarofim@reaction-eng.com
EN,EV
Silcox, Geoffrey D. (Associate Chair) .. (801) 581-8820
geoff@eng.utah.edu
EN,EV
Smith, Philip J. (Chair) .... (801) 585-3129
philip.smith@utah.edu
EN,MO
Wendt, Jost O. L. .... (801) 581-5763
wendt@eng.utah.edu
EN,EV

Associate Professors

Eddings, Eric .... (801) 585-3931
eddings@che.utah.edu
EN,EV
Krahenbuhl, Melinda (Research) ...... (801) 581-4188
mpk@nuclear.utah.edu
EN,SE
Magda, Jules J. .... (801) 581-7536
jj.magda@m.cc.utah.edu
BM,PO,SU
Skliar, Mikhail .... (801) 581-6918
mikhail.skliar@utah.edu
BM,PR
Trujillo, Edward M. .... (801) 581-4460
edward.trujillo@utah.edu
BM,BT,EV,MO

Tyler, Bonnie .... (801) 587-9696
bonniet@che.utah.edu
BT,EV
Whitty, Kevin .... (801) 585-9388
whitty@eng.utah.edu
EN,EV

Assistant Professors

Roper, D. Keith .... (801) 581-6915
kroper@eng.utah.edu
BM,BT,NT
Sutherland, James C. .... (801) 581-6915
EN,MO

Accredited by: ABET

Degrees granted 2005-2006:
B.S.: 33   M.S.: 8   Ph.D.: 12

Graduate advisor: Leda Mareth

Undergraduate advisor: Leda Mareth

Student organization: AIChE
Advisor: Edward M. Trujillo

Department reports to: Richard B. Brown (801) 581-6911, Dean of Engineering

Placement service: Lisa Christensen

Virginia

Hampton University

Chemical Engineering Department
Hampton, VA 23668

University phone	(757) 727-5000
Department phone	(757) 727-5288
Department FAX	(757) 727-5189

Professors

Adeyiga, Adeyinka A. .... (757) 727-5289
    adeyinka.adeyiga@hamptonu.edu
Akyurtlu, Ates (Chair) .... (757) 727-5599
    ates.akyurtlu@hamptonu.edu
Akyurtlu, Jale F. .... (757) 727-5589
    jale.akyurtlu@hamptonu.edu
Morgan, Morris H. III .... (757) 727-5583
    morris.morgan@hamptonu.edu

Associate Professors

Chegini, Amir .... (757) 727-5817
    amir.chegini@hamptonu.edu

Accredited by: ABET

Degrees granted 2004-2005:
    B.S.: 2

Undergraduate advisor: Department Chair

Student organization: AIChE
Advisor: Dr. Ates Akyurtlu

Department reports to: Eric Sheppard., Dean, School of Engineering and Technology

Placement service: Bessie Willis (757) 727-5331

University of Virginia

Department of Chemical Engineering
School of Engineering and Applied Science
102 Engineers' Way
P.O. Box 400741
Charlottesville, VA 22904-4741
Deliveries: Chemical Engineering Building Room 117, McCormick Road

University phone	(434) 924-0311
Department phone	(434) 924-7778
Department FAX	(434) 982-2658

Professors

Carta, Giorgio .... (434) 924-6281
    gc@virginia.edu
Davis, Robert J. (Chair) .... (434) 924-6284
    rjd4f@virginia.edu
Fernandez, Erik J. .... (434) 924-1351
    ejf3c@virginia.edu
Ford, Roseanne M. .... (434) 924-6283
    rmf3f@virginia.edu
Gaden, Elmer L. Jr. (Emeritus) .... (434) 924-7778
    elg@virginia.edu
Gainer, John L. (Emeritus) .... (434) 924-7778
    jlg@virginia.edu
Hudson, J. L. .... (434) 924-6275
    hudson@virginia.edu
Kirwan, Donald J. .... (434) 924-6278
    djk@virginia.edu
Laurencin, Cato .... (434) 923-0250
    ctl3f@virginia.edu
Lilleleht, Lembit (Emeritus) .... (434) 924-7778
    lul@virginia.edu
Neurock, Matthew .... (434) 924-6248
    mn4n@virginia.edu
O'Connell, John P. .... (434) 924-3428
    jpo2x@virginia.edu

Associate Professors

Aronson, Mark .... (434) 924-6276
    mta2z@virginia.edu

Assistant Professors

Green, David L. .... (434) 924-1302
    dlg9s@virginia.edu
McIntosh, Steven .... (434) 924-7778
Oberhauser, James P. .... (434) 924-7974
    jpo8a@virginia.edu

Other Faculty

Espino, Ramon .... (434) 924-6279
    re4n@virginia.edu

Accredited by: ABET

Degrees granted 2005-2006:
    B.S.: 22   M.S.: 9   Ph.D.: 5

Graduate advisor: Erik Fernandez

Undergraduate advisor: Giorgio Carta

Student organization: AIChE
Advisor: John O'Connell

Department reports to: James H. Aylor, Interim Dean, School of Engineering and Applied Sciences

Placement service: C. J. Livesay, Assistant Dean and Director, Center for Engineering Career Development

Virginia Commonwealth University

Department of Chemical and Life Science Engineering
601 West Main St.
Richmond, Va 23284

University phone .... (804) 828-0100
University FAX .... (804) 828-3846
Department phone .... (804) 828-7789
Department FAX .... (804) 828-3846

Professors

Jamison, Russell (Dean) .... (804) 828-0190
    rjamison@vcu.edu
    BM
McGee, Henry (Emeritus) .... (804) 828-3636
    hmcgee@vcu.edu
    TP
McHugh, Mark .... (804) 828-7031
    mmchugh@vcu.edu
    PO,TH
Peters, Michael (Chair) .... (804) 828-7789
    mpeters@vcu.edu
    BM,BT,SU,TP
Wynne, Kenneth .... (804) 828-9303
    kjwynne@vcu.edu
    AM,BM,PO,SU

Associate Professors

Huvard, Gary .... (804) 827-7000
    gshuvard@vcu.edu
    PO,RE

Assistant Professors

Fong, Stephen .... (804) 828-7789
    sfong@vcu.edu
    BM,BT
Rao, Raj .... (804) 828-7789
    rrao@vcu.edu
    BM,BT

Accredited by: ABET/AIChE

Degrees granted 2005-2006:
    B.S.: 15   M.S.: 6   Ph.D.: 3

Graduate advisor: Dr. Kenneth Wynne

Undergraduate advisor: Dr. Gary Huvard

Student organization: AIChE and AIMBE
Advisor: Dr. Gary Huvard

Department reports to: Dean Robert J. Mattauch

Placement service: Ms. Jennifer Rivers

Virginia Polytechnic Institute and State University/Virginia Tech

Chemical Engineering Department
0211
Blacksburg, VA 24061
Deliveries: 133 Randolph Hall

University phone .... (540) 231-6000
Department phone .... (540) 231-6631
Department FAX .... (540) 231-5022

Professors
Baird, Donald G. .... (540) 231-5998
dbaird@vt.edu
Durrill, Preston L. (Adjunct) .... (540) 231-6774
pdurrill@radford.edu
Hassler, John (Visiting) .... (540) 231-3775
jhassler@vt.edu
Kiran, Erdogan .... (540) 231-1375
ekiran@vt.edu
Liu, Y. A. .... (540) 231-7800
design@vt.edu
Oyama, S. Ted .... (540) 231-5309
oyama@vt.edu
Rony, Peter R. (Emeritus)
rony@vt.edu
Squires, Arthur M. (Emeritus) .... (540) 231-5972
verasqu@vt.edu
Sullivan, Joseph (Emeritus)
jtsull@vt.edu
Walz, John (Head) .... (540) 231-4213
jywalz@vt.edu
Wilkes, Garth L. (Emeritus) .... (540) 231-5498
gwilkes@vt.edu
Wills, George B. (Emeritus)

Associate Professors
Cox, David F. .... (540) 231-6829
dfcox@vt.edu
Davis, Richey M. .... (540) 231-4578
rmdavis@vt.edu
Marand, Eva .... (540) 231-8231
emarand@vt.edu
Michelsen, Donald L. (Emeritus)
Mischke, Roland A. (Emeritus)

Assistant Professors
Goldstein, Aaron .... (540) 231-3674
goldst@vt.edu
Martin, Stephen
Sum, Amadeu .... (540) 231-7869
asum@vt.edu

Accredited by: ABET

Degrees granted 2005-2006:
    B.S.: 41   M.S.: 2   Ph.D.: 7

Graduate advisor: Amadeu Sum

Undergraduate advisor: D. F. Cox

Student organization: AIChE
Advisor: Y. A. Liu

Department reports to: Richard Benson, Dean of Engineering

Placement service: J. H. Malone

Washington

University of Washington

Chemical Engineering Department
Box 351750
Seattle, WA 98195-1750
Deliveries: Benson Hall, Room 105

University phone .... (206) 543-2100
Department phone .... (206) 543-2250
Department FAX .... (206) 543-3778

Professors

Allan, G. Graham (Joint) .... (206) 543-1491
create@u.washington.edu
AM,EN,PO,RE

Babb, Albert L. (Emeritus)
mercury@u.washington.edu
BM,EN,MO

Baneyx, François .... (206) 685-7659
baneyx@u.washington.edu
AM,BT,NT

Berg, John C. .... (206) 543-2029
berg@cheme.washington.edu
EV,NT,PO,SU

Bowen, J. Ray (Emeritus) .... (206) 616-8128
bowen@engr.washington.edu
EN

Bryers, James D. (Adjunct) .... (206) 221-5876
jbryers@u.washington.edu
AM,BT,NT

Campbell, Charles T. (Adjunct) .... (206) 616-6085
campbell@chem.washington.edu
EV,MO,NT,RE,SU

Castner, David G. .... (206) 543-8094
castner@nb.engr.washington.edu
AM,BM,SU

David, Morton M. (Emeritus)

Davis, E. James (Emeritus) .... (206) 543-0298
jdavis@cheme.washington.edu
EV

Finlayson, Bruce A. (Emeritus) .... (206) 685-1634
finlayson@cheme.washington.edu
MO,PR

Garlid, Kermit L. (Emeritus) .... (206) 543-4807
kgarlid@u.washington.edu
EN

Gustafson, Richard (Adjunct) .... (206) 543-2790
pulp@u.washington.edu
PO,PR,RE

Heideger, William J. (Emeritus)

Hodgson, Kevin (Joint) .... (206) 543-7346
hodgson@u.washington.edu
SU

Hoffman, Allan S. (Joint) .... (206) 543-9423
hoffman@u.washington.edu
BM

Horbett, Thomas A. (Joint) .... (206) 685-1392
horbett@uweb.engr.washington.edu
BM,PO,SU

Jenekhe, Samson A. (Joint) .... (206) 543-5525
jenekhe@u.washington.edu
AM,EN,ME,NT,PO

Jiang, Shaoyi .... (206) 616-6509
sjiang@u.washington.edu
BT,MO,NT,SU

Johanson, Lennart N. (Emeritus)

Lidstrom, Mary E. (Joint) .... (206) 616-5282
lidstrom@u.washington.edu
BT,EV

McKean, Wm. T. (Joint) .... (206) 543-1626
wmckean@u.washington.edu

Pilat, Michael J. (Adjunct) .... (206) 543-4789
mpilat@u.washington.edu
EV

Ratner, Buddy D. (Joint) .... (206) 685-1005
ratner@uweb.engr.washington.edu
AM,BM,BT,NT,PO,SU

Ricker, N. Lawrence .... (206) 543-8786
ricker@u.washington.edu
MO,PR

Schwartz, Daniel T. .... (206) 685-4815
dts@u.washington.edu
EN,ME,NT

Sleicher, Charles A. (Emeritus) .... (206) 543-4807
charles@sleicher.net

Stuve, Eric M. (Chair) .... (206) 543-2253
stuve@u.washington.edu
EN,RE,SU

Woodruff, Gene L. (Emeritus)
woodruff@u.washington.edu
EN

Xia, Younan (Adjunct) .... (206) 543-1767
xia@chem.washington.edu
NT

Yager, Paul (Adjunct) .... (206) 543-6126
yagerp@u.washington.edu
BM,BT

Associate Professors

Adler, Stuart .... (206) 543-2131
stuadler@u.washington.edu
AM,EN,EV,MO,SU

Holt, Bradley R. .... (206) 543-0554
holt@cheme.washington.edu
PR

Krieger-Brockett, Barbara (Emeritus) .. (206) 543-2216
krieger@cheme.washington.edu
BT,EN,EV,RE

Overney, Rene M. .... (206) 543-4353
roverney@u.washington.edu
AM,NT,PO

Sarikaya, Mehmet (Adjunct) .... (206) 543-0724
sarikaya@u.washington.edu
AM,BM,BT,MO,NT,SU

Assistant Professors

132 United States

Pun, Suzie (Adjunct) .... (206) 685-3488
spun@u.washington.edu
BM,BT,NT

Shen, Hong .... (206) 543-5961
hs24@u.washington.edu
BM,BT,NT

Other Faculty

Baratuci, William B. .... (206) 543-2271
baratuci@u.washington.edu

Accredited by: ABET

Degrees granted 2005-2006:

    B.S.: 55   M.S.: 8   Ph.D.: 11

Graduate advisor: N. Lawrence Ricker

Undergraduate advisor: Devota Madrano

Student organization: AIChE
Advisor: John C. Berg

Department reports to: Matthew O'Donnell (206) 543-0340, Dean of Engineering

Placement service: Susan Terry (206) 543-0535, Director

Washington State University

School of Chemical Engineering and Bioengineering
P.O. Box 642710
Pullman, WA 99164-2710
Deliveries: 118 Dana Hall, Spokane St.

University phone	(509) 335-3564
Department phone	(509) 335-4332
Department FAX	(509) 335-4806

Professors

Campbell, Kenneth .... (509) 335-8011
    cvselkbc@vetmed.wsu.edu
    BM,BT
Davis, Denny C. .... (509) 335-7993
    davis@wsu.edu
Ivory, Cornelius F. .... (509) 335-7716
    ivory@mail.wsu.edu
    BT,MO,SE
Lee, James M. .... (509) 335-5252
    jmlee@wsu.edu
    BT,EN,MO
Liddell, KNona C. .... (509) 335-3710
    liddell@che.wsu.edu
    AM,EV
Miller, Reid C. (Emeritus) .... (509) 335-4001
    millerrc@wsu.edu
    TH
Petersen, James N. .... (509) 335-49141
    jn_petersen@wsu.edu
    BT,EV,MO,RE
Thomson, William J. (Emeritus) ...... (509) 335-8580
    thomson@che.wsu.edu
    RE
Van Wie, Bernard J. .... (509) 335-4103
    bvanwie@che.wsu.edu
    BM,BT,EN,EV,ME,NT
Zollars, Richard L. (Director) .... (509) 335-4332
    rzollars@che.wsu.edu
    MO,PO,RE,SU

Assistant Professors

Abu-Lail, Nehal
    BM,BT,EV,NT,PO,SU
Beyenal, Haluk
    BT,EN,EV,SU
Brewer, Lawrence
    BM,BT,NT
Ha, Su .... (509) 335-3786
    suha@wsu.edu
    BT,EN,RE
Lin, David .... (509) 335-7534
    davidlin@wsu.edu
    BM,BT
Vasavada, Anita .... (509) 335-7533
    vasavada@wsu.edu
    BM

Accredited by: ABET

Degrees granted 2005-2006:
    B.S.: 18   M.S.: 9
Student organization: AIChE
Advisor: Richard L. Zollars

Department reports to: Candis Claiborn, Interim Dean, College of Engineering and Architecture

Placement service: Alton Jamison

West Virginia

West Virginia University

Department of Chemical Engineering
P.O. Box 6102
Morgantown, WV 26506-6102
Deliveries: 403 ESB, Evansdale Drive

University phone . . . . (304) 293-0111
Department phone . . . . (304) 293-2111
Department FAX . . . . (304) 293-4139

Professors

Bailie, Richard C. (Emeritus)
Cho, Eung H. . . . . (304) 293-2111x2433
Eung.Cho@mail.wvu.edu
EN,EV,SE,SU
Cilento, Eugene V. (Dean) . . . . . . (304) 293-4821x2237
Gene.Cilento@mail.wvu.edu
BM,BT,MO
Dadyburjor, Dady B. (Chair) . . . . (304) 293-2111x2411
Dady.Dadyburjor@mail.wvu.edu
AM,EN,RE
Galli, Alfred F. (Emeritus)
Gupta, Rakesh K. . . . . (304) 293-2111x2427
Rakesh.Gupta@mail.wvu.edu
AM,EV,MO,NT,PO
Kono, Hisashi O. (Emeritus) . . . . . (304) 293-2111x2421
Hisashi.Kono@mail.wvu.edu
Kugler, Edwin L. . . . . (304) 293-2111x2414
Edwin.Kugler@mail.wvu.edu
EN,NT,RE
Stiller, Alfred H. . . . . (304) 293-2111x2408
Alfred.Stiller@mail.wvu.edu
EN,RE
Turton, Richard . . . . (304) 293-2111x2415
Richard.Turton@mail.wvu.edu
EN,EV,MO,PR,RE
Yang, Ray Y. K. . . . . (304) 293-2111x2419
Ray.Yang@mail.wvu.edu
AM,BT,EV,MO,NT,RE
Zondlo, John W. . . . . (304) 293-2111x2409
John.Zondlo@mail.wvu.edu
AM,EN,EV

Associate Professors

Shaeiwitz, Joseph A. . . . . (304) 293-2111x2410
Joseph.Shaeiwitz@mail.wvu.edu
Stinespring, Charter D. . . . . (304) 293-2111x2425
Charter.Stinespring@mail.wvu.edu
AM,EN,ME,NT,SU

Assistant Professors

Anderson, Brian J. . . . . (304) 293-2111 x2435
Brian.Anderson@mail.wvu.edu
BM,EN,EV,MO
Klinke, David J. . . . . (304) 293-2111 x2432
David.Klinke@mail.wvu.edu
BM,MO,RE

Other Faculty

Doraiswamy, Deepak (Adjunct)
Henry, Joseph D. (Adjunct)
Jaffe, Charles M. (Adjunct) . . . . . . (304) 293-3435x4425
Charles.Jaffe@mail.wvu.edu
Keller, George (Adjunct)
Liang, Ruifeng . . . . (304) 293-2111x2623
Ruifeng.Liang@mail.wvu.edu
AM,PO
Olson, Fred (Adjunct)
Wallace, William E. (Adjunct)
Wildi, Robert (Adjunct)
Zhang, Wu . . . . (304) 293-2111x2422
Wu.Zhang@mail.wvu.edu
AM,EN,EV,MO,NT,PO,PR,SE,SU

Accredited by: ABET

Degrees granted 2005-2006:
    B.S.: 15   M.S.: 9   Ph.D.: 2

Graduate advisor: Charter D. Stinespring

Undergraduate advisor: Joseph A. Shaeiwitz

Student organization: AIChE
Advisor: Dady Dadyburjor

Department reports to: Dean, College of Engineering and Mineral Resources

Placement service: David Stewart, Interim Director (304) 293-2221

West Virginia University Institute of Technology

Chemical Engineering Department
405 Fayette Pike
Montgomery, WV 25136

University phone 304-442-3161
University FAX 304-442-1006
Department phone 304-442-3296
Department FAX 304-442-3164

Professors

Crum, Edward (Emeritus)
Doner, David 304-442-3376
　　　　　　　　　　ddoner@wvutech.edu
　　　　　　　　　　EN,RE,SE
Minnick, Michael 304-442-3378
　　　　　　　　　　mminnick@wvutech.edu
　　　　　　　　　　PR

Associate Professors

Thomas, Garth (Chair) 304-442-3377
　　　　　　　　　　garth.thomas@mail.wvu.edu
　　　　　　　　　　EN,MO,PR,SE

Assistant Professors

Wang, Jin 304-442-3163
　　　　　　　　　　jin.wang@mail.wvu.edu
　　　　　　　　　　MO,PR

Accredited by: North Central, ABET

Degrees granted 2005-2006:
　　B.S.: 6

Student organization: AIChE
Advisor: Garth E Thomas Jr.

Department reports to: Dean M. Sathyamoorthy

Placement service: Cantrell Miller

Wisconsin

University of Wisconsin-Madison

Department of Chemical and Biological Engineering
1415 Engineering Drive
Madison, WI 53706-1691

University phone .... (608) 262-1234
Department phone .... (608) 262-1092
Department FAX .... (608) 262-5434

Professors

Abbott, Nicholas L. (Associate Chair) . (608) 265-5278
abbott@engr.wisc.edu
BT,NT,PO,SU
Bird, R. Byron (Emeritus) .... (608) 262-5920
bird@engr.wisc.edu
PO,RE,TP
Chapman, Thomas W. (Emeritus) ..... (608) 263-1979
chapman@engr.wisc.edu
de Pablo, Juan J. .... (608) 262-7727
depablo@engr.wisc.edu
BT,MO,PO,TH
Dumesic, James A. .... (608) 262-1095
dumesic@engr.wisc.edu
EN,RE,SU
Graham, Michael D. (Chair) .... (608) 265-3780
graham@engr.wisc.edu
MO,TP
Hill, Charles G. Jr. (Emeritus) .... (608) 263-4593
hill@engr.wisc.edu
RE,SE
Klingenberg, Daniel J. .... (608) 262-8932
klingenberg@engr.wisc.edu
AM,SU,TP
Kuech, Thomas F. .... (608) 263-2922
kuech@engr.wisc.edu
AM,ME
Langer, Stanley H. (Emeritus) .... (608) 262-1190
langer@engr.wisc.edu
EN,EV,RE
Lightfoot, Edwin N. (Emeritus) .... (608) 262-6934
lightfoot@engr.wisc.edu
BT,SE,TP
Murphy, Regina M. (Associate Chair) .. (608) 262-1587
murphy@engr.wisc.edu
BM,BT
Nealey, Paul F. .... (608) 265-8171
nealey@engr.wisc.edu
AM,BM,BT,PO,SU
Rawlings, James B. .... (608) 263-5859
jbraw@che.wisc.edu
MO,PR
Ray, W. Harmon (Emeritus) .... (608) 263-4732
ray@engr.wisc.edu
MO,PO,PR,RE
Sather, Glenn A. (Emeritus) .... (608) 262-1092
gasather@facstaff.wisc.edu

Yin, John .... (608) 265-3779
yin@engr.wisc.edu
BT,MO

Associate Professors

Mavrikakis, Manos .... (608) 262-9053
manos@engr.wisc.edu
AM,EN,MO,RE,SU
Palecek, Sean P. .... (608) 262-8931
palecek@engr.wisc.edu
BM,BT
Root, Thatcher W. .... (608) 262-8999
thatcher@engr.wisc.edu
BT,EV,RE,SE
Swaney, Ross E. .... (608) 262-3641
swaney@engr.wisc.edu
MO,PR

Assistant Professors

Lynn, David M. .... (608) 262-1086
dlynn@engr.wisc.edu
AM,BT,PO
Maravelias, Christos T. .... (608) 265-9026
maravelias@engr.wisc.edu
BT,MO,PR
Shusta, Eric V. .... (608) 262-1092
shusta@engr.wisc.edu
BM,BT

Accredited by: ABET

Degrees granted 2005-2006:
    B.S.: 78   M.S.: 5   Ph.D.: 18

Graduate advisor: Eric V. Shusta

Undergraduate advisor: Regina M. Murphy

Student organization: AIChE
Advisor: David Lynn/Christos Maravelias

Department reports to: Paul Peercy, Dean of Engineering

Placement service: Sandra L. Arnn (608) 262-3471, Director

Wyoming

University of Wyoming

Department of Chemical and Petroleum Engineering
Dept 3295, 1000 E University Avenue
Laramie, Wyoming 82071
Deliveries: 16th and Gibbon Street

University phone	(307) 766-5160
University FAX	(307)-766-4042
Department phone	(307) 766-2500
Department FAX	(307) 766-6777

Professors

Ackerman, John (Adjunct) .... (307) 766-2566
wyoming1@uwyo.edu
NT
Banaszak, Michael (Adjunct)
Cha, Chang Yul (Emeritus) .... (307) 766-2500
ccha@chacorporation.com
Deans, Harry A. (Emeritus)
Evers, John F. (Emeritus)
Harris, H. Gordon .... (307) 766-6558
harrishg@uwyo.edu
EN
Haynes, Henry W. (Emeritus) .... (307) 766-2500
haynes@uwyo.edu
Mason, Geoffrey (Adjunct)
Morrow, Norman R. .... (307) 766-2838
morrownr@uwyo.edu
EN
Radosz, Maciej .... (307) 766-2500
radosz@uwyo.edu
PO,SE
Sharma, Mrityunjai P. .... (307) 766-6317
sharma@uwyo.edu
EN,EV
Takamura, Koichi (Adjunct)
Towler, Brian F. (Chair) .... (307) 766-2189
towler@uwyo.edu
EN

Associate Professors

Bell, David A. .... (307) 766-5769
davebell@uwyo.edu
SU

Assistant Professors

Adidharma, Hertanto .... (307) 766-2909
adidharm@uwyo.edu
TH
Alvarado, Vladimir .... (307) 766-6464
vladimir_11@msn.com
EN
Argyle, Morris .... (307) 766-2500
mdargyle@uwyo.edu
RE,SE
Johnson, Patrick .... (307) 766-6524
pjohns27@uwyo.edu
BM,NT
LaForce, Tara .... (307) 766-3278
tlaforce@uwyo.edu
EN
Shen, Youqing .... (307) 766-2500
BT,PO

Other Faculty

Wo, Shaochang (Research) .... (307) 766-2780
swo@uwyo.edu

Accredited by: ABET

Degrees granted 2005-2006:
  B.S.: 19   Ph.D.: 3

Graduate advisor: Faculty

Undergraduate advisor: Faculty

Student organization: AIChE
Advisor: Morris D. Argyle

Department reports to: O. A. (Gus) Plumb, Dean of Engineering

Placement service: Jo Chytka, Director

Argentina

Instituto Tecnologico de Buenos Aires

Departamento de Ingenieria Quimica
Av. E. Madero 399
1106 Buenos Aires

University phone	(54-11)63934800
Department phone	(54-11)63934860
Department FAX	(54-11)63934861

Professors

Cao, Eduardo G.
   EN,SE
Costa, Jose L.
Kammerer, Jorge R.
Marques, Dardo (Head) .... (54-11)63934800 ext 4862
   dmarques@itba.edu.ar
   PR
Paya, Miguel A. .... (54-11)63934800 ext 5820
   mpaya@itba.edu.ar
Pizarro, Ramon A.
   BT
Pontiggia, Claudio J.
Soler, Susana M.
Zuazo, Beatriz N.

Associate Professors

Peralta, Carmen E.
Pirk, Tomas
Rausei, Diego N.
   dnrausei@itba.edu.ar
   RE

Assistant Professors

Aude Luppi, Vergenie .... (54-11)63934860 ext 5821
   vaude@itba.edu.ar
   EV
Bertini, Liliana M.
Coppari, Norberto R.
Fernandez, Beatriz E.
   MO,PR
Fidalgo, Maria M. .... (54-11) 63934800 ext5821
   mfidalgo@itba.edu.ar
   EV,NT
Gagey, Susana M.
Iglesias, Graciela S.
Iñon, Fernando A.
Popik, Pablo R.
   EN
Sanchez, Lidia P.
Testa Fernandez, Juan J.

Accredited by: CONEAU

Degrees granted 2005-2006:
   B.S.: 11

Undergraduate advisor: Dardo Marques

Department reports to: Ing. Jose Luis Roces (Vicerrector)

Universidad Nacional de la Plata

Facultad de Ingenieria
Departmento Ingenieria Quimica
1 esq. 47-1900 La Plata

Department phone	54 0221 423-6680
Department FAX	54 0221 425-0122

Professors

Barreto, Guillermo F.
   barreto@quimica.unlp.edu.ar
Elsner, Cecilia I.
   cielsner@volta.ing.unlp.edu.ar
Ferretti, Osmar A.
   ferretti@ing.unlp.edu.ar
Iglesias, Omar A.
   oaiglesi@volta.ing.unlp.edu.ar
Martinez, Osvaldo M.
   ommartin@volta.ing.unlp.edu.ar
Mascheroni, Rodolfo H.
   rhmasche@ing.unlp.edu.ar
Navarro, Agustin F. (Director)
   anavarro@ing.unlp.edu.ar
Pessacq, Raul A.
   rpessacq@volta.ing.unlp.edu.ar
Siri, Guillerno J.
   gsiri@quimica.unlp.edu.ar
Zaritzky, Noemi E.
   zaritzky@volta.ing.unlp.edu.ar

Associate Professors

Mazza, German D.
   gmazza@ing.unlp.edu.ar

Assistant Professors

Bevilacqua, Alicia E.
   aebevila@volta.ing.unlp.edu.ar
Caminos, Rolando A.
   caminos@ing.unlp.edu.ar
D. Sorbo, Carlos L.
   lelio_ds@ciudad.com.ar
Gervasi, Claudio A.
   gervasi@inifta.unlp.edu.ar
Giner, Sergio A.
   saginer@ing.unlp.edu.ar
Graieb, Jorge A.
   jagraieb@interar1.com.ar
Pereira Duarte, Susana I.
   spereira@ing.unlp.edu.ar
Salvadori, Viviana O.
   vosalvad@ing.unlp.edu.ar
Santana, Ines M.
   isantan@ing.unlp.edu.ar

Accredited by:

Department reports to: Ing.Agustín Navarro Directory

Universidad Nacional del Litoral

Inst. de Desarrollo Tecnol. para la Industria Quimica - INTEC
Guemes 3450
S3000GLN - Santa Fe - Santa Fe

University phone	(54-342) 4571110
University FAX	(54-342) 4571111
Department phone	(54-342) 4511079
Department FAX	(54-342) 4511079

Professors

Alfano, Orlando M. — alfano@intec.unl.edu.ar
Baltanas, Miguel A. — tderliq@ceride.gov.ar
Campanella, Enrique A. — tquique@ceride.gov.ar
Cassano, Alberto E. — acassano@ceride.gov.ar
Cerda, Jaime — jcerda@intec.unl.edu.ar
Chiovetta, Mario G. (Director) .... 4511079 — mchiove@intec.unl.edu.ar
Costanza, Vicente — tsinoli@ceride.gov.ar
Deiber, Julio A. — treoflu@ceride.gov.ar
Grau, Ricardo J. — cqfina@ceride.gov.ar
Gugliotta, Luis M. — lgug@intec.unl.edu.ar
Henning, Gabriela — ghenning@intec.unl.edu.ar
Irazoqui, Horacio A. — hirazo@ceride.gov.ar
Isla, Miguel A. — misla@ceride.gov.ar
Luna, Julio — jluna@ceride.gov.ar
Marchetti, Jacinto L. — jlmarch@ceride.gov.ar
Meira, Gregorio R. — gmeira@ceride.gov.ar
Rubiolo, Amelia C. — tceidal@ceride.gov.ar
Saita, Fernando A. — fasaita@ceride.gov.ar

Associate Professors

Bortolozzi, Raul A. — rabor@ceride.gov.ar
Brandi, Rodolfo J. — rbrandi@ceride.gov.ar
Cabrera, María I. — cqfina@ceride.gov.ar
Estenoz, Diana A. — destenoz@ceride.gov.ar
Giavedoni, Maria D. — madelia@ceride.gov.ar
Martin, Carlos A. — cmartin@ceride.gov.ar
Pozzo, Roberto L. — rpozzo@intec.unl.edu.ar
Vega, Jorge R. — jvega@ceride.gov.ar
Zorrilla, Susana E. — zorrilla@intec.unl.edu.ar

Assistant Professors

Berli, Claudio L. — cberli@ceride.gov.ar
Dondo, Rodolfo G. — rdondo@ceride.gov.ar
Murguía, Marcelo C. — mmurguia@ceride.gov.ar

Accredited by:

Degrees granted 2004-2005:

Ph.D.: 2

Department reports to: Mario Barletta, Rector de la Universidad Nacional del Litoral

Universidad Nacional del Salta

Facultad de Ingenieria
Buenos Aires 177
4400 Salta
Deliveries: Lorgio MERCADO FUENTES

University phone 54 387 432-0563
University FAX 54 387 425-5535
Department phone 54 387 425-5420
Department FAX 54 387 425-5351

Professors

Armada, Margarita (Chair) .... 425 5362 [†]
 — armadam@unsa.edu.ar
Borla, Ricardo J. (Chair) 425 5351
 — rborla@epassaporte.com
Cuevas, Carlos M. (Chair) 425 5362
 — ccuevas@unsa.edu.ar
Flores, Horacio R. (Chair) 254 5412
 — hrflores@unsa.edu.ar
Gonzo, Elio E. (Chair) 425 5410
 — gonzo@unsa.edu.ar
Gottifredi, Juan C. (Emeritus) 425 5410
 — gottifre@unsa.edu.ar
Mercado, Lorgio (Associate Dean) 425 5343
 — lmercado@unsa.edu.ar
Pocoví, Rubens E. (Emeritus) 425 5412
 — inbemi@unsa.edu.ar
Quiroga, Oscar D. (Emeritus) 425 5409
 — oquiroga@unsa.edu.ar

Associate Professors

Abán, Francisco E. (Chair) 425 5351
 — faban@unsa.edu.ar
Capretto, María E. (Chair) 425 5326
 — castillo@unsa.edu.ar
Destéfanis, Hugo A. (Chair) 425 5391
 — hdestefa@unsa.edu.ar
Macoritto, Alberto M. (Chair) 425 5350
 — amacori@unsa.edu.ar
Romero, Luis César (Chair) 425 5343
 — lcromero@unsa.edu.ar

[†] Faculty numbers have prefix 54 387.

Accredited by: CONEAU

Degrees granted 2004-2005:
 B.S.: 10   Ph.D.: 2

Undergraduate advisor: Carlos Cuevas

Student organization: Centro Estudiantes de Ingeniería

Department reports to: Lorgio Mercado Fuentes, Facultad de Ingenieria UNSa

Placement service: Facultad de Ingeniería, Universidad Nacional de Salta

Universidad Nacional del Sur

Departamento de Ingeniería Química
Avenida Alem 1253
8000 Bahía Blanca
Buenos Aires

University phone 54 291 459-5101
Department phone 54 291 459-5170
Department FAX 54 291 459-5171

Professors

Brignole, Esteban A. (Emeritus) .... 4861700-231 [†]
　ebrignole@plapiqui.edu.ar
　MO,PR,SE,TH
Capiati, Numa J. 4861700-205
　ncapiati@plapiqui.edu.ar
　AM,PO
Crapiste, Guillermo Héctor 4861700-203
　gcrapiste@plapiqui.edu.ar
　BT,MO,TP
Damiani, Daniel E. 4861700-211
　ddamiani@plapiqui.edu.ar
　RE,SU
Lozano, Jorge E. 4595170-212
　jlozano@plapiqui.edu.ar
　BT,PR,SU
Mayer, Carlos E. (Honorary) 4595170-3610
　cmayer@criba.edu.ar
Porras, José A. 4538605
　jporras@criba.edu.ar
　RE
Urbicain, Martín J. (Honorary) 4861700-201
　urbicain@criba.edu.ar
　HT,PR
Valles, Enrique M. 4861700-228
　valles@plapiqui.edu.ar
　AM,NT,PO

Associate Professors

Bandoni, José A. (Head) 4861700-254
　abandoni@plapiqui.edu.ar
　BM,BT,EN,MO,PR
Bottini, Susana B. 4861700-213
　sbottini@plapiqui.edu.ar
　SE,TH
Duarte, Marta M. 4595170-3618
　mduarte@criba.edu.ar
　EN,RE
Echarte, Roberto E. 4861700-248
　recharte@plapiqui.edu.ar
　HT,PR
Eliceche, Ana M. 4861700-252
　meliceche@plapiqui.edu.ar
　EN,EV,MO,PR,SE
Errazu, Alberto F. 4861700-269
　aerrazu@plapiqui.edu.ar
　EV,RE
Salinas, Daniel R. 4595170-3614
　dsalinas@criba.edu.ar
　AM,NT,SU
Villar, Marcelo A. 4861700-212
　mvillar@plapiqui.edu.ar
　NT,PO

Assistant Professors

Aduriz, Hugo R. 4861700-260
　raduriz@plapiqui.edu.ar
　EV,RE

Barbosa, Silvia E. 4861700-271
　sbarbosa@plapiqui.edu.ar
　NT,PO,SU
Borio, Daniel O. 4861700-208
　dborio@plapiqui.edu.ar
　MO,RE
Brandolin, Adriana 4861700-272
　abrandolin@plapiqui.edu.ar
　MO,PO,PR,RE
Bucalá, Verónica 4861700-265
　vbucala@plapiqui.edu.ar
　RE
Carelli, Amalia A. 4861700-238
　acarelli@plapiqui.edu.ar
　BT
Diaz, María S. 4861700-243
　sdiaz@plapiqui.edu.ar
　EV,MO,PR
García, Silvana G. 4595100-3514
　sgarcia@criba.edu.ar
　AM,NT,SU
Quinzani, Lidia M. 4861700-282
　poquinza@criba.edu.ar
　AM,NT,PO
Saidman, Silvana B. 4595100-3615
　ssaidman@criba.edu.ar
　AM,BT,PO,SU
Sanchez, Mabel C. 4861700-249
　msanchez@plapiqui.edu.ar
　MO,PR,SE
Sarmoria, Claudia 4861700-273
　csarmoria@plapiqui.edu.ar
　MO,PO,PR,RE
Schbib, Noemí S. 4861700-267
　sschbib@plapiqui.edu.ar
　MO,PR,RE
Tonelli, Stella M. 4861700-251
　stonelli@plapiqui.edu.ar
　EV,MO

[†] Phone code 54 291 precedes each number.

Accredited by:

Degrees granted 2005-2006:
　B.S.: 27　　M.S.: 1　　Ph.D.: 12

Graduate advisor: Lidia M. Quinzani

Undergraduate advisor: Adriana Brandolin

Department reports to: Dr. José Alberto Bandoni - Head

Australia

University of Adelaide

School of Chemical Engineering
University of Adelaide
Adelaide SA 5005

University phone 61 8 8303 4455
Department phone 61 8 8303 5455
Department FAX 61 8 8303 4373

Professors

Agnew, John A. (Emeritus) 8303 4118
　jagnew@chemeng.adelaide.edu.au

King, Keith D. (Chair) 8303 5448
    kking@chemeng.adelaide.edu.au

Associate Professors

Alwahabi, Zeyad T. 8303 3768
    zalwahab@chemeng.adelaide.edu.au
Ashman, Peter J. 8303 5072
    pashman@chemeng.adelaide.edu.au
Colby, Chris B. 8303 5846
    cbcolby@chemeng.adelaide.edu.au
Davey, Ken R. 8303 5457
    kdavey@chemeng.adelaide.edu.au
Lewis, David M. 8303 5503
    dlewis@chemeng.adelaide.edu.au
Mullinger, Peter J. 8303 5085
    pmulling@chemeng.adelaide.edu.au
Nguyen, Q. D. 8303 5456
    dnguyen@chemeng.adelaide.edu.au
O'Neill, Brian K. 8303 4647
    boneill@chemeng.adelaide.edu.au

Assistant Professors

Ngothai, Yung 8303 5445
    yngothai@chemeng.adelaide.edu.au

Accredited by: Institute of Engineers (Australia), IChemE (UK)

Degrees granted 2004-2005:
    B.S.: 44   M.S.: 1   Ph.D.: 3

Graduate advisor: Dr Peter Ashman

Undergraduate advisor: Dr Yung Ngothai

Student organization: Ms Claire Barber

Department reports to: Professor Peter Dowd, Executive Dean of Engineering

Placement service: Ms Amanda Phillis, Careers Administrator

Curtin University of Technology

Department of Chemical Engineering
GPO Box U1987
Perth WA 6845
Deliveries: Kent Street, Bentley, WA 6102

University phone .... (618) 9266-9266
Department phone .... (618) 9266-7581
Department FAX .... (618) 9266-3554

Professors

Smith, Terence N. (Emeritus)
Tade, Moses O. (Head) 618 9266 7704
    tadem@che.curtin.edu.au
Zhang, Dong-ke 618 9266 7581
    dkzhang@che.curtin.edu.au

Associate Professors

Ang, Ha M. 618 9266 7894
    angm@che.curtin.edu.au
Ray, Martyn 618 9266 7702
    M.Ray@exschange.curtin.edu.au

Assistant Professors

Balliu, Nicoleta 618 9266 2683
    N.Balliu@curtin.edu.au
Li, Qin 618 9266 7704
Lou, Xia 618 9266 1682
    X.Lou@curtin.edu.au
Pareek, Vishnu 618 9266 4687
    pareekv@vesta.curtin.edu.au
Vuthaluru, Hari B. 618 9266 4685
    haribabu@che.curtin.edu.au
Wang, Shaobin 618 9266 3776
    wangshao@vesta.curtin.edu.au
Wu, Hongwei 618 9266 7592
    Wuh@vesta.curtin.edu.au

Accredited by: Institution of Engineers Australia, Institution of Chemical Engineers (UK)

Degrees granted 2004-2005:
    B.S.: 69   M.S.: 1

Graduate advisor: H.M. Ang

Undergraduate advisor: H Vuthaluru

Student organization: ACES

Department reports to: Professor Peter Lee, Executive Dean, Engineering and Science

University of Melbourne

Department of Chemical and Biomolecular Engineering
Victoria 3010
Deliveries: Monash Road, Parkville 3052

University phone .... (61 3) 8344-4000
Department phone .... (61 3) 8344 6631
Department FAX .... (61 3) 8344 4153

Professors

Boger, David V. (Director) 8344-7440
    dvboger@unimelb.edu.au
    AM,EV,SU
Caruso, Frank 8344-3461
    fcaruso@unimelb.edu.au
    AM,BT,NT,SU
Davidson, Malcolm R. 8344-6615
    m.davidson@unimelb.edu.au
    MO,PR,TP
Ducker, William 8344 3430
    wducker@unimelb.edu.au
    AM,EN,NT
Dunstan, David E. 8344-8261
    davided@unimelb.edu.au
    BM,BT,NT,PR
Gray, Neil B. (Emeritus) 8344-6639
    n.gray@unimelb.edu.au
    AM,MO,RE
Healy, Thomas W (Emeritus) 8344-7624
    tomhealy@unimelb.edu.au
    AM,BT,EV,PR,SU
Pamment, Neville B. 8344-6627
    nbp@unimelb.edu.au
    BM,BT
Reuter, Marcus 8344 6780
    mreuter@unimelb.edu.au
    EV,MO,PR
Scales, Peter J (Head) 8344-6480
    peterjs@unimelb.edu.au
    AM,EV,SU

Shallcross, David C. 8344-6614
dcshal@unimelb.edu.au
PR
Siemon, S. R. (Emeritus)
Solomon, David H. 8344-8200
davids@unimelb.edu.au
PO
Stevens, Geoffrey W. (Director) 8344-6621
gstevens@unimelb.edu.au
AM,EN,EV,PR,SU
van Deventer, Jannie S.J. (Dean) 8344-6619
jannie@unimelb.edu.au
AM,PR
Wood, David G. (Emeritus) 8344-3800
dgwood@unimelb.edu.au

Associate Professors

Connor, Michael A 8344-6875
maconnor@unimelb.edu.au
EN,EV
Dagastine, Raymond R 8344 4704
rrd@unimelb.edu.au
AM,MO,NT,SU
Franks, George 8344 9020
gvfranks@unimelb.edu.au
BT,EV,PR,SU
Kentish, Sandra E. 8344-6682
sandraek@unimelb.edu.au
BT,EN,EV,HT,PR
O'Connor, Andrea J. 8344-8962
andreajo@unimelb.edu.au
BM,BT,NT
Qiao, Greg G. 8344-3683
gregghq@unimelb.edu.au
AM,BT,EN,NT,PO
Yeow, Y. L. 8344-6613
yly@unimelb.edu.au
MO,PO

Assistant Professors

Tirtaatmadja, Viyada 8344-4676
viyadat@unimelb.edu.au

Accredited by: IChemE (UK), Institute of Engineers (Australia)

Degrees granted 2005-2006:
  B.S.: 65   Ph.D.: 20

Graduate advisor: A/Prof David Shallcross

Undergraduate advisor: Dr Michael Connor

Student organization: MUCESS
Advisor: Mr Mike Parksinson

Department reports to: Jannie van Deventer 8344-6619, Dean of Engineering

Monash University

Chemical Engineering Department
PO Box 36
Clayton, Victoria 3800
Deliveries: Building 35, Room 226, Wellington Road

University phone 61 3 990 54000
Department phone 61 3 990 53450
Department FAX 61 3 990 59649

Professors

Brisk, Mike L. (Emeritus) .... 990 52592 [†]
mike.brisk@eng.monash.edu.au
Johnston, Bob E. (Emeritus) 990 52592
R.E.Johnston@eng.monash.edu.au
Lawson, Frank (Honorary) 990 53426
frank.lawson@eng.monash.edu.au
Mathews, Joseph F. (Honorary) 990 53425
joe.mathews@eng.monash.edu.au
Potter, Owen E. (Emeritus) 990 53420
Prince, Ian G. 990 53449
ian.prince@eng.monash.edu.au
Rhodes, Martin J. (Head) 990 53445
martin.rhodes@eng.monash.edu.au
Seale, Alan (Adjunct) 990 59620
alan.seale@eng.monash.edu.au
Sridhar, Tamarapu (Dean) 990 53427
tam.sridhar@eng.monash.edu.au
Tiu, Carlos (Honorary) 990 53423
carlos.tiu@eng.edu.au
Toner, Mark (Adjunct) 990 59620
mark.toner@eng.monash.edu.au

Associate Professors

Andrews, John R.G. 990 53428
john.andrews@eng.monash.edu.au
Batchelor, Warren J. 990 53452
warren.batchelor@eng.monash.edu.au
Hoadley, Andrew F. 990 53421
andrew.hoadley@eng.monash.edu.au
Jagadeeshan, Ravi P. 990 53274
ravi.jagadeeshan@eng.monash.edu.au
Jeffrey, Matthew I. 990 51873
matthew.jeffrey@eng.monash.edu.au
Li, Chun-Zhu 990 59623
chun-zhu.li@eng.monash.edu.au
Nguyen, Kien L. 990 53429
loi.nguyen@eng.monash.edu.au
Olbrich, W. E. 990 53436
erich.olbrich@eng.monash.edu.au
Parker, Ian H. 990 55078
ian.parker@eng.monash.edu.au
Shen, Wei 990 53447
wei.shen@eng.monash.edu.au
Webley, Paul A. 990 51874
paul.webley@eng.monash.edu.au

[†] Numbers have 61 3 prefix.

Accredited by: IChemE (U.K.), Inst. Engrs. (Aust.), RACI

Degrees granted 2003-2004:
  B.S.: 25   M.S.: 1   Ph.D.: 8

Graduate advisor: Dr Paul A Webley

Undergraduate advisor: Prof Martin J Rhodes

Student organization: SMUCE
Advisor: Dr W. Erich Olbrich

Department reports to: Prof Tamarupu Sridhar, Dean of Engineering

Placement service: Ms Gilda Moss, 61 3 990 54170

University of Newcastle

Chemical Engineering Department
New South Wales, NSW 2308

University phone	(02) 4921-5000
Department phone	(02) 4921-6180
Department FAX	(02) 4921-6920

Professors

Dlugogorski, Bogdan .... (02) 4921-6176
    Bogdan.Dlugogorski@newcastle.edu.au
    RE
Evans, Geoff .... (02) 4921-5897
    Geoffrey.Evans@newcastle.edu.au
    EV
Galvin, Kevin .... (02) 4921-6194
    Kevin.Galvin@newcastle.edu.au
    SE
Jameson, Graeme J. .... (02) 4921-6181
    Graeme.Jameson@newcastle.edu.au
    SU
Kennedy, Eric (Head) .... (02) 4921-6177
    Eric.Kennedy@newcastle.edu.au
    RE
Wall, Terry F. (Joint) .... (02) 4921-6179
    Terry.Wall@newcastle.edu.au

Associate Professors

Lucas, John .... (02) 4921 6193
    John.Lucas@newcastle.edu.au
    PR
Moghtaderi, Behdad .... (02) 4921-6183
    Behdad.Moghtaderi@newcastle.edu.au
Nguyen, Anh .... (02) 4921 6189
    Anh.Nguyen@newcastle.edu.au
    NT

Accredited by: Institution of Engineers (Australia), IChemE (UK)

Degrees granted 2005-2006:

    Ph.D.: 3

Graduate advisor: B Dlugogorski

Undergraduate advisor: B Moghtaderi

Department reports to: J Carter (02) 4921-6025, Pro Vice-Chancellor

Placement service: Susan Eade

University of Queensland

Division of Chemical Engineering
St. Lucia, Brisbane
Queensland 4072
Deliveries: Building 74 College Road

University phone	61 7 3365 1111
Department phone	61 7 3365 3708
Department FAX	61 7 3365 4199

Professors

Bhatia, Suresh 3365 4263
    sureshb@cheque.uq.edu.au
Cameron, Ian T. (Head) 3365 4261
    ianc@donald.cheque.uq.edu.au
Do, Duong D. 3365 4154
    duongd@cheque.uq.edu.au
James, David 3365 3708
    davidj@cheque.uq.edu.au
Kavanagh, Lydia 3365 7517
    lydiak@cheque.uq.edu.au
Keller, Jurg 3365 4727
    j.keller@cheque.uq.edu.au
Litster, Jim D. 3365 3616
    j.litster@cheque.uq.edu.au
Lu, G. Q. (Max) 3365 3735
    maxlu@cheque.uq.edu.au
Middelberg, Anton (Director) 3366 8784
    a.middelberg@uq.edu.au
Rudolph, Victor 3365 4171
    victorr@cheque.uq.edu.au
White, E.T. (Ted) (Emeritus) 3365 4153
    tedw@cheque.uq.edu.au

Associate Professors

Bell, Peter 3365 3801
    p.bell@cheque.uq.edu.au
Clarke, Bill 3365 6464
    billc@cheque.uq.edu.au
Cooper-White, Justin 3346 8715
    j.cooperwhite@uq.edu.au
Crosthwaite, Caroline 3365 4264
    carolc@cheque.uq.edu.au
Halley, Peter 3365 4158
    p.halley@cheque.uq.edu.au
Howes, T. 3365 4262
    tonyh@cheque.uq.edu.au
Lant, Paul 3365 4728
    paull@cheque.uq.edu.au
Nielsen, Lars 3365 6960
    larsn@cheque.uq.edu.au
Wiles, Robert J. (Gus) (Hon) 3365 3708
    rjwiles@pocketmail.com.au

Assistant Professors

da Costa, Joe 3365 4218
    joedac@cheque.uq.edu.au
Hardin, Matt 3365 3708
    m.hardin@auckland.ac.nz
Martin, Darren 3365 4152
    darrenm@cheque.uq.edu.au
Masserotto, Paul 3365 4152
    paulm@cheque.uq.edu.au
Nicholson, Timothy 3365 4081
    t.m.nicholson@uq.edu.au
O'Brien, Kate 3365 3534
    k.obrien@mailbox.uq.edu.au
Reid, Steve 3365 4001
    stever@cheque.uq.edu.au
Sopade, Peter 3365 3931
    p.sopade@cheque.uq.edua.u
Wang, F-Y. 3365 4552
    fuyang@cheque.uq.edu.au
Zhu, Zhonghua (John) 3365 3528
    johnz@cheque.uq.edu.au
Zou, Jin
    j.zou@uq.edu.au

Accredited by: Institute of Engineers (Australia), IChemE (UK)

Degrees granted 2004-2005:
    B.S.: 52   M.S.: 4   Ph.D.: 14

Graduate advisor: Tony Howes

Undergraduate advisor: Paul Lant

Department reports to: Jim Litster, Head, School of Engineering

Placement service: Marilyn Wilckens

Royal Melbourne Institute of Technology

School of Civil and Chemical Engineering
GPO Box 2476V
Melbourne, Victoria 3001

University phone .... (61 3) 9925-2000
Department phone .... (61-3) 9925-2080
Department FAX .... (61 3) 9639 0138

Professors

Bhattacharya, Sati N. 9925-2086
    satinath.bhattacharya@rmit.edu.au
Roddick, Felicity (Head) 9925-2080
    felicity.roddick@rmit.edu.au

Associate Professors

Jollands, Margaret 9925-2089
    margaret.jollands@rmit.edu.au
Swinbourne, Doug 9925-2201
    douglas.swinbourne@rmit.edu.au

Accredited by: Institution of Engineers (Australia), IChemE (UK)

Undergraduate advisor: Assoc Prof Roger Hadgraft

Department reports to: Prof. Daine Alcorn, Pro-Vice Chancellor

Placement service: Joanne Tyler 61 3 9925-2078, Careers and Appointments

University of Sydney

Chemical Engineering Department - J01
The University of Sydney
New South Wales 2006

University phone .... (02) 9351-2222
Department phone .... (02) 9351-2455
Department FAX .... (02) 9351-2854

Professors

Coster, Hans G. L. (Visiting) 9351 2256
    hcoster@chem.eng.usyd.edu.au
Glasser, David (Visiting) 9351 3001
    glasser@chem.eng.usyd.edu.au
Haynes, Brian S. 9351-3435
    haynes@chem.eng.usyd.edu.au
Petrie, Jim G. 9351-4115
    petrie@chem.eng.usyd.edu.au
Prince, Rolf G. H. (Emeritus) 9351-2354
    prince@chem.eng.usyd.edu.au
Romagnoli, Jose A (Chair) 9351-4794
    jose@chem.eng.usyd.edu.au

Associate Professors

Barton, Geoff W. (Head) 9351-3780
    barton@chem.eng.usyd.edu.au
Langrish, Tim A. G. 9351-4568
    timl@chem.eng.usyd.edu.au

Assistant Professors

Fletcher, David F. (Adjunct) 9351-4147
    davidf@chem.eng.usyd.edu.au
Gomes, Vincent G. 9351-4868
    vgomes@chem.eng.usyd.edu.au
Harris, Andrew 9351 2926
    aharris@chem.eng.usyd.edu.au
Kavannah, John M. 9306-9442
    kavanagh@chem.eng.usyd.edu.au
See, Howard T 9351-3832
    howards@chem.eng.usyd.edu.au
Valix, Marjorie 9351-4995
    mvalix@chem.eng.usyd.edu.au

Accredited by: IChemE (UK), Institute of Engineers (Australia)

Graduate advisor: T A G Langrish

Undergraduate advisor: H T See

Student organization: S U C E A

Department reports to: Dean of Engineering

Placement service: The Secretary, Appointments Board

Bangladesh

Bangladesh University of Engineering and Technology

Chemical Engineering Department
BUET, Dhaka 1000
Bangladesh

University phone 880-2-9665650-80
University FAX 880-2-9665622
Department phone 880-2-9665609

Professors

Ahmed, Nooruddin .... 7347 [†]
    nahmed@che.buet.ac.bd
Ali, M. S. 7309
    ali@buet.ac.bd
Begum, Dil A. 7346
    afroza@buet.ac.bd
Hossain, Ijaz (Head) 7329
    pmrebuet@bangla.net
Mahmud, Iqbal (Emeritus) 7341
    imahmed@bangla.net
Quader, A.K.M.A. 7308
    quader@buet.ac.bd

Associate Professors

Islam, M. S. 7559
    seraj@buet.ac.bd
Khan, Sirajul H. 7326
    shkhan@buet.ac.bd
Mondal, Harendra N. 7245
    hmondal@buet.ac.bd

Assistant Professors

Kazi, M F. K. 7894

Syeda, Sultana R. 7893
syedasrazia@che.buet.ac.bd

† Numbers are extensions of the university number.

Accredited by:

Degrees granted 2003-2004:
    B.S.: 62   M.S.: 2

Graduate advisor: Feroz Kabir

Undergraduate advisor: Syeda Sultana Razia

Student organization: Chemical Engineering Association Advisor: M. Sabder Ali

Department reports to: A S w Kurny, Dean of Engineering

Placement service: M. Fazlul Bari, Director of Students Welfare

Belgium

University of Gent

Laboratorium voor Petrochemische Techniek
Krijgslaan 281 S5
9000 Gent
Deliveries: Krijgslaan 281, Building S5 B-9000

Department phone 32 9 264 45 16
Department FAX 32 9 264 49 99

Professors
Marin, Guy B. (Director)
Yablonsky, Gregory (Visiting)

Associate Professors
Dumez, Francis.
Heynderickx, Geraldine
Olea, Maria (Visiting)
van den Berg, Henk

Assistant Professors
De Schepper, Sandra
Reyniers, Marie-Francoise
Thybaut, Joris

Accredited by:

Degrees granted 2004-2005:
    B.S.: 25   M.S.: 15   Ph.D.: 3

Department reports to: D. De Zutter, Dean, Faculty of Applied Science

Katholieke Uiniversiteit Leuven

Department of Chemical Engineering
W. de Croylaan 46
B-3001 Heverlee (Leuven)
Belgium

Department phone 32 16 32 26 76
Department FAX 32 16 32 29 91

Professors
Creemers, Claude 32 16 322346
    Claude.Creemers@CIT.kuleuven.ac.be
Mewis, Jan (Emeritus)
    jan.mewis@cit.kuleuven.ac.be
Moldenaers, Paula 32 16 322359
    Paula.Moldenaers@CIT.kuleuven.ac.be
Van Impe, Jan 32 16 321466
    Jan.VanImpe@cit.kuleuven.ac.be
Vandecasteele, Carlo (Chair) 32 16 322727
    Carlo.Vandecasteele@cit.kuleuven.ac.be

Associate Professors
Degrève, Jan 32 16 322367
    Jan.Degreve@CIT.kuleuven.ac.be
Vermant, Jan 32 16 32 23 55
    jan.vermant@cit.kuleuven.ac.be

Assistant Professors
van puyvelde, Peter 32 16 322357
    peter.vanpuyvelde@cit.kuleuven.ac;be
VanderBruggen, Bart 32 16 322340
    Bart.VanderBruggen@CIT.kuleuven.ac.be

Accredited by: Flemisch interuniversitary council

Degrees granted 2003-2004:
    B.S.: 35   M.S.: 27   Ph.D.: 11

Graduate advisor:

Student organization: VTK : www.vtk.be

Placement service: Jan.Degreve@CIT.kuleuven.ac.be

University of Liege

Group of Applied Chemistry
and Chemical Engineering
Institut de Chimie (B6)
4000 Sart Tilman—Liege

Department phone .... (32) 4 366 3541
Department FAX .... (32) 4 366 3545

Professors
Crine, M. .... (32) 4 366 3556
    M.Crine@ulg.ac.be
    BM,BT,SE
Germain, A. H. .... (32) 4 366 3547
    Albert.Germain@ulg.ac.be
    EN,EV,RE
Heyen, G. .... (32) 4 366 3527
    G.Heyen@ulg.ac.be
    EN,MO,PR
Kalitventzeff, B. (Emeritus) .... (32) 4 366 3521
    B.Kalitventzeff@ulg.ac.be
L'Homme, G. A. (Emeritus) .... (32) 4 366 3541
    Guy.LHomme@ulg.ac.be
    EV
Lecloux, A.
Marchot, P. .... (32) 4 366 3513
    Pierre.Marchot@ulg.ac.be
    BM,MO,RE,SE
Pirard, J. P. .... (32) 4 366 3558
    Jean-Paul.Pirard@ulg.ac.be
    AM,MO,NT,RE,SU

Assistant Professors
Heinrichs, B. .... (32) 4 366 35 05
B.Heinrichs@ulg.ac.be
AM,RE,SU
Toye, D. .... (32) 4 366 35 09
Dominique.Toye@ulg.ac.be
BM,BT,MO,RE,SE

Accredited by:

Degrees granted 2005-2006:
           M.S.: 25   Ph.D.: 5

Student organization: AEES

Faculté Polytechnique de Mons

56 rue de l'Epargne
Mail Code 7000
Mons, Belgium

Professors
Delvosalle, Christian (Director)
Thomas, Diane

Accredited by:

Brazil

Universidade Estadual de Campinas (UNICAMP)

School of Chemical Engineering - FEQ
Av. Albert Einstein, 500
Cidade Universitaria, C.P. 6066
13081-970, Campinas SP
University phone ....(55) 019-788 3900
Department FAX ....(55) 019 788 3910

Professors
Bittencourt, Edison
    edsonbit@dtp.feq.unicamp.br
    PO
Francesconi, Artur Z.
    francesconi@desq.feq.unicamp.br
    SE
Maciel Filho, Rubens
    maciel@feq.unicamp.br
    BT,PO,RE
Maciel, Maria R. W.
    wolf@feq.unicamp.br
    SE
Mei, Lúcia H. I.
    lumeibit@turing.unicamp.br
    PO
Mendes, Mario de J.
    mendes@desq.feq.unicamp.br
    RE
Mori, Milton
    mori@feq.unicamp.br
    MO,PR
Rocha Pereira, J. A. F. da
    pereira@desq.unicamp.br
    PR
Rocha, Sandra C.dos S.
    rocha@feq.unicamp.br
    PR
Santana, Cesar C.
    santana@feq.unicamp.br
    PR

Associate Professors
Bueno, Sônia M. A.
    sonia@feq.unicamp.br
    BT
Campos, Joao S. de C.
    siniezio@dtp.feq.unicamp.br
    PO
Cremasco, Marco A.
    cremasco@feq.unicamp.br
    PR
Cruz, Sandra L. da
    cruz@desq.unicamp.br
    PR
Franco, Telma T.
    franco@feq.unicamp.br
    BT,SE
Fratini, Ana M. F.
    fattini@desq.feq.unicamp.br
    PR
Guirardello, Reginaldo
    guira@feq.unicamp.br
    MO,PR
Jordão, Elisabete
    bete@desq.feq.unicamp.br
    RE
Kieckbusch, Theo G.
    theo@feq.unicamp.br
    PR
Krähenbühl, Maria A.
    mak@feq.unicamp.br
    SE
Lona, Liliane M.F.
    liliane@feq.unicamp.br
    PO
Miranda, Everson A.
    everson@feq.unicamp.br
    BT
Moraes, Ângela M.
    ammoraes@feq.unicamp.br
    BT
Ravagnani, Sergio P.
    ravag@dtp.feq.unicamp.br
    PO
Ravagnani, Teresa M. K.
    kakuta@desq.feq.unicamp.br
    PR
Santana, Maria H. A.
    lena@feq.unicamp.br
Silva, Maria A.
    cida@feq.unicamp.br
Silva, Meuris G. C. da
    meuris@bla.feq.unicamp.br
Tannous, Kátia
    katia@feq.unicamp.br
Taranto, Osvaldir P. (Associate Dean)
    val@feq.unicamp.br
Zemp, Roger J.
    zemp@desq.feq.unicamp.br

Assistant Professors
Abreu, Charlles R. A.
    MO

Aznar, Martin
    maznar@feq.unicamp.br
    SE
Bartoli, Julio R.
    bartoli@feq.unicamp.br
    PO
Beppu, Marisa M.
    beppu@feq.unicamp.br
    BM
Bufo, Moacir J.
    bufo@feq.unicamp.br
    EV
Cobo, Antonio J. G.
    acobo@bla.feq.unicamp.br
    RE
Costa, Aline C.
    BT
Lisboa, Antonio C. L.
    lisboa.feq.unicamp.br
    PR
Morales, Ana R.
    morales@dtp.feq.unicamp.br
    PO
Oliveira, Wagner S.
    oliveira@dtp.feq.unicamp.br
    PO
Peres, Leila
    lperes@dtp.feq.unicamp.br
    PO
Silva, Flávio V.
    flavio@feq.unicamp.br
    PR
Sprung, Renato
    sprung@bla.feq.unicamp.br
Tomaz, Edson
    etomaz@feq.unicamp.br
Valenca, Gustavo P.
    gustavo@feq.unicamp.br

Accredited by: MEC, CAPES

Degrees granted 2005-2006:
    B.S.: 125   M.S.: 53   Ph.D.: 32

Graduate advisor: Liliane Lona

Undergraduate advisor: Roger Zemp

Placement service: Silvana Fenga neves

Universidade Estadual de Maringa

Departamento de Engenharia Quimica
Av. Colombo, 5790, BL D-90
CEP 87020-900 - MARINGA
University phone . . . . (55 44) 261 4000
Department phone . . . . (55 44) 261 44755
Department FAX . . . . (55 44) 263 3440

Professors
de Moraes, Flavio F. . . . . 261 4754 [†]
    flavio@deq.uem.br
Pereira, Nehemias C. (Head) 261 4760
    nehemias@deq.uem.br
Zanin, Gisella M. (Dean) 261 4754
    gisellazanin@deq.uem.br

Associate Professors
Bergamasco, Rosangela 261 4763
    rosangela@deq.uem.br
dos Santos, Onelia A. A. (Associate Head) 261 4756
    onelia@deq.uem.br
Gimenes, Marcelino L. 261 4761
    marcelino@deq.uem.br
Machado, Nadia R. C. F. 261 4747
    nadia@deq.uem.br
Mendes, Elisabete S. (Dean) 261 4756
    elisabete@deq.uem.br
Neitzel, Ivo 261 4761
    ibneitzel@deq.uem.br
Ravagnani, Mauro A. 261 4746
    ravag@deq.uem.br
Tavares, Celia R. G. 261 4746
    celia@deq.uem.br

Assistant Professors
Andrade, Cid M. G. 261 4752
    cid@deq.uem.br
Arroyo, Pedro A. 261 4758
    arroyo@deq.uem.br
Barros, Maria A. D. 261 4758
    angelica@deq.uem.br
Canassa, Edson M. 261 4752
    edson@deq.uem.br
Cardozo Filho, Lucio 261 4749
    lucio@deq.uem.br
Cossich, Eneida S. 261 4748
    eneida@deq.uem.br
Costa, Alexando M. de S. 2614755
    costa@deq.uem.br
da Rocha Pietrobon, Carmen L. 261 4762
    carmen@deq.uem.br
de Barros Jr, Carlos 261 4780
    carlos@deq.uem.br
de Barros, Sueli T. D. (Associate Dean) 261 4762
    sueli@deq.uem.br
Gianotto, Valter R. 261 4753
    valter@deq.uem.br
Higa, Márcio 2614755
    higa@deq.uem.br
Jorge, Luiz M. M. 261 4747
    lmmj@deq.uem.br
Mafra, Luciana I. 261 4749
    luciana@deq.uem.br
Motta Lima, Oswaldo C. da 261 4759
    oswaldo@deq.uem.br
Muller, José M. 261 4759
    jmu@deq.uem.br
Olivo, Jose E. 261 4753
    olivo@deq.uem.br
Paraiso, Paulo R. 261 4752
    paulo@deq.uem.br
Povh, Nancy P. 261 4748
    nancy@deq.uem.br
Ugri, Miriam C. B. A. 261 4749
    miriam@deq.uem.br

[†] Faculty numbers have prefix 55 44.

Accredited by: Brazilian Ministry of Education

Degrees granted 2003-2004:
    B.S.: 73   M.S.: 13   Ph.D.: 4

Graduate advisor: Gisella Maria Zanin

Undergraduate advisor: Oswaldo Curty da Motta Lima

Student organization: CAEQ

Advisor: Oswaldo Curty da Motta Lima

Department reports to: Nehemias Curvelo Pereira

Placement service: Jose Eduardo Olivo

Universidade Federal Do Rio De Janeiro

Escola de Quimica
Centro de Tecnologia-Bloco E
Cidade Universitaria
21949-900, Rio de Janeiro, RJ

University phone	(5521) 2562-7037
University FAX	(5521) 2562-7567
Department phone	(5521) 2562-7040
Department FAX	(5521) 2562-7567

Professors

Coutinho, Jorge de A. (Emeritus)
Gentil, Vicente (Emeritus) .... (5521) 2562-7595
Martelli, Hebe H. L. (Emeritus)
Mascarenhas, Bernardo J. G. (Emeritus)
Moritz, Vitalis (Emeritus)
Perlingeiro, Carlos A. G. (Emeritus) .... (5521) 2562-7532
    per@eq.ufrj.br
Rajagopal, Krishnaswamy .... (5521) 2562-7654
    raja@eq.ufrj.br
Schmal, Martin
    schmal@peq.coppe.ufrj.br
Seidl, Peter .... (5521) 2562-7586
    pseidl@eq.ufrj.br
Telles, Affonso C.S da S. .... (5521) 2562-7653
    affonso@eq.ufrj.br
Valdman, Belkis (Director) .... (5521) 2562-7037
    diretora@eq.ufrj.br

Associate Professors

Aguiar, Eduardo F. de S.
    falabella@cenpes.petrobras.com
Almeida, Valéria C. de .... (5521) 2562-7595
    valeria@eq.ufrj.br
Antunes, Adelaide M. de S. .... (5521) 2562-7426
    adelaide@eq.ufrj.br
Aranda, Donato A. G. .... (5521) 2562-7657
    donato@eq.ufrj.br
Araujo, Ofelia de Q. F. .... (5521) 2562-7637
    ofelia@eq.ufrj.br
Barreto, Daniel W. .... (5521) 2562-7475
    dbarreto@eq.ufrj.br
Borschiver, Suzana .... (5521) 2562-7426
    suzana@eq.ufrj.br
Brasil, Simone L. D.C. .... (5521) 2562-7642
    simone@eq.ufrj.br
Calado, Verônica M. de A. .... (5521) 2562-7533
    calado@eq.ufrj.br
Cammarota, Magali C. .... (5521) 2562-7568
    christe@eq.ufrj.br
Carvalho, Denize D. de. .... (5521) 2562-7566
    denize@eq.ufrj.br
Castier, Marcelo .... (5521) 2562-7607
    castier@eq.ufrj.br
Castro, Mário Sérgio O. de .... (5521) 2562-7584
    msoc@eq.ufrj.br
Chrisman, Erika C. A. N. .... (5521) 2562-7582
    enunes@eq.ufrj.br
Coelho, Maria A. Z. .... (5521) 2562-7572
    alice@eq.ufrj.br
Couto, Maria A. P.G. .... (5521) 2562-7577
    gimenes@eq.ufrj.br
Cunha, Osvaldo G. C. da .... (5521) 2562-7641
    osvaldo@eq.ufrj.br
d'Avila, Luiz A. .... (5521) 2562-7041
    davila@eq.ufrj.br
Dweck, Jo .... (5521) 2562-7594
    dweck@eq.ufrj.br
Folly, Rossana O. M. .... (5521) 2562-7652
    rossana@eq.ufrj.br
Guimarães, Maria J. de O. .... (5521) 2562-7597
    zeze@eq.ufrj.br
Leite, Selma G. F. .... (5521) 2562-7578
    selma@eq.ufrj.br
Leão, Maria H. M. R. .... (5521) 2562-7580
    mhrl@eq.ufrj.br
Maia, Maria C. A. .... (5521) 2562-7576
    antun@eq.ufrj.br
Margarit, Isabel C. P. .... (5521) 2562-8550
    margarit@eq.ufrj.br
Martins, José Vítor B. .... (5521) 2562-7610
    vitor@eq.ufrj.br
Medeiros, José L. de .... (5521) 2562-7535
    jlm@eq.ufrj.br
Medronho, Ricardo de A. .... (5521) 2562-7635
    medronho@eq.ufrj.br
Milfont Jr., Wilson de N. .... (5521) 2562-7585
    milfont@eq.ufrj.br
Mothé, Cheila G. .... (5521) 2562-7587
    cheila@eq.ufrj.br
Nicolaiewsky, Eliôni M. de A. .... (5521) 2562-7634
    elioni@eq.ufrj.br
Penteado Filho, Alberto de F. .... (5521) 2562-7537
    penteado@eq.ufrj.br
Pereira Jr., Nei .... (5521) 2562-7639
    nei@eq.ufrj.br
Pessoa, Fernando L. P. .... (5521) 2562-7603
    pessoa@eq.ufrj.br
Peçanha, Ricardo P. .... (5521) 2562-7633
    pecanha@eq.ufrj.br
Queiroz, Eduardo M. .... (5521) 2562-7603
    mach@eq.ufrj.br
Salgado, Andréa M. .... (5521) 2562-7579
    andrea@eq.ufrj.br
Servulo, Eliana F. C. .... (5521) 2562-7600
    servulo@eq.ufrj.br
Silva, Monica A. P. da .... (5521) 2562-7606
    monica@eq.ufrj.br
Souza Jr., Maurício B. de .... (5521) 2562-7636
    mbsj@eq.ufrj.br
Souza, Mariana de M. V.M. .... (5521) 2562-7640
    mmattos@eq.ufrj.br
Tavares, Frederico W. .... (5521) 2562-7650
    tavares@eq.ufrj.br
Valle, Maria L. M. .... (5521) 2562-7581
    murta@eq.ufrj.br
Yokoyama, Lidia .... (5521) 2562-7640
    lidia@eq.ufrj.br
Zakon, Abraham .... (5521) 2562-7643
    zakon@eq.ufrj.br

Assistant Professors

Alhadeff, Eliana M. .... (5521) 2562-7647
    ema@eq.ufrj.br
Andrade, Hubmaier L. B. de .... (5521) 2206-9229
    hubmaier.andrade@br.bureauveritas.com

Andrade, José E. P.de .... (5521) 2562-7717
    joseedu@eq.ufrj.br
Awerianow, Cláudia de M. J. .... (5521) 2598-1715
    jardim@eq.ufrj.br
Campello, Ana E. F. .... (5521) 2562-7638
    campello@eq.ufrj.br
Gomes, Alexandre C. L. .... (5521) 2562-7574
    aleiras@eq.ufrj.br
Silva, Sílvia M. C. da .... (5521) 2562-7651
    sebrao@eq.ufrj.br
Vieira, Pedro A. P. .... (5521) 2562-7536
    papv@eq.ufrj.br

Accredited by: ABEQ, ABQ, CRQ, MEC-SESU, CAPES

Degrees granted 2003-2004:
    B.S.: 80   M.S.: 40   Ph.D.: 20

Undergraduate advisor: Prof. Eduardo Mach Queiroz

Student organization: DAEQ

Department reports to: Profa. Belkis Valdman

Placement service: Prof. Pedro Antonio Peixoto Vieira

Universidade de Sao Paulo USP

Escola Politecnica da USP
Department of Chemical Engineering
Av. Prof. Luciano Gualberto, travessa 3, n. 380
05508-900 Sao Paulo, SP, Brasil
Department phone 55 11 3091-2236
Department FAX 55 11 3031-3020

Professors

Buchler, Pedro M. .... 3091-2225 [†]
    pbuchler@usp.br
    EV
Facciotti, Maria C. R. (Head) 3091-2234
    mcrfacci@usp.br
    BT
Giudici, Reinaldo 3091-2254
    rgiudici@usp.br
    MO,PO,PR,RE
Guardani, Roberto 3091-2277
    guardani@usp.br
    EV,PR,SE
Nascimento, Cláudio A. O. 3091-2216
    oller@usp.br
    EV,MO,PR
Odloak, Darci (Associate Head) 3091-2261
    odloak@usp.br
    PR

Associate Professors

Kilikian, Beatriz V. 3091-2232
    kilikian@usp.br
    BT
LeRoux, Galo A.C. 3091-1170
    galoroux@usp.br
    MO,PR
Neves, Jose M. 3091-2280
    jmango@ipt.br
Pinto, Jose M. 3091-2237
    jompinto@usp.br
    PR

Silva, Gil A. 3091-2213
    ganderis@usp.br
    EV
Tadini, Carmen C. 3091-2258
    catadini@usp.br
Terron, Luiz R. 3091-2240
    lrterron@usp.br

Assistant Professors

Antunha, Andre G. 3091-2244
    aantunha@usp.br
Aoki, Idalina V. 3091-2274
    idavaoki@usp.br
Camacho, Jose L. P. 3091-2243
    jlpcam@usp.br
Gombert, Andreas K. 3091-2229
    andreas.gombert@poli.usp.br
    BT
Guedes, Isabel C. 3091-2223
    icguedes@usp.br
Gut, Jorge A.W. 3091-2253
    jorgewgut@usp.br
    PR
Loureiro, Luiz V. 3091-2237
    valcov@usp.br
    PR
Maeda, Masazi 3091-2233
    masmaeda@usp.br
Matai, Patricia 3091-2224
    pmatai@usp.br
Melo, Hercilio G. 3091-2231
    hgdemelo@usp.br
Neiva, Augusto C. 3091-2228
    augusto.neiva@poli.usp.br
Pacheco, Claudio R. F. 3091-2265
    crfp@usp.br
    SE
Paiva, Jose L. 3091-2262
    jolpaiva@usp.br
    SE
Park, Song W. 3091-2271
    sonwpark@usp.br
    PR
Pessoa, Pedro A. 3091-1106
    pedro.pessoa@poli.usp.br
    SE
Salvagnini, Wilson M. 3091-2285
    jackwil@usp.br
Song, Tah W. 3091-2245
    tahwsong@usp.br
    SE
Taqueda, Maria E. S. 3091-2252
    mtaqueda@usp.br
    SE
Teixeira, Antonio C. S.C. 3091-2263
    acscteix@usp.br
    EV,MO,RE
Tonso, Aldo 3091-2283
    aldtonso@usp.br
    BT

[†] Dial with prefix 55 11.

Accredited by:

Student organization: AEQ

Universidade Federal de Uberlandia

Federal University of Uberlândia
School of Chemical Engineering
Av. João Naves de Ávila, 2121
P. O. Box 593
38400-902 Uberlândia-MG/Brazil

University phone 55 34 3239-4292
Department phone 55 34 3239-4292
Department FAX 55 34 3239-4188/4249

Professors

Damasceno, João J. R. (Director) . . . . ext. 213/235 [†]
   jjrdamasceno@ufu.br
de Araújo, Euclides H. . . . . ext. 234
   euclides@ufu.br
Limaverde, José R. . . . . ext. 206
   jrlimaverde@ufu.br
Ribeiro, Eloízio J. . . . . ext. 239
   ejriberio@ufu.br

Associate Professors

Ataíde, Carlos H. . . . . ext. 211
   chataide@ufu.br
Burjaili, Mauro M. . . . . ext. 236
   mmburjaili@ufu.br
Cardoso, Vicelma L. . . . . ext. 220
   vicelma@ufu.br
Coelho, Márcia G. . . . . ext. 232
   mgcoelho@ufu.br
de Assis, Adilson J. . . . . ext. 233
   ajassis@ufu.br
de Souza Barrozo, Marcos A. . . . . ext. 203
   masbarrozo@ufu.br
Franco, Moilton R. . . . . ext. 218
   moilton@ufu.br
Henrique, Humberto M. . . . . ext. 217
   humberto@ufu.br
Hori, Carla E. . . . . ext. 212
   cehori@ufu.br
Murata, Valéria V. . . . . ext. 229
   valeria@ufu.br
Nascimento, Alvimar F. . . . . ext. 228
   alvimar@ufu.br
Oliveira Lopes, Luís C. . . . . ext. 230
   lcol@ufu.br
Oliveira, Daniel T. . . . . ext. 237
   tostes@ufu.br
Romanielo, Lucienne L. . . . . ext. 226
   lucienne@ufu.br
Soares, Ricardo R. . . . . ext. 209
   rrsoares@ufu.br

Assistant Professors

Sesso, Maria das G. V. . . . . ext. 236
   mgvas@ufu.br

[†] Faculty numbers are extensions of 55 34-3239-4189

Accredited by: Brazilian Ministry of Education

Placement service: copev@ufu.br

Canada

University of British Columbia

Department of Chemical and Biological Engineering
2360 East Mall
Vancouver, British Columbia
V6T 1Z3
Deliveries: Mr. Horace Lam

University phone . . . . (604) 822-2211
Department phone . . . . (604) 822-3238
Department FAX . . . . (604) 822-6003

Professors

Bennington, Chad . . . . (604) 822-8573
   cpjb@chml.ubc.ca
   EN,EV,RE
Bowen, Bruce D. . . . . (604) 822-3198
   bowen@chml.ubc.ca
Branion, Richard (Emeritus) . . . . (604) 822-8752
   branion@chml.ubc.ca
Duff, Sheldon . . . . (604) 822-9485
   sduff@chml.ubc.ca
   BT,EN,EV
Englezos, Peter . . . . (604) 822-6184
   englezos@chml.ubc.ca
   EN,EV,PR
Epstein, Norman (Emeritus) . . . . (604) 822-2815
Grace, John R. . . . . (604) 822-3121
   jgrace@chml.ubc.ca
   EN,EV,MO,RE
Hatzikiriakos, Savvas (Associate Dean) . . . . . (604) 822-3107
   hatzikir@chml.ubc.ca
   AM,PO
Haynes, Charles . . . . (604) 822-5136
   israels@chml.ubc.ca
   BT,SE,SU
Kerekes, Richard J. (Emeritus) . . . . (604) 224-8561
   kerekes@chml.ubc.ca
Lim, C. J. . . . . (604) 822-4871
   cjlim@chml.ubc.ca
   EN,EV,MO,PR,RE,SE
Oloman, Colin W. (Emeritus) . . . . (604) 822-4345
   coloman@intergate.ca
   EN,EV,RE
Pinder, Kenneth L. (Emeritus) . . . . (604) 822-2583
   klpinder@chml.ubc.ca
   BM,BT,EV,RE
Piret, James M. . . . . (604) 822-5835
   jpiret@chml.ubc.ca
   BM,BT,MO,PR,RE
Smith, Kevin J. (Head) . . . . (604) 822-3601
   kjs@interchange.ubc.ca
   AM,EN,RE,SU
Watkinson, A. P. (Emeritus) . . . . (604) 822-2741
   apw@chml.ubc.ca
Wilkinson, David . . . . (604) 822-4888
   dwilkinson@chml.ubc.ca
   AM,EN,EV,NT,RE,SU

Associate Professors

Baldwin, Susan . . . . (604) 822-1973
   sbaldwin@chml.ubc.ca
   BM,BT,EV,MO,RE
Bi, Xiaotao (Associate Head) . . . . (604) 822-4408
   xbi@chml.ubc.ca
   EN,EV,MO,PR,RE

Feng, James .... (604) 822-8875
jfeng@chml.ubc.ca
AM,BM,MO,PO,SU
Kwok, K. E. .... (604) 822-3238
ezra@chml.ubc.ca
BM,MO,PR
Lau, Anthony .... (604) 822-3476
aklau@chml.ubc.ca
EN,EV
Martinez, Mark .... (604) 822-2693
martinez@chml.ubc.ca
MO,PR,RE
Mohseni, Madjid (Associate Head) .... (604) 822-0047
mmohseni@chml.ubc.ca
BT,EV
Petrell, Royanne .... (604) 822-3475
rpetrell@chml.ubc.ca
AM,BM,EV,ME,MO
Taghipour, Fariborz .... (604) 822-1902
fariborz@chml.ubc.ca
EN,EV,MO,RE

Assistant Professors
Ellis, Naoko .... (604) 822-1243
nellis@chml.ubc.ca
EN,EV,MO,RE
Gyenge, Elod .... (604) 822-3217
egyenge@chml.ubc.ca
NT,SU

Accredited by: Canadian Engineering Accreditation Board

Degrees granted 2005-2006:
    B.S.: 53   M.S.: 14   Ph.D.: 4

Graduate advisor: S. Baldwin, E. Gyenge, A. Lau

Undergraduate advisor: S. Duff, R. Petrell, M. Martinez, J. Piret

Student organization: AIChE and CSChE
Advisor: R. J. Petrell

Department reports to: M. Isaacson(604) 822-6413, Dean of Applied Science

Dalhousie University

Chemical Engineering Department
P.O. Box 1000
Halifax, Nova Scotia B3J 2X4
Deliveries: 1360 Barrington Street, Room F201, B3J 1Z1
University phone........................ (902) 494-2211
Department phone .... (902) 494-3953
Department FAX .... (902) 420-7639

Professors
Al Taweel, Adel M. .... (902) 494-3992
Al.Taweel@dal.ca
Amyotte, Paul R. .... (902) 494-3976
Paul.Amyotte@dal.ca
Chen, Bin-hwa (Emeritus) .... (902) 494-3953
Fels, Mort .... (902) 494-3254
Mort.Fels@dal.ca
Gomaa, Hassan (Adjunct) .... (902) 494-3953
Gupta, Yash P. (Head) .... (902) 494-3948
Yash.Gupta@dal.ca
Mintz, Kenneth (Adjunct) .... (902) 494-3953

Pegg, Michael J. .... (902) 494-3225
Michael.Pegg@dal.ca
Thibault, Paul (Adjunct) .... (902) 494-3953
Woo, Stephen S. (Adjunct) .... (902) 494-3953

Associate Professors
Ghanem, Amyl .... (902) 494-3225
Amyl.Ghanem@dal.ca
Kuzak, Stephen G. .... (902) 494-3253
Stephen.Kuzak@dal.ca

Assistant Professors
Yuet, Pak K. .... (902) 494-3213
Pak.Yuet@dal.ca

Accredited by: Canadian Engineering Accreditation Board

Degrees granted 2003-2004:
    B.S.: 32   M.S.: 6   Ph.D.: 1

Graduate advisor: A.M. Al Taweel

Undergraduate advisor: S. G. Kuzak

Student organization: CSChE
Advisor: S. G. Kuzak

Department reports to: W. Caley (902) 494-6055, Dean of Engineering

Placement service: A. M. Coolen

University of Alberta

Department of Chemical and Materials Engineering
University of Alberta
Edmonton, Alberta, T6G 2G6
Deliveries: Room 536, 116 St. and 92 Ave.
University phone........................ (780) 492-3111
University FAX .... (780)492-7172
Department phone .... (780) 492-3321
Department FAX .... (780) 492-2881

Professors
Burrell, Robert .... (780) 492-8111
rburrell@ualberta.ca
Choi, Phillip .... (780) 492-9018
phillip.choi@ualberta.ca
Chuang, Karl T. (Emeritus) .... (780) 492-4676
karlt.chuang@ualberta.ca
Dalla Lana, Ivo G. (Emeritus) .... (780) 492-3391
dalla.lana@ualberta.ca
Eadie, Reginald L. .... (780) 492-2858
reg.eadie@ualberta.ca
Elliott, Janet A. W. .... (780) 492-7963
janet.elliott@ualberta.ca
Etsell, Thomas H. .... (780) 492-5594
tom.etsell@ualberta.ca
Fisher, D. G. (Emeritus) .... (780) 492-3301
grant.fisher@ualberta.ca
Forbes, J. F. (Chair) .... (780) 492-0873
fraser.forbes@ualberta.ca
Gray, Murray R. .... (780) 492-7965
murray.gray@ualberta.ca
Gupta, Rajender .... (780) 492-6861
rajender.gupta@ualberta.ca
Hayes, Robert E. .... (780) 492-3571
bob.hayes@ualberta.ca
Henein, Hani .... (780) 492-7304
hani.henein@ualberta.ca

Huang, Biao .... (780) 492-9016
    biao.huang@ualberta.ca
Ivey, Douglas G. .... (780) 492-2957
    doug.ivey@ualberta.ca
Kresta, Suzanne M. .... (780) 492-9221
    suzanne.kresta@ualberta.ca
Kuznicki, Steven M. .... (780) 492-8819
    kuznicki@ualberta.ca
Li, Dongyang .... (780) 492-6750
    dongyang.li@ualberta.ca
Liu, Qi .... (780) 492-8628
    qi.liu@ualberta.ca
Luo, Jingli .... (780) 492-2232
    jingli.luo@ualberta.ca
Lynch, David T. (Dean) .... (780) 492-3596
    dave.lynch@ualberta.ca
Masliyah, Jacob H. .... (780) 492-4673
    jacob.masliyah@ualberta.ca
Mather, Alan E. (Emeritus) .... (780) 492-3957
    alan.mather@ualberta.ca
McCaffrey, William .... (780) 492-6733
    william.mccaffrey@ualberta.ca
Nandakumar, Kumar .... (780) 492-5810
    kumar.nandakumar@ualberta.ca
Otto, Fred D. (Emeritus) .... (780) 492-5963
    fred.otto@ualberta.ca
Patchett, Barry M. (Emeritus) .... (780) 492-2604
    barry.patchett@ualberta.ca
Ryan, James T. (Emeritus)
    jim.ryan@ualberta.ca
Shah, Sirish L. .... (780) 492-5162
    sirish.shah@ualberta.ca
Shaw, John .... (780) 492-8236
    jmshaw@ualberta.ca
Sundararaj, Uttandaraman .... (780) 492-1044
    u.sundararaj@ualberta.ca
Wanke, Sieghard E. .... (780) 492-3817
    sieg.wanke@ualberta.ca
Wayman, Michael L. (Emeritus) .... (780) 492-3418
    mike.wayman@ualberta.ca
Williams, Michael C. (Emeritus) ..... (780) 492-3962
    mike.williams@ualberta.ca
Wood, Reginald K. (Emeritus) .... (780) 492-5963
    reg.wood@ualberta.ca
Xu, Zhenghe .... (780) 492-7667
    zhenghe.xu@ualberta.ca

Associate Professors

Chen, Weixing .... (780) 492-7706
    weixing.chen@ualberta.ca
Sanders, Sean .... (780) 492-0981
    sean.sanders@ualberta.ca
Uludag, Hasan .... (780) 492-0988
    hasan.uludag@ualberta.ca
Yeung, Anthony .... (780) 492-3669
    tony.yeung@ualberta.ca

Assistant Professors

Ben-Zvi, Amos .... (780) 492-7651
    abenzvi@ualberta.ca
    MO,RE
Lee, Jong M. .... (780) 492-8092
    jongmin.lee@ualberta.ca
    PR
Mitlin, David .... (780) 492-1542
    dmitlin@ualberta.ca

Accredited by: Canadian Engineering Accreditation Board

Degrees granted 2005-2006:
    B.S.: 160   M.S.: 14   Ph.D.: 6
Undergraduate advisor: Qi Liu

Student organization: CSChE and AIChE
Advisor: Anthony Yeung

Department reports to: J. Fraser Forbes, Department Chair

Placement service: Wendy L. Coffin (780) 492-4291

University of Calgary

Department of Chemical and Petroleum Engineering
2500 University Drive N.W.
Calgary, Alberta, T2N 1N4

Department phone .... (403) 220-5751
Department FAX .... (403) 284-4852

Professors

Aguilera, Roberto (Adjunct)
Badakhshan, Amir (Emeritus)
Behie, Leo A. .... (403) 220-6692
    behie@ucalgary.ca
Bennion, Douglas W. (Emeritus)
Berruti, Franco (Adjunct)
Bishnoi, P. R. (Emeritus) .... (403) 220-6695
    bishnoi@ucalgary.ca
Bolkan, Yasemin (Adjunct)
Card, Colin (Adjunct)
Chakma, Amitabha (Adjunct)
Chung, Keng (Adjunct)
Donnelly, John K. (Adjunct)
Exall, Douglas (Adjunct)
Farouq Ali, S. M. (Honorary)
Goobie, Lorraine (Adjunct)
Gregory, Garry A. (Adjunct)
Harding, Thomas (Head)
Heidemann, Robert A. (Emeritus) ..... (403) 220-8755
    heideman@ucalgary.ca
Jeje, Ayodeji A. (Associate Head) ... (403) 220-5753
    jeje@ucalgary.ca
Kantzas, Apostolos .... (403) 220-5752
    akantzas@ucalgary.ca
Keith, David .... (403) 220-6154
    keith@ucalgary.ca
Maini, Brij B. .... (403) 220-8777
    bmaini@ucalgary.ca
Mattar, Luis (Adjunct)
McGee, Bruce (Adjunct)
Mehrotra, Anil K. .... (403) 220-7406
    mehrotra@ucalgary.ca
Mehta, S.A. (Raj) .... (403) 220-4804
    mehta@ucalgary.ca
Mohtadi, Matt F. (Emeritus)
Moore, R. G. .... (403) 220-5750
    moore@ucalgary.ca
Mungan, Nick (Honorary)
Murray, Alan (Adjunct)
Nghiem, Long (Adjunct)
Pereira, Pedro .... (403) 220-4799
    ppereira@ucalgary.ca
Potter, Ian (Adjunct)
Pruden, Barry B. (Emeritus)
Satyro, M. (Adjunct)
Schramm, Laurie (Adjunct)
Settari, A. (Tony) .... (403) 220-7133
    asettari@ucalgary.ca

Stanislav, Jaroslav F. (Emeritus)
Svrcek, William Y. (Associate Head) ...... (403) 220-5755
svrcek@ucalgary.ca
Thusoo, Autar (Adjunct)
Tollefson, Eric L. (Emeritus)
Towson, Donald (Adjunct)
Trebble, Mark A. .... (403) 220-4823
trebble@ucalgary.ca
Wichert, Edward (Adjunct)
Wong, Dale (Adjunct)
Younger, A. (Andy) (Adjunct)
Zanzotto, Ludo .... (403) 220-8918
lzanzott@ucalgary.ca

Associate Professors

Adegbesan, Kenny (Adjunct)
Aikman, Michael (Adjunct)
Azaiez, Jalel .... (403) 220-7526
azaiez@ucalgary.ca
Baker, Richard (Adjunct)
Bellehumeur, Celine T. .... (403) 220-8804
cbellehu@ucalgary.ca
Monnery, Wayne D. (Adjunct)
Pooladi-Darvish, Mehran .... (403) 220-8779
pooladi@ucalgary.ca
Yarranton, Harvey W. .... (403) 220-6529
hyarrant@ucalgary.ca
Young, Brent R. .... (403) 220-8751
byoung@ucalgary.ca

Assistant Professors

Abedi, Jalal .... (403) 220-5594
jabedi@ucalgary.ca
Clarke, Mathew .... (403) 220-7341
maclarke@ucalgary.ca
De Visscher, Alex .... (403) 220-4739
adevissc@ucalgary.ca
Farrell, Patrick J. (Adjunct)
Gates, Ian .... (403) 220-5752
ian.gates@ucalgary.ca
Hill, Josephine .... (403) 210-9488
jhill@ucalgary.ca
Husein, Maen .... (403) 220-6691
mhusein@ucalgary.ca
Kallos, Michael S. .... (403) 220-7447
mskallos@ucalgary.ca
Sen, Arindom .... (403) 210-9452
asen@ucalgary.ca
Srinivasan, Sanjay (Adjunct)

Accredited by: Canadian Engineering Accreditation Board

Degrees granted 2004-2005:
    B.S.: 65   M.S.: 28   Ph.D.: 2

Undergraduate advisor: Dr. A. Jeje

Department reports to: S. C. Wirasinghe, Dean (403) 220-5731 Faculty of Engineering

Ecole Polytechnique-University of Montreal

Department of Chemical Engineering
P.O. Box 6079, Station Centre-ville
Montreal, Quebec H3C 3A7
Deliveries: 2500 Chemin Polytechnique, Montreal H3T 1J4
University phone .... (514) 340-4711
Department phone .... (514) 340-4613
Department FAX .... (514) 340-4159

Professors

Ajersch, Frank .... 4533 [†]
frank.ajersch@polymtl.ca
Bala, Srinivasan 7472
srinivasan.bala@polymtl.ca
Bale, Christopher 4535
christopher.bale@polymtl.ca
Bertrand, Francois 5773
francois.bertrand@polymtl.ca
Buschmann, Michael D. 4931
mike@grbb.polymtl.ca
Carreau, Pierre 4924
pierre.carreau@polymtl.ca
Chaouki, Jamal 4034
jamal.chaouki@polymtl.ca
Chartrand, Patrice 4089
patrice.chartrand@polymtl.ca
De Crescenzo, Gregory 7428
gregory.decrescenzo@polymtl.ca
Deschenes, Louise 5974
louise.deschesnes@polymtl.ca
Favis, Basil D. 4527
basil.favis@polymtl.ca
Guy, Christophe (Dean) 4526
christophe.guy@polymtl.ca
Klvana, Danilo 4927
danilo.klvana@polymtl.ca
Lafleur, Pierre G. (Dean) 4618
pierre.lafleur@polymtl.ca
Legros, Robert (Head) 4922
robert.legros@polymtl.ca
Patience, Gregory S. 3439
gregory-s.patience@polymtl.ca
Pelton, Arthur 4531
arthur.pelton@polymtl.ca
Perrier, Michel 4130
michel.perrier@polymtl.ca
Samson, Rejean 4898
Samson@biopro.polymtl.ca
Tanguy, Philippe A. 4017
philippe.tanguy@polymtl.ca

Associate Professors

Stuart, Paul 4384
paul.stuart@polymtl.ca

Assistant Professors

Dubois, Charles 4893
charles.dubois@polymtl.ca
Heuzey, Marie-Claude 5930
marie-claude.heuzey@polymtl.ca
Jolicoeur, Mario 4525
mario.jolicoeur@polymtl.ca

[†] Faculty numbers are extensions of (514) 340-4711.

Accredited by: Engineering Institute of Canada

Degrees granted 2004-2005:
B.S.: 117   M.S.: 27   Ph.D.: 6

Graduate advisor: Louise Deschênes

Undergraduate advisor: Charles Dubois

Student organization: CEGCh

Department reports to: Robert L. Papineau, Dean of Engineering

Placement service: Maryse Deschênes (514) 340-4374

Lakehead University

Chemical Engineering Department
955 Oliver Road
Thunder Bay, Ontario P7B 5E1
University phone .... (807) 343-8110
Department phone .... (807) 343-8170
Department FAX .... (807) 343-8928

Professors

Garred, Laurie J. (Emeritus) .... (807) 343-8415
    Laurie.Garred@lakeheadu.ca
    BM
Gilbert, Allan F. (Chair) .... (807) 343-8583
    Allan.Gilbert@lakeheadu.ca
    MO,PR
Nirdosh, Inderjit .... (807) 343-8343
    INirdosh@gale.lakeheadu.ca
    PR,SE
Puttagunta, V. R. (Emeritus)
    Venugopala.Puttagunta@lakeheadu.ca
    EN

Associate Professors

Catalan, Lionel .... (807) 343-8573
    Lionel.Catalan@lakeheadu.ca
    EV
Cooper, David .... (807) 343-8697
    David.Cooper@lakeheadu.ca
    EV
Liao, Baoqiang .... (807) 343-8437
    baoqiang.liao@lakeheadu.ca
    BT,EV

Assistant Professors

Xu, Chunbao (Charles) .... (807) 343-8761
    cxu@lakeheadu.ca
    EN,RE

Accredited by: CEAB

Degrees granted 2005-2006:
    B.S.: 9

Graduate advisor: L. Catalan

Undergraduate advisor: A. Gilbert

Student organization: CSChE
Advisor: I. Nirdosh

Department reports to: H.T. Saliba, (807) 343-8509, Dean, Faculty of Engineering

Placement service: John DeGiacomo, Manager, Student Placement and Cooperative Education Centre

Universite Laval

Chemical Engineering Department
Pouliot Building (3550)
Faculte des Sciences et de Genie
Cite Universitaire
Quebec, Quebec, G1K 7P4
University phone .... (418) 656-2131
Department phone .... (418) 656-3375
Department FAX .... (418) 656-5993

Professors

Bousmina, Mostapha .... (418) 656-2769
    mosto.bousmina@gch.ulaval.ca
Cholette, Albert (Emeritus)
Garnier, Alain .... (418) 656-3106
    alain.garnier@gch.ulaval.ca
Grandjean, Bernard (Chair) .... (418) 656-2859
    bernard.grandjean@gch.ulaval.ca
Kaliaguine, Serge .... (418) 656-2708
    serge.kaliaguine@gch.ulaval.ca
Larachi, Faical .... (418) 656-3566
    faical.larachi@gch.ulaval.ca
LeDuy, Anh .... (418) 656-2634
    anh.leduy@gch.ulaval.ca
Ramalho, Rubens S. (Emeritus)
Rodrigue, Denis (Associate Chair) .... (418) 656-2903
    denis.rodrigue@gch.ulaval.ca
Roy, Christian .... (418) 656-7406
    christian.roy@gch.ulaval.ca

Assistant Professors

Do, Trong-On .... (418) 656-3774
    Trong-On.Do@gch.ulaval.ca
Duchesne, Carl .... (418) 656 5184
    carl.duchesne@gch.ulaval.ca
Mighri, Frej .... (418) 656-2241
    Frej.Mighri@gch.ulaval.ca

Accredited by: Canadian Engineering Accreditation Board

Degrees granted 2004-2005:
    B.S.: 37   M.S.: 4   Ph.D.: 1

Undergraduate advisor: Denis Rodrigue

Student organization: CSChE
Advisor: Denis Rodrigue

Department reports to: Jean Serodes (418) 656-2354, Dean, Faculty of Science and Engineering

Placement service: Micheline Grenier (418) 656-6877

McGill University

Department of Chemical Engineering
M.H. Wong Building
Room 3060, Floor 3A
3610 University Street
Montreal, Quebec, Canada
H3A 2B2
University phone .... (514) 398-4455
Department phone .... (514) 398-4494
Department FAX .... (514) 398-6678

Professors

Cooper, David G. .... (514) 398-4278
  dcoope5@po-box.mcgill.ca
Dealy, John M. .... (514) 398-4264
  jdealy@po-box.mcgill.ca
Douglas, W. J. M. (Emeritus) .... (514) 398-6186
  mdougl2@po-box.mcgill.ca
Kamal, Musa R. .... (514) 398-4262
  mkamal@po-box.mcgill.ca
Munz, Richard J. (Chair) .... (514) 398-4277
  rmunz@po-box.mcgill.ca
Rey, Alejandro D. .... (514) 398-4196
  arey@po-box.mcgill.ca
Vera, Juan H. (Emeritus) .... (514) 398-4274
  jvera@po-box.mcgill.ca
Weber, Martin E. (Emeritus) .... (514) 398-4269
  mweber7@po-box.mcgill.ca

Associate Professors

Berk, Dimitrios .... (514) 398-4271
  dberk@po-box.mcgill.ca
Meunier, Jean-Luc .... (514) 398-8331
  jmeuni1@po-box.mcgill.ca
Simandl, Jana .... (514) 398-8308
  jsiman@po-box.mcgill.ca

Assistant Professors

Omanovic, Sasha .... (514) 398-4273
  TBA

Accredited by: Canadian Engineering Accreditation Board

Degrees granted 2004-2005:
  B.S.: 52   M.S.: 12   Ph.D.: 7

Graduate advisor: Professor Sylvain Coulombe

Undergraduate advisor: Jean-Luc Meunier

Student organization: CSChE
Advisor: Reghan Hill

Department reports to: Dr. Christoff Pierre, Dean of Engineering

Placement service: Gregg Blachford, Director

McMaster University

Chemical Engineering Department
Hamilton, Ontario L8S 4L7

University phone .... (905) 525-9140
Department phone .... (905) 525-9140x24957
Department FAX .... (905) 521-1350

Professors

Baird, Malcolm H. (Emeritus) .... x24945 †
  bairdmhi@mcmaster.ca
  RE
Crowe, Cameron M. (Emeritus) x24947
  crowe@mcmaster.ca
  PR
Dickson, James x24948
  dickson@mcmaster.ca
  SE
Feuerstein, Irwin A. (Emeritus) x23118
  feuerst@mcmaster.ca
  BM
Hamielec, Alvin E. (Emeritus) x24950
  hamielec@mcmaster.ca
  PO
Hrymak, Andrew N. (Chair) x23136
  hrymak@mcmaster.ca
  PO
Loutfy, Rafik x26616
  loutfyr
MacGregor, John F. x24951
  macgreg@mcmaster.ca
  PR
Mahalec, Vladimir x26386
  mahalec@mcmaster.ca
  PR
Marlin, Thomas E. x27125
  marlint@mcmaster.ca
  PR
Pelton, Robert H. x27045
  peltonrh@mcmaster.ca
  NT,SU
Shemilt, Leslie W. (Emeritus) x24735
  shemiltl@mcmaster.ca
Taylor, Paul A. x24952
  taylorpa@mcmaster.ca
  PR
Vlachopoulos, John x24954
  vlachopj@mcmaster.ca
  PO
Wood, Philip E. x24755
  woodpe@mcmaster.ca
  HT
Woods, Donald R. (Emeritus) x27339
  woodsdr@mcmaster.ca
Wright, Joseph D. (Adjunct) .... (514) 630-4102
  wright@paprican.ca
Zhu, Shiping x24962
  zhuship@mcmaster.ca
  NT,PO

Associate Professors

Jones, Lyndon (Adjunct) .... (519) 888 4567x5030
  lwjones@uwaterloo.ca
  BM
Kourti, Theodora (Adjunct) x27035
  kourtit@mcmaster.ca
  PR
Potter, David x23442
  potterd@mcmaster.ca
  PO
Sheardown, Heather x24794
  sheardow@mcmaster.ca
  BM
Swartz, Christopher x27945
  swartzc@mcmaster.ca
  PR

Assistant Professors

Filipe, Carlos x27278
  filipec@mcmaster.ca
  BM,BT,EV
Ghosh, Raja x27415
  rghosh@mcmaster.ca
  BM,BT
Jones, Kim x26333
  kjones@mcmaster.ca
  BM
Kostanski, Kris (Adjunct) x27018
  kostans@mcmaster.ca
  PO

156 Canada

Mhaskar, Prashant x23273
mhaskar@mcmaster.ca
PR
Quinn, Shannon (Adjunct) .... (905) 548-7200x6648
Shannon_quinn@dofasco.ca
PR
Thompson, Michael R. x23213
mthomps@mcmaster.ca
PO

† Faculty numbers are extensions of (905) 525-9140.

Accredited by: Canadian Engineering Accreditation Board

Degrees granted 2005-2006:
    B.S.: 52   M.S.: 9   Ph.D.: 2

Student organization: CSChE
Advisor: J.M. Dickson

Department reports to: M. Elbestawi (905) 525-9140x24288, Dean, Faculty of Engineering

Placement service: Anne Markey (905) 525-9140x22571

University of New Brunswick

Chemical Engineering Department
Post Office Box 4400
Fredericton, New Brunswick
E3B 5A3
Deliveries: 15 Dineen Drive, Head Hall, Room B-14

University phone....................... (506) 453-4666
Department phone .... (506) 453-4520
Department FAX .... (506) 453-3591

Professors

Bendrich, Guida .... (506) 453-5019
bendrich@unb.ca
Chaplin, Robin A. .... (506) 453-4656
rchaplin@unb.ca
Couturier, Michel F. (Associate Dean) ..... (506) 453-4690
cout@unb.ca
Eic, Mladen .... (506) 453-4689
meic@unb.ca
Lister, Derek H. (Chair) .... (506) 453-3299
dlister@unb.ca
Ni, Yonghao (Director) .... (506) 453-4547
yonghao@unb.ca
Picot, Jules J. C. (Emeritus) .... (506) 453-3542
jjcp@unb.ca

Associate Professors

Li, Kecheng .... (506) 451-6861
kecheng@unb.ca
Lowry, Brian .... (506) 453-4691
bjl@unb.ca
Singh, Kripa .... (506) 453-5108
singhk@unb.ca
Xiao, Huining .... (506) 453-3532
hxiao@unb.ca
Zheng, Ying .... (506) 447-3329
yzheng@unb.ca

Assistant Professors

Cook, William .... (506) 452-6318
wcook@unb.ca
Romero-Zeron, Laura .... (506) 447-3356
laurarz@unb.ca

Accredited by: Canadian Engineering Accreditation Board

Degrees granted 2004-2005:
    B.S.: 21   M.S.: 3   Ph.D.: 3

Graduate advisor: K. Li

Undergraduate advisor: B. Lowry

Student organization: CSChE
Advisor: B. Lowry

Department reports to: Dr. D. Coleman, Dean of Engineering

Placement service: Lois Clowater, Student Employment Service

Queen's University

Department of Chemical Engineering
Dupuis Hall
Kingston, ON
K7L 3N6
Deliveries: Dupuis Hall, Room 201

University phone....................... (613) 533-6000
Department phone .... (613) 533-2765
Department FAX .... (613) 533-6637

Professors

Bacon, David W. (Emeritus) .... (613) 533-2781
bacond@post.queensu.ca
Becker, Henry A. (Emeritus) .... (613) 533-2767
beckerha@chee.queensu.ca
Code, Russel K. (Emeritus)
Cunningham, Michael F. .... (613) 533-2782
michael.cunningham@chee.queensu.ca
PO
Daugulis, Andrew J. .... (613) 533-2784
andrew.daugulis@chee.queensu.ca
RE
Downie, John (Emeritus)
downiej@post.queensu.ca
Grandmaison, Edward W. .... (613) 533-2771
grandmai@chee.queensu.ca
Harris, Thomas J. (Dean) .... (613) 533-2056
harrist@post.queensu.ca
PR
Hsu, Cheng C. (Emeritus) .... (613) 533-2786
hsuc@chee.queensu.sa
McAuley, Kimberley B. .... (613) 533-2768
kim.mcauley@chee.queensu.ca
PO
McLellan, P. J. (Head) .... (613) 533-6343
james.mclellan@chee.queensu.ca
PR
Neufeld, Ronald J .... (613) 533-2785
ron.neufeld@chee.queensu.ca
BM
Wojciechowski, Bohdan W. (Emeritus)
wojciech@post.queensu.ca

Associate Professors

Amsden, Brian G. .... (613) 533-3093
amsden@chee.queensu.ca
BM
Guay, Martin .... (613) 533-2788
guaym@chee.queensu.ca
PR

Hutchinson, Robin A. (613) 533-3097
robin.hutchinson@chee.queensu.ca
PO
Parent, J. S. . . . . (613) 533-6266
parent@chee.queensu.ca
PO
Ramsay, Juliana A. . . . . (613) 533-2770
ramsayj@chee.queensu.ca
BT

Assistant Professors

Docoslis, Aris . . . . (613) 533-6949
docoslis@chee.queensu.ca
NT
Karan, Kunal . . . . (613) 533-3095
karan@chee.queensu.ca
EN
Kontopoulou, Marianna . . . . (613) 533-3079
marianna.kontopoulou@chee.queensu.ca
PO
Waldman, Stephen . . . . (613) 533-2896
waldman@me.queensu.ca
BM

Accredited by: Canadian Engineering Accreditation Board

Degrees granted 2005-2006:
B.S.: 108   M.S.: 13   Ph.D.: 3

Graduate advisor: Dr. J. Scott Parent

Undergraduate advisor: Dr. E.W. Grandmaison

Department reports to: Thomas J. Harris, Dean of the Faculty of Applied Science

Placement service: Paul Smith, Career Planning and Placement

Royal Military College of Canada

Chemistry and Chemical Engineering Department
P. O. Box 17000 Station Forces
Kingston, Ontario, K7K 7B4

Department phone . . . . (613) 541-6000x6271
Department FAX . . . . (613) 542-9489

Professors

Bates, P. J. . . . . (613) 541-6609x6609
bates-p@rmc.ca
Bennett, Leslie G. I. . . . . (613) 541-6000x6614
bennett-l@rmc.ca
Bonin, Hugues W. . . . . (613) 541-6000x6613
bonin-h@rmc.ca
Bui, V. T. . . . . (613) 541-6000x6621
bui-v@rmc.ca
Creber, K.A.M. (Head) . . . . (613) 541-6000x6272
creber@rmc.ca
Lewis, Brent J. . . . . (613) 541-6000x6611
lewis-b@rmc.ca
Mann, Ronald F. (Emeritus) . . . . (613) 541-6000x6055
mann-r@rmc.ca
Roberge, Pierre . . . . (613) 541-6000x6485
roberge-p@rmc.ca
Thompson, William T. . . . . (613) 541-6000x6081
thompson-w@rmc.ca
Weir, Ronald D. (613) 541-6612x6612
weir-r@rmc.ca

Associate Professors

Andrews, William S. . . . . (613) 541-6000x6052
andrews-w@rmc.ca
Peppley, Brant A . . . . (613) 541-6000x6702
peppley-b@rmc.ca

Assistant Professors

Cunningham, Nicolas . . . . (613) 541-6000x6610
Nicolas.Cunningham@rmc.ca
Thurgood, Christopher . . . . (613) 541-6000x6981
thurgood-c@rmc.ca

Accredited by: Canadian Engineering Accreditation Board

Student organization: CSChE
Advisor: H. W. Bonin

Department reports to: J. C. Amphlett, Head, Department of Chemistry and Chemical Engineering

Ryerson University

Department of Chemical Engineering
350 Victoria Street
Toronto, Ontario, Canada M5B 2K3

University phone . . . . (416) 979-5000
Department phone . . . . (416) 979-5157
Department FAX . . . . (416) 979-5083

Professors

Cuenca, Manuel A. . . . . ext. 6346 †
mcuenca@ryerson.ca
BT,EV,PR
Lohi, Ali (Chair) . . . . ext. 7028
alohi@ryerson.ca
EN,MO,PR,RE
Turcotte, Ginette . . . . ext. 7312
gturcott@ryerson.ca
BT,EV,PR

Associate Professors

Chan, Philip (Associate Chair) . . . . ext. 6960
p4chan@ryerson.ca
AM,MO,PO
Dhib, Ramdhane . . . . ext. 6343
rdhib@ryerson.ca
MO,PO,PR,RE
Doan, Huu (Associate Chair) . . . . ext. 6341
hdoan@ryerson.ca
EV,PR,SE
Dutton, Roshni . . . . ext. 4081
roshni.dutton@ryerson.ca
BM,BT,EV
Mehrvar, Mehrab . . . . ext. 6555
mmehrvar@ryerson.ca
BT,EV,MO,PR
Upreti, Simant R. . . . . ext. 6344
supreti@ryerson.ca
MO,PO,PR
Wu, Jiangning . . . . ext. 6549
j3wu@ryerson.ca
BT,EV,RE

Assistant Professors

Dahman, Yaser . . . . ext. 4080
ydahman@ryerson.ca
AM,BT,PO

Ein-Mozaffari, Farhad .... ext. 4251
fmozaffa@ryerson.ca
MO,PR

[†] Faculty numbers are extensions of (416) 979-5000

Accredited by: Canadian Engineering Accreditation Board

Degrees granted 2005-2006:
    B.S.: 36   M.S.: 15

Graduate advisor: Dr. Philip Chan

Undergraduate advisor: Dr. Huu Doan

Student organization: CSChE
Advisor: Dr. Huu Doan

Department reports to: Dr. Stalin Boctor (416) 979-5000, ext. 7224, Dean, Faculty of Engineering, Architecture and Science

Placement service: Dr. John Easton (416) 979-5000, ext. 6589, Director, Office of Co-operative Education

University of Saskatchewan

Department of Chemical Engineering
College of Engineering
57 Campus Drive
Saskatoon, Saskatchewan S7N 5A9

University phone .... (306) 966-4342
Department phone .... (306) 966-4760
Department FAX .... (306) 966-4777

Professors
Bakhshi, Narendra N. (Emeritus) .... (306) 966-4763
bakhshi@engr.usask.ca
EN,RE
Dalai, Ajay .... (306) 966-4771
dalai@engr.usask.ca
EN,RE
Hill, Gordon .... (306) 966-4765
hill@engr.usask.ca
BT
Macdonald, Douglas G. (Emeritus)
Peng, D.-Y. .... (306) 966-4767
pengd@engr.usask.ca
TH
Postlethwaite, John (Emeritus)
Pugsley, Todd (Head) .... (306) 966-4761
pugsley@engr.usask.ca
PR

Associate Professors
Evitts, Richard .... (306) 966-4766
Richard_W_Evitts@engr.usask.ca
PR
Lin, Yen-Han .... (306) 966-4764
linyen@engr.usask.ca
BM,BT
Nemati, Mehdi .... (306) 966-4769
Mehdi.Nemati@usask.ca
BT
Phoenix, Aaron .... (306) 966-4190
phoenix@engr.usask.ca
MO,TH
Sumner, Robert .... (306) 966-4775
sumner@engr.usask.ca
TP

Assistant Professors
Niu, Hui (Catherine) .... (306) 966-2174
catherine.niu@usask.ca
BT
Wang, Hui .... (306) 966-2685
hui.wang@usask.ca
EN

Accredited by: Canadian Engineering Accreditation Board

Degrees granted 2005-2006:
    M.S.: 7   Ph.D.: 3

Graduate advisor: D.-Y. Peng

Undergraduate advisor: R. Sumner

Student organization: CSChE
Advisor: A. Phoenix

Department reports to: C. Lague, Dean of Engineering

Placement service: Canada Manpower Centre

Universite de Sherbrooke

Chemical Engineering Department
Faculty of Engineering
2500, Blvd Universite
Sherbrooke, Quebec J1K2R1

University phone .... (819) 821-8000
Department phone .... (819) 821-7171
Department FAX .... (819) 821-7955

Professors
Abatzoglou, Nicolas .... (819) 821-7904 [†]
Nicolas.Abatzoglou@USherbrooke.ca
EN,EV,PR,RE
Boulos, Maher I. .... (819) 821-7168
Maher.Boulos@USherbrooke.ca
AM,MO,NT,PR
Broadbent, Arthur D. (Adjunct) x62172
Arthur.Broadbent@USherbrooke.ca
Chornet, Esteban .... (819) 821-7170
Esteban.Chornet@USherbrooke.ca
EN,EV,PR,RE
Gitzhofer, Francois .... (819) 821-7841
Francois.Gitzhofer@USherbrooke.ca
AM,EN,NT,PR
Gravelle, Denis x62177
Denis.Gravelle@USherbrooke.ca
MO,RE
Heitz, Michèle (Associate Dean) x62827
Michele.Heitz@USherbrooke.ca
BT,EN,EV
Jones, J. P. x62165
Peter.Jones@USherbrooke.ca
EV
Marcos, Bernard x62166
Bernard.Marcos@USherbrooke.ca
MO
Proulx, Pierre x62173
Pierre.Proulx@USherbrooke.ca
MO,NT,RE
Soucy, Gervais (Chair) x62167
Gervais.Soucy@USherbrooke.ca
AM,EV,NT,PR

Thérien, Normand (Adjunct) x62172
Normand.Therien@USherbrooke.ca
EV,MO

Associate Professors

Faucheux, Nathalie x61343
Nathalie.Faucheux@USherbrooke.ca
BM,BT,NT,SU
Jurewicz, Jerzy .... (819) 821-7178
Jerzy.Jurewicz@USherbrooke.ca
AM,EN,NT
Vermette, Patrick x62826
Patrick.Vermette@USherbrooke.ca
AM,BM,BT,NT,PO,SU

Assistant Professors

Sirois, Joël x62169
Joel.Sirois@USherbrooke.ca
BT,MO,PR,RE

† Extensions have prefix (819) 821-8000.

Accredited by: Canadian Engineering Accreditation Board

Degrees granted 2005-2006:
    B.S.: 46   M.S.: 4   Ph.D.: 2

Graduate advisor: Denis Gravelle

Undergraduate advisor: Denis Gravelle

Student organization: Societe canadienne de genie chimique
Advisor: Nicolas Abatzoglou

Department reports to: Gérard Lachiver, Dean (819) 821-7111

Placement service: Renald Mercier, Director, Service des stages et du placement

University of Toronto

Department of Chemical Engineering and Applied Chemistry
200 College Street
Toronto, Ontario, Canada, M5S 3E5

University phone........................ (416) 978-2011
Department phone .... (416) 978-3063
Department FAX .... (416) 978-8605

Professors

Allen, D. G. (Associate Chair) .... (416) 978-8517
allendg@chem-eng.utoronto.ca
Balke, Stephen T. (Emeritus) .... (416) 978-7495
balke@chem-eng.utoronto.ca
PO
Barham, David (Emeritus) .... (416) 978-3063
Basmadjian, Diran (Emeritus) .... (416) 978-5621
Bidleman, Terry F. (Adjunct) .... (416) 739-5730
terry.bidleman@ec.gc.ca
Boocock, David G.B. (Emeritus) .... (416) 978-4020
boocock@chem-eng.utoronto.ca
EN
Brook, Jeffrey (Adjunct) .... (416) 978-3063
Burgess, William H. (Emeritus) .... (416) 978-5827
Chaffey, Charles E. (Emeritus) .... (416) 978-3067
chaffey@chem-eng.utoronto.ca
Charles, Michael E. (Emeritus) .... (416) 978-5890
charles@chem-eng.utoronto.ca

Cheng, Yu-Ling .... (416) 978-5500
ylc@chem-eng.utoronto.ca
Cluett, William R. .... (416) 978-6697
cluett@chem-eng.utoronto.ca
Colcleugh, David W. (Adjunct) .... (416) 978-3063
Cormack, Donald E. .... (416) 978-4074
cormack@chem-eng.utoronto.ca
EN,MO
Cox, Brian (Joint) .... (416) 978-2127
cox@ecf.utoronto.ca
Davies, John E. (Joint) .... (416) 978-1471
davies@ecf.utoronto.ca
Diamond, Miriam L. (Joint) .... (416) 978-1749
diamond@geog.utoronto.ca
EV
Diosady, Levente L. .... (416) 978-4137
diosady@chem-eng.utoronto.ca
BT,EV,PR,RE,SE
Edwards, Elizabeth A. .... (416) 978-3506
edwards@chem-eng.utoronto.ca
Evans, Greg J. (Associate Dean) .... (416) 978-1821
evans@chem-eng.utoronto.ca
Foulkes, Frank R. (Emeritus) .... (416) 978-6432
foulkes@chem-eng.utoronto.ca
Goodfellow, Howard D. (Adjunct) .... (416) 978-3063
Grace, Thomas M. (Adjunct) .... (416) 978-3063
Graydon, William F. (Emeritus) .... (416) 978-3063
Grynpas, Marc D. (Joint) .... (416) 586-4464
grynpas@mshri.on.ca
Hummel, Richard L. (Emeritus) .... (416) 978-3063
James, David F. (Joint) .... (416) 978-3049
james@mie.utoronto.ca
PO
Jervis, Robert E. (Emeritus) .... (416) 978-3071
robert.jervis@utoronto.ca
Jones, Andrew (Adjunct) .... (416) 978-3063
Kawaji, Masahiro .... (416) 978-3064
kawaji@chem-eng.utoronto.ca
Kirk, Donald W. .... (416) 978-7406
kirk@chem-eng.utoronto.ca
Kortschot, Mark T. (Associate Chair) ..... (416) 978-8926
kortsch@chem-eng.utoronto.ca
AM,PO,SU
Kuhn, David C.S. (Adjunct) .... (416) 978-3065
kuhn@chem-eng.utoronto.ca
Luus, Rein (Emeritus) .... (416) 978-5200
luus@chem-eng.utoronto.ca
Lyne, M. B. (Adjunct)
Mackay, Donald (Emeritus) .... (416) 978-3063
Mims, Charles A. .... (416) 978-4575
mims@chem-eng.utoronto.ca
Missen, Ronald W. (Emeritus) .... (416) 978-3063
Napier, Douglas H. (Emeritus) .... (416) 978-3063
Newman, Roger C. .... (416) 946-0604
newman@chem-eng.utoronto.ca
Ojha, Matadial (Joint) .... (416) 978-7584
Papangelakis, Vladimiros G. .... (416) 978-1093
papange@chem-eng.utoronto.ca
EV,MO,PR,RE
Paradi, Joseph C. .... (416) 978-3714
paradi@mie.utoronto.ca
MO,PR
Phillips, M. J. (Emeritus) .... (416) 978-5872
mjp@chem-eng.utoronto.ca
RE
Piggott, Michael R. (Emeritus) .... (416) 978-4745
michael.piggot@ecf.utoronto.ca

Reeve, Douglas W. (Chair) .... (416) 946-7939
    reeve@chem-eng.utoronto.ca
Rizvi, Syed S.H. (Adjunct) .... (607) 255-7913
    ssr3@cornell.edu
Sandler, Samuel (Emeritus) .... (416) 978-3063
Santerre, Paul (Joint) .... (416) 979-4903 x4341
    paul.santerre@utoronto.ca
    BM,PO,SU
Saville, Bradley A. .... (416) 978-7745
    saville@chem-eng.utoronto.ca
    BT,EN
Sefton, Michael V. .... (416) 978-3088
    sefton@chem-eng.utoronto.ca
    BM,PO
Shoichet, Molly S. .... (416) 978-1460
    molly@chem-eng.utoronto.ca
    AM,BM,BT,NT,PO,SU
Smith, James W. (Emeritus) .... (416) 978-4906
    smithjw@chem-eng.utoronto.ca
    EN,EV
Smith, William R. (Adjunct) .... (416) 978-3063
Spinner, Irving H. (Emeritus) .... (416) 978-3063
Stanford, William L. (Joint) .... (416) 946-8379
    william.stanford@utoronto.ca
    BM
Szabo, Paul (Adjunct) .... (416) 978-6985
    paul.szabo@chem-eng.utoronto.ca
Tran, Honghi .... (416) 978-8585
    tranhn@chem-eng.utoronto.ca
Trass, Olev (Emeritus) .... (416) 978-6901
    trass@chem-eng.utoronto.ca
    EN,SE
Tremaine, Peter (Adjunct) .... (519) 823-4120
    tremain2@uoguelph.ca
Utigard, Torstein A. (Joint) .... (416) 978-6267
    utigard@ecf.utoronto.ca
Wania, Frank (Joint) .... (416) 287-7225
    frank.wania@utoronto.ca
Winnik, Mitchel A. (Joint) .... (416) 978-6495
    mwinnik@chem.utoronto.ca
Woo, Stephen S (Adjunct) .... (905) 940-1859
    sswoo@rogers.com
Woodhams, Raymond T. (Emeritus) ...... (416) 978-3063
Woodhouse, Kimberly A .... (416) 978-3060
    kas@chem-eng.utoronto.ca
    BM

Associate Professors
Beller, Harry R. (Adjunct) .... (925) 422-0081
    beller2@llnl.gov
Cooper, Paul A. (Joint) .... (416) 946-5078
    p.cooper@utoronto.ca
Farnood, Ramin R. .... (416) 946-7525
    farnood@chem-eng.utoronto.ca
    AM,EV,MO,SU
Fulthorpe, Roberta (Joint) .... (416) 287-7221
    fulthorpe@scar.utoronto.ca
Jia, Charles Q. .... (416) 946-3097
    cqjia@chem-eng.utoronto.ca
    EN,EV,SU
Kumacheva, Eugenia (Joint) .... (416) 978-3576
    ekumache@alchemy.chem.utoronto.ca
Liss, Steven N. (Adjunct) .... (416) 979-5000 x6357
    sliss@ryerson.ca
McKague, Bruce (Adjunct) .... (416) 978-7297
    mckague@chem-eng.utoronto.ca
Oshinowo, Toks (Adjunct) .... (416) 978-6948
    toks@chem-eng.utoronto.ca

Sain, Mohini (Joint) .... (416) 946-3191
    m.sain@utoronto.ca
Sayad, Saed (Adjunct) .... (416) 978-3063
Sodhi, Rana (Adjunct) .... (416) 978-5381
    sodhi@chem-eng.utoronto.ca
Thomson, Murray J. (Joint) .... (416) 978-1827
    thomson@mie.utoronto.ca
Yip, Chris M. .... (416) 978-7853
    christopher.yip@utoronto.ca
Zandstra, Peter (Joint) .... (416) 978-8888
    peter.zandstra@utoronto.ca
    BM,BT,MO,NT,PR,RE

Assistant Professors
Acosta, Edgar J. .... (416) 946-0742
    acosta@chem-eng.utoronto.ca
    BM,EV,MO,SE,SU
Allen, Christine (Joint) .... (416) 946-8594
    callen@adm.utoronto.ca
    BT,EN,EV,PR,RE
Audet, Julie (Joint) .... (416) 946-0209
    julie.audet@utoronto.ca
Bender, Timothy P. .... (416) 978-6140
    tim.bender@utoronto.ca
    AM,NT,PO
Chan, Warren C.W. .... (416) 946-8416
    warren.chan@utoronto.ca
    NT
Gong, Sunling (Adjunct) .... (416) 739-5749
    Sunling.Gong@ec.gc.ca
Mahadevan, Radhakrishnan .... (416) 946-0996
    mahadevan@chem-eng.utoronto.ca
    BT,MO
Master, Emma R. .... (416)946-7861
    emaster@chem-eng.utoronto.ca
    BT
Radisic, Milica .... (416) 946-5295
    milica@chem-eng.utoronto.ca
Yan, Ning (Joint) .... (416) 946-8070
    ning.yan@utoronto.ca

Accredited by: Canadian Engineering Accreditation Board

Degrees granted 2005-2006:
    B.S.: 71   M.S.: 49   Ph.D.: 9

Graduate advisor: D.G. Allen

Undergraduate advisor: M.T. Kortschot

Department reports to: C. Amon, Dean, Faculty of Applied Science and Engineering

Placement service: Ms. Rivi Frankle

University of Waterloo

Chemical Engineering Department
200 University Ave. W.
Waterloo, Ontario, N2L 3G1
University phone .... (519) 888-4567
Department phone .... (519) 888-4567x32404
Department FAX .... (519) 746-4979

Professors
Anderson, William A. (Associate Chair) .... x35011 [†]
    wanderson@uwaterloo.ca
    BM,EV,RE

Budman, Hector (Associate Chair) x36980
hbudman@uwaterloo.ca
BT,PR
Chatzis, Ioannis x33306
ichatzis@uwaterloo.ca
EN,EV,NT,SU,TP
Douglas, Peter L. (Associate Dean) x32913
pdouglas@uwaterloo.ca
EV,MO,PR
Duever, Thomas (Chair) x32540
tduever@uwaterloo.ca
AM,MO,PO,PR
Fahidy, Thomas (Emeritus) .... x 32092
tfahidy@uwaterloo.ca
MO,TP
Huang, Robert (Emeritus) x33409
ryhuang@uwaterloo.ca
AM,SE,SU
Hudgins, Robert R. (Emeritus) x32092
rhudgins@uwaterloo.ca
EN,RE
Legge, Raymond L. x36728
rllegge@uwaterloo.ca
AM,BT,EV,NT
Moo-Young, Murray (Emeritus) x84006
mooyoung@uwaterloo.ca
BT,EN,EV,RE,SE,SU,TP
Ng, Flora T. T. x33979
fttng@uwaterloo.ca
EN,RE,SE
Pal, Rajinder x32985
rpal@uwaterloo.ca
AM,PO,SE,SU,TP
Penlidis, Alexander x36634
penlidis@uwaterloo.ca
AM,MO,PO,PR,RE
Pritzker, Mark x32542
pritzker@uwaterloo.ca
AM,EN,SU
Rempel, G. L. x32702
grempel@uwaterloo.ca
AM,BM,EN,EV,HT,MO,NT,PO,RE,SE,TP
Scharer, Jeno M. (Emeritus) x32703
jscharer@uwaterloo.ca
BM,BT
Silveston, Peter (Emeritus) x32092
silvesto@uwaterloo.ca
EN,RE
Soares, Joao B. P. x33436
jsoares@uwaterloo.ca
MO,PO
Tzoganakis, Costas x33442
ctzogan@uwaterloo.ca
AM,MO,PO,TP

Associate Professors
Chen, Pu x35586
p4chen@uwaterloo.ca
BM,BT,MO,NT,PO,SE,SU,TH,TP
Chou, Chih-Hsiung 33310
cpchou@uwaterloo.ca
BT,EV
Croiset, Eric x36472
ecroiset@uwaterloo.ca
EN,EV,MO,RE
Feng, Xianshe x36555
xfeng@uwaterloo.ca
EN,EV,PO,SE,SU

Ioannidis, Marios x32914
mioannidis@uwaterloo.ca
EN,EV,MO,TP
Jervis, Eric x33928
ejervis@uwaterloo.ca
BM,BT,MO,TP
Moresoli, Christine x35254
cmoresoli@uwaterloo.ca
BT
Pan, Qinman x37111
qpan@uwaterloo.ca
AM,BM,EN,NT,PO,RE,SE,TP

Assistant Professors
Elkamel, Ali x37157
aelkamel@uwaterloo.ca
EN,EV,MO,PR
Epling, William .... x 37157
wepling@uwaterloo.ca
EN,EV,RE,SU
Folwer, Michael x33415
mfowler@uwaterloo.ca
AM,EN,RE,TP
Henneke, Dale x37833
henneke@uwaterloo.ca
NT,RE
Simon, Leonardo x33301
lsimon@uwaterloo.ca
AM,NT,PO

[†] Faculty numbers are extensions of (519) 888-4567.

Accredited by: Canadian Engineering Accreditation Board

Degrees granted 2005-2006:
  B.S.: 85   M.S.: 33   Ph.D.: 19

Graduate advisor: Patricia Anderson

Undergraduate advisor: William A. Anderson

Student organization: CSChE
Advisor: T.A. Duever

Department reports to: A. Sedra, Dean of Engineering

Placement service: J. F. Westlake

University of Western Ontario

Chemical and Biochemical Engineering
London, Ontario N6A 5B9
University phone .... (519) 661-2111
Department phone .... (519) 661-2131
Department FAX .... (519) 661-3498

Professors
Bassi, Amarjeet S. .... ext. 88324 [†]
abassi@eng.uwo.ca
Bergougnou, Maurice (Emeritus) .... (519) 661-2143
Berruti, Franco (Dean) .... (519) 661-2128
berruti@eng.uwo.ca
Briens, Cedric .... (519) 661-2145
cbriens@eng.uwo.ca
de Lasa, Hugo .... (519) 661-2144
hdelasa@eng.uwo.ca
Jutan, Arthur .... ext. 88322
ajutan@uwo.ca

Margaritis, Argyrios (Emeritus) .... (519) 661-2146
    amarg@uwo.ca
Ray, Ajay .... ext. 81279
    array@eng.uwo.ca
Rohani, Sohrab (Chair) .... (519) 661-4116
    rohani@eng.uwo.ca
Zhu, Jesse .... (519) 661-3807
    jzhu@eng.uwo.ca

Associate Professors

Karamanev, Dimitre .... ext. 88230
    dkaramanev@eng.uwo.ca
Nakhla, George .... (519) 661-5470
    nakhla@eng.uwo.ca
Prakash, Anand .... ext. 88528
    aprakash@eng.uwo.ca
Ray, Mita .... ext. 81273
    mray@eng.uwo.ca
Wan, WanKei .... ext. 88440
    wkwan@eng.uwo.ca

Assistant Professors

Barghi, Shahzad .... ext. 81275
    sbarghi2@eng.uwo.ca
Briens, Lauren .... ext. 88849
    lbriens@eng.uwo.ca
Charpentier, Paul .... (519) 661-3466
    pcharpentier@eng.uwo.ca
Mequanint, Kibert .... ext. 88573
    kmequani@eng.uwo.ca
Rizkalla, Amin .... ext. 82212
    arizkalla@eng.uwo.ca

† Numbers are extensions of (519) 661-2111.

Accredited by: Canadian Engineering Accreditation Board, APEO Canadian Society, ChE/CIC

Graduate advisor: A. Bassi

Undergraduate advisor: A. Bassi

Student organization: CSChE
Advisor: L. Briens

Department reports to: F. Berruti (519) 661-2128, Dean of Engineering

Placement service: Lesley Mounteer, Career Development Officer

Chile

Pontificia Universidad Catolica de Chile

Departamento de Ingeniería Química y Bioprocesos
Vicuña Mackenna 4860,
C.P. 6904411
Santiago
Chile

University phone .... (562) 354-4198
Department phone .... (562) 354-4254
Department FAX .... (562) 354-5803

Professors

Agosin, Eduardo (Associate Dean) .... 354-4253 †
    agosin@ing.puc.cl
Aguilera, José M. 354-4256
    jmaguile@ing.puc.cl

Associate Professors

De la Barra, Fernando 354-4237
del Valle, José M. 354-4418
    delvalle@ing.puc.cl
Jorquera, Héctor I. 354-4421
    jorquera@ing.puc.cl
    EV,MO
Olivares, Marcela 354-4234
    maoliva@ing.puc.cl
Pérez, José R. 354-4258
    perez@ing.puc.cl
    BT,MO,PR
San Martín, Ricardo 354-4263
    sanmarti@ing.puc.cl
Schmidt, Cristian 354-4237

Assistant Professors

Bouchon, Pedro 354-4927
    pbouchon@ing.puc.cl
Sáez, César A. 354-4257
    csaez@ing.puc.cl

† Numbers above have prefix 562.

Accredited by: CONAP, ABET

Degrees granted 2005-2006:
    B.S.: 30   M.S.: 8   Ph.D.: 3

Graduate advisor: Ricardo Perez

Undergraduate advisor: Cesar Saez

Department reports to: Hernan de Solminihac, Dean, School of Engineering

Placement service: William Young, Assistant Director, Students Affairs Office, (562) 686-4200

Universidad de Santiago de Chile

Facultad de Ingenieria
Depto. Ingenieria Quimica
Avda. Libertador Bernardo
O'Higgins 3363
P.O. Box 10233
Santiago

University phone 56 2 681-1100
Department phone 56 2 681-2398
Department FAX 56 2 681-1422

Professors

Alvarez, Ivan (Head)
Correa, Horacio
Romo, Claudio
Vega, Ricardo

Associate Professors

Bravo, Luis
Blasco, Ramon
Carvajal, Cynthia
Castro, Jose
Correa, Duberlis

Herrera, Luis
Levy, Isaac
Moyano, Pedro
Retamales, Lautaro
Reyes, Alejandro
Thoma, Enrico
Vega, Rolando

Assistant Professors

Contreras, Elsa
Cubillos, Francisco
Diaz, Georgina
Marquardt, Fritz-Hans
Palominos, Pedro

Accredited by: Ministry of Education, Colegio de Ing. de Chile A.G.

Department reports to: Bernd Schulz, Decano, Facultad de Ingenieria

Placement service: Elsa Contreras

Peoples Republic of China

Beijing University of Chemical Technology

Chemical Engineering Department
15 KeiSanhuan East Road
ChaoYang District
Beijing 100029

University phone 886-10 6421-2091
Department phone 86-10-64434754
Department FAX 86-10-64423610

Professors

Chen, Biaohua . . . . 64429057 [†]
    chenbh@mail.buct.edu.cn
Guo, Fen 64448808
    guof@mail.buct.edu.cn
Guo, Kai 64448808
    guok@163bj.com
Hai, Reti 64427356
    hjzhx@mail.buct.edu.cn
Li, Chengyue 64436787
    licy@mail.buct.edu.cn
Li, Chunxi 64444912
    Licx@mail.buct.edu.cn
Liu, Hui 64433695
    hliu@mail.buct.edu.cn
Ma, Runyu 64448919
    r.ma@mail.buct.edu.cn
Wang, Jianhong 64411065
    wjhmaster@263.net
Wang, Wenchuan 64444905
    wangwc@mail.buct.edu.cn
Zhang, Zeting 64434775
    zhangzt@mail.buct.edu.cn
Zheng, Danxing 64416406
    dxzh@mail.buct.edu.cn
Zhong, Chongli 64419862
    zhongcl@mail.buct.edu.cn

Associate Professors

Chen, Xiaochun 64434783
    chenxc@mail.buct.edu.cn
Feng, Lui 64427356
    chenql@btamail.net.cn
Gao, Zhengming 64418267
    gaozm@mail.buct.edu.cn
Huang, Xiongbin 64418267
    huangxb@mail.buct.edu.cn
Li, Jianwei 64433695
    lijw@mail.buct.edu.cn
Li, Qunsheng 64449695
    qsli@mail.buct.edu.cn
Li, Xiujin 64427356
    lxiujin@hotmail.com

[†] Faculty numbers preceded by 86-10

Accredited by: State Education Commission of China

Department reports to: Xiangzhi Wu, Dean

Tsinghua University

Department of Chemical Engineering
Beijing, 100084, China

University phone . . . . (86) 10 62782035
Department phone . . . . (86) 10 62784523
Department FAX . . . . (86) 10 62770304

Professors

Cao, Zhuan 62784120
    cza-dce@tsinghua.edu.cn
Chen, Bingzhen 62784129
    dcecbz@tsinghua.edu.cn
Chen, Cuixian 62783747
    chencx@tsinghua.edu.cn
Chen, Jian 62789195
    cj-dce@tsinghua.edu.cn
Dai, Youyuan 62784153
    daiyy@tsinghua.edu.cn
Ding, Fuxin 62785410
    dingfx@tsinghua.edu.cn
Duan, Zhanting 62784175
    duanzt@tsinghua.edu.cn
Fei, Weiyang 62784185
    fwy-dce@tsinghua.edu.cn
Gao, Guanghua 62786448
    gaogh@tsinghua.edu.cn
He, Xiaorong 62783913
    hexr@tsinghua.edu.cn
Hu, Ping 62787617
    hspinghu@tsinghua.edu.cn
Hu, Xianhua 62784263
    hxh-dce@tsinghua.edu.cn
Jin, Yong 62784294
    jiny@tsinghua.edu.cn
Li, Jiding 62785929
    lijiding@tsinghua.edu.cn
Li, Yourun 62771039
    liyr@tsinghua.edu.cn
Lin, Zhanglin
    zhanglinlin@tsinghua.edu.cn
Liu, Dehua 62772825
    dhliu@tsinghua.edu.cn
Liu, Deshan 62784263
Liu, Zheng (Chair) 62788858
    liuzheng@tsinghua.edu.cn

Luo, Guangsheng 62788545
    gsluo@tsinghua.edu.cn
Shen, Jingzhu 62784479
    shenjz@tsinghua.edu.cn
Wang, Dezheng 62794404
    wangdezheng@tsinghua.edu.cn
Wang, Jiading 62785145
    wjd-dce@tsinghua..edu.cn
Wang, Jinfu 62789233
    wangjfu@tsinghua.edu.cn
Wang, Xiaogong (Associate Chair) 62787673
    wxg-dce@tsinghua.edu.cn
Wang, Xiaolin 62773732
    xl-wang@tsinghua.edu.cn
Wang, Yundong 62782748
    wangyd@chemeng.tsinghua.edu.cn
Wang, Zhanwen 62786622
    wangzw@tsinghua.edu.cn
Wei, Fei 62783626
    wf-dce@tsinghua.edu.cn
Xie, Xuming 62786785
    xxm-dce@tsinghua.edu.cn
Xing, Xinhui 62785514
    xing@chemeng.tsinghua.edu.cn
Yang, Jichu 62786176
    yjc-dce@tsinghua.edu.cn
Yu, Jian 62788201
    yuj@tsinghua.edu.cn
Zhou, Rongqi 62784431
    zhourq@tsinghua.edu.cn
Zhou, Xiao 62783483
    zhoux@tsinghua.edu.cn
Zhu, Shenlin 62787911
    zhusl@tsinghua.edu.cn

Associate Professors

Chen, Fuming 62785922
    chenfm@tsinghua.edu.cn
Gao, Yanfang 62786448
    yfgao@tsinghua.edu.cn
Guo, Baohua 62786645
    bhguo@tsinghua.edu.cn
Guo, Qingfeng 62786303
    zhp-dms@tsinghua.edu.cn
Guo, Zhigang 62785930
    guozhig@tsinghua.edu.cn
Han, Minghan 62773374
    hanmh@tsinghua.edu.cn
Hao, Jihua 62773082
    haojh@tsinghua.edu.cn
Hu, Shanying 62771249
    hxr-dce@tsinghua.edu.cn
Kan, Chengyou 62794191
    kancy@tsinghua.edu.cn
Li, Qiang 62787870
    liqiang@tsinghua.edu.cn
Lian, Yanqing 82865079
    yqlian@tsinghua.edu.cn
Lin, Aiguang 62786090
    linaig@tsinghua.edu.cn
Liu, Fuzhen 62786096
    lfz-dce@tsinghua.edu.cn
Liu, Ruizhi 62784470
Luo, Guohua 62773733
    luoguoh@tsinghua.edu.cn
Qin, Wei 62789197
    qinw@chemeng.tsinghua.edu.cn
Shen, Jinyu 62787604
    shenjy@tsinghua.edu.cn

Shi, Dianwen (Associate Chair) 62786483
Sun, Dengwen 62785105
Tang, Liming 62771539
    tanglm@tsinghua.edu.cn
Tang, Zhigang 62773572
    zhg-tang@tsinghua.edu.cn
Wang, Baoguo 62783869
    bgwang@tsinghua.edu.cn
Wang, Tao 62789196
    taowang@tsinghua.edu.cn
Wang, Tingjie 62789232
    wangtj@tsinghua.edu.cn
Xiang, Lan 62329289
    lxiang@chemeng.tsinghua.edu.cn
Yu, Lixin 62789238
    yulixin@tsinghua.edu.cn
Yu, Yangxin 62773533
    yuyx@tsinghua.edu.cn
Zhang, Liping 62754895
Zhao, Hong (Associate Chair) 62789236
    zhh@chemeng.tsinghua.edu.cn

Department reports to: Zheng Liu, Chairman

Colombia

Universidad Industrial de Santander

Escuela de Ingenieria Quimica
P.O. Box 678
Bucaramanga

University phone .... (976) 456141
Department phone .... (976) 456141x434
Department FAX .... (976) 352554

Professors

Acevedo, Leonardo (Chair) .... x451 [†]
Alvarez, Mario x434
Barrera, Alvaro (Emeritus) x451
Centeno, Aristobulo x434
Correa, Rodrigo x434
del Ferrada, Pedro x434
Gonzalez, Cesar (Emeritus) x434
Guerra, Carlos F. (Emeritus) x246
Hernandez, Herzen x434
Kafarov, Vjacheslav x434
Lavente, Dionisio x434
Pulido, Jorge E. x434
Ramirez, Alvaro x451
Retamoso, Clemente (Emeritus) x434
Salazar, Ramiro x434
Sanchez, Luis D. x434
Santaelle, Jose F. x434

Associate Professors

Barajas, Crisostomo x434
Castillo, Edgar x434
Escalante, Humberto x434
del Giraldo, Sonia x434
Idarraga, Luis M. x434
Martinez, Ramiro x434
Valencia, Hugo A. x433

[†] Dial (976) 459647 and ask for extension.

Department reports to: Orlando Aguirre, Dean, Ph-Ch. Sci. Div.

Placement service: Clara Helena Gomez (x121)

Croatia

University of Zagreb

Faculty of Chemical Engineering and Technology
Marulicev trg 19
10 000 Zagreb

University phone	385 1 45 64 111
Department phone	385 1 45 97 281
Department FAX	385 1 45 97 260

Professors

Bozicevic, Juraj 45 97 131
  jbozic@marie.fkit.hr
Budin, Rajka 45 97 138
  rbudin@marie.fkit.hr
Cerjan-Stefanovic, Stefica 45 97 210
  scerjan@pierre.fkit.hr
Duic, Ljerka 45 97 141
  lduic@marie.fkit.hr
Glasnovic, Antun (Associate Dean) 45 97 221
  aglasnov@pierre.fkit.hr
Gomzi, Zoran 45 97 105
  zgomzi@marie.fkit.hr
Hraste, Marin 45 97 220
  mhraste@pierre.fkit.hr
Janovic, Zvonimir 45 97 125
  zjanov@marie.fkit.hr
Jelencic, Jasenka (Dean) 45 97 281
  jjelen@marie.fkit.hr
Karminski Zamola, Grace 45 97 215
  gzamola@pierre.fkit.hr
Kastelan-Macan, Marija 45 97 211
  mkastela@jagor.srce.hr
Koprivanac, Natalija 45 97 124
  nkopri@marie.fkit.hr
Kovacevic, Vera 48 46 378
  vkovac@marie.fkit.hr
Kunst, Branko 45 97 232
  kunstb@pierre.fkit.hr
Lopac, Vjera 45 97 106
  vlopac@marie.fkit.hr
Matusinovic, Tomislav 45 97 218
  tmatusin@pierre.fkit.hr
Metikos-Hukovic, Mirjana 45 97 140
  mmetik@marie.fkit.hr
Mintas, Mladen 45 97 214
  mmintas@pierre.fkit.hr
Rek, Vesna 48 28 476
  vrek@marie.fkit.hr
Sindler, Marija 45 97 246
  msindler@marie.fkit.hr
Soljic, Zvonimir 45 97 204
  zsoljic@pierre.fkit.hr
Vasic-Racki, ura 45 97 104
  dvracki@marie.fkit.hr
Zrncevic, Stanka 45 97 102
  szrnce@marie.fkit.hr

Associate Professors

Ivankovic, Hrvoje 45 97 228
  hivan@pierre.fkit.hr
Ivankovic, Marica 45 97 230
  mivan@pierre.fkit.hr
Sertic-Bionda, Katica 45 97 129
  kserti@marie.fkit.hr
Sipos, Laszlo 45 97 290
  laszlo.sipos@zg.tel.hr
Stupnisek-Lisac, Ema (Associate Dean) 45 97 281
  elisac@marie.fkit.hr

Assistant Professors

Agic, Ante 48 33 850
  aagic@marie.fkit.hr
Bajza, Zeljko 48 33 850
  zeljko@zg.tel.hr
Briski, Felicita 45 97 269
  fbriski@pierre.fkit.hr
Dananic, Vladimir 45 97 107
  vdanan@marie.fkit.hr
Gusic, Ivica 45 97 266
  igusic@pierre.fkit.hr
Hrnjak-Murgic, Zlata 45 97 122
  zhrnjak@marie.fkit.hr
Kurajica, Stanislav 45 97 226
  stankok@pierre.fkit.hr
Mance, Ana D. 45 97 243
  admance@pierre.fkit.hr
Matijasevic, Ljubica 45 97 101
  ljmatij@marie.fkit.hr
Mestrovic-Markovinovic, Antonija 45 97 113
  amarko@marie.fkit.hr
Metes, Azra 45 97 122
  ametes@marie.fkit.hr
Milardovic, Stjepan 45 97 286
  smilard@pierre.fkit.hr
Papic, Sanja 45 97 122
  spapic@marie.fkit.hr
Rogosic, Marko 45 97 299
  mrogosic@pierre.fkit.hr
Tomasic, Vesna 45 97 103
  vtomas@marie.fkit.hr
Volovsek, Vesna 45 97 135
  volovsek@marie.fkit.hr

Czech Republic

University of Pardubice

Chemical Engineering Department
nam. Cs. Legii 565
532 10 Pardubice

Department phone	420 466037140
Department FAX	420 466037068

Professors

Lecjaks, Zdenek (Emeritus)
Machac, Ivan 420 466 037 131
  Ivan.Machac@upce.cz
  TP
Mikulasek, Petr 420 466 037 130
  Petr.Mikulasek@upce.cz
  SE

Associate Professors

Cakl, Jiri 420 466 037 128
  Jiri.Cakl@upce.cz
  SE

Palaty, Zdenek (Head) 420 466 037 503
Zdenek.Palaty@upce.cz
MO,SE

Assistant Professors

Dolecek, Petr 420 466 037 129
Petr.Dolecek@upce.cz
MO,SE,TP
Jirankova, Hana 420 466 037 134
Hana.Jirankova@upce.cz
SE
Siska, Bedrich 420 466 037 126
Bedrich.Siska@upce.cz
MO,RE,SE
Velikovska, Pavlina 420 466 037 134
Pavlina.Velikovska@upce.cz
SE
Zakova, Alena 420 466 037 135
Alena.Zakova@upce.cz
SE

Accredited by:

Degrees granted 2005-2006:
B.S.: 2   M.S.: 11   Ph.D.: 1
Department reports to: Jiri Malek, President of University

Finland

Abo Akademi University

Faculty of Chemical Engineering
Biskopsgatan 8
FIN-20500
Abo

University phone........................... 358 2 21531
Department FAX 358 2 2154967

Professors

Back, Ralph-Johan
Fagervik, Kaj
Fardim, Pedro
Holmbom, Bjarne
Hupa, Mikko
Häggblom, Kurt-Erik
Ivaska, Ari
Lewenstam, Andrzej
Lilius, Johan
Murzin, Dmitry
Salmi, Tapio (Dean)
Saxén, Henrik
Toivakka, Martti
Toivonen, Hannu
Westerholm, Jan
Westerlund, Tapio
Wikström, Kim
Wilen, Carl-Eric
Zevenhoven, Ron

Accredited by:

Degrees granted 2004-2005:
M.S.: 59   Ph.D.: 14
Department reports to: Tapio Salmi

Helsinki University of Technology

Department of Chemical Technology
P.O.Box 6100
FIN-02015 HUT
Deliveries: Kemistintie 1 M

University phone............................ 358 9 4511
Department phone 358 9 4511
Department FAX 358 9 462373

Professors

Aittamaa, Juhani 4512 630
juhani.aittamaa@hut.fi
MO,PR,SE
Hurme, Markku 4512 632
markku.hurme@hut.fi
EN,EV,PR
Jamsa-Jounela, Sirkka-Liisa 4512 631
sirkka-liisa.jamsa-jounela@hut.fi
PR
Jokela, Reija 4512 531
reija.jokela@hut.fi
Karppinen, Maarit 4512 602
maarit.karppinen@tkk.fi
AM,NT
Kontturi, Kyösti 4512 575
kyosti.kontturi@hut.fi
Koskinen, Ari 4512 526
ari.koskinen@hut.fi
Krause, Outi (Associate Dean) 4512 613
outi.krause@hut.fi
NT,RE
Kulmala, Sakari 4512 601
sakari.kulmala@hut.fi
Laakso, Simo 4512 550
simo.laakso@hut.fi
BT
Leisola, Matti (Head) 4512 546
matti.leisola@hut.fi
BT
Niinistö, Lauri 4512 600
lauri.niinisto@hut.fi
Nordstrom, Katrina 4512 549
katrina.nordstrom@hut.fi
BT
Seppala, Jukka 4512 614
jukka.seppala@hut.fi
AM,PO

Accredited by:

Degrees granted 2005-2006:
M.S.: 75   Ph.D.: 20
Department reports to: Markku Hurme

University of Oulu

Department of Process and Environmental Engineering
P.O. Box 4300
FIN-90014

University phone......................... 358-8-5531011
Department phone 358-8-5532301
Department FAX 358-8-5532304

Professors

Harkki, Jouko
Keiski, Riitta

Kortela, Urpo
Lakso, Esko
Leiviska, Kauko (Chair)
Neubauer, Peter
Niinimaki, Jouko
Pohjola, Veikko
Sillanpää, Mika
Ylinen, Raimo

France

Ecole Nationale Superieure des Industries Chimiques (ENSIC)

Institut National Polytechnique de Lorraine (INPL)
Chemical Engineering Department - ENSIC
1, rue Grandville - B.P. 20451 -
54001 Nancy Cedex
Department phone .... (33) 383 17 50 00
Department FAX .... (33) 383 35 08 11

Professors
Bouchy, Michel .... (33) 383 17 51 37
  Michel.Bouchy@ensic.inpl-nancy.fr
Choplin, Lionel (Associate Chair) .... (33) 383 17 50 10
  Lionel.Choplin@ensic.inpl-nancy.fr
Corriou, Jean-Pierre .... (33) 383 17 52 13
  Jean-Pierre.Corriou@ensic.inpl-nancy.fr
Dirand, Michel (Chair) .... (33) 383 17 50 13
  Michel.Dirand@ensic.inpl-nancy.fr
Favre, Eric .... (33) 383 17 53 90
  Eric.Favre@ensic.inpl-nancy.fr
Houzelot, Jean-Léon .... (33) 383 17 52 35
  Jean-Leon.Houzelot@ensic.inpl-nancy.fr
Jamart, Brigitte .... (33) 383 17 52 79
  Brigitte.Jamart@ensic.inpl-nancy.fr
Jonquières, Anne .... (33) 383 17 50 29
  Anne.Jonquieres@ensic.inpl-nancy.fr
Latifi, Abderrazak .... (33) 383 17 52 34
  Abderrazak.Latifi@ensic.inpl-nancy.fr
Li, Huai-Zhi .... (33) 383 17 51 09
  Huai-Zhi.Li@ensic.inpl-nancy.fr
Marchal-Heussler, Laurent .... (33) 383 17 51 60
  Laurent.Marchal-Heussler@ensic.inpl-nancy.fr
Matlosz, Michael (Associate Chair) ...... (33) 383 17 52 57
  Michael.Matlosz@ensic.inpl-nancy.fr
Midoux, Noël .... (33) 383 17 52 50
  Noel.Midoux@ensic.inpl-nancy.fr
Molleyre, François .... (33) 383 17 50 71
  François.Molleyre@ensic.inpl-nancy.fr
Pla, Fernand .... (33) 383 17 50 49
  Fernand.Pla@ensic.inpl-nancy.fr
Plasari, Edouard .... (33) 383 17 50 99
  Edouard.Plasari@ensic.inpl-nancy.fr
Roizard, Christine (Associate Chair) ..... (33) 383 17 53 06
  Christine.Roizard@ensic.inpl-nancy.fr
Saatdjian, Esteban .... (33) 383 17 52 53
  Esteban.Saatdjian@ensic.inpl-nancy.fr
Sardin, Michel .... (33) 383 17 52 54
  Michel.Sardin@ensic.inpl-nancy.fr
Scacchi, Gérard .... (33) 383 17 50 83
  Gerard.Scacchi@ensic.inpl-nancy.fr
Schuffenecker, Louis .... (33) 383 59 59 90
  Louis.Schuffenecker@inpl-nancy.fr
Solimando, Roland .... (33) 383 17 53 06
  Roland.Solimando@ensic.inpl-nancy.fr

Associate Professors
Arrault, Axelle .... (33) 383 17 52 14
  Axelle.Arrault@ensic.inpl-nancy.fr
Bourdet, Jean-Bernard .... (33) 383 17 50 40
  Jean-Bernard.Bourdet@ensic.inpl-nancy.fr
Bouroukba, Mohammed .... (33) 383 17 51 32
  Mohammed.Bouroukba@ensic.inpl-nancy.fr
Burkle, Valérie
  Valerie.Burkle@ensic.inpl-nancy.fr
Castel, Christophe .... (33) 383 17 51 40
  Christophe.Castel@ensic.inpl-nancy.fr
Chachuat, Benoît
  Benoit.Chachuat@ensic.inpl-nancy.fr
Clément, Robert .... (33) 383 17 50 29
  Robert.Clement@ensic.inpl-nancy.fr
Commenge, Jean-Marc .... (33) 383 17 51 80
  Jean-marc.Commenge@ensic.inpl-nancy.fr
Devin, Isabelle
  Isabelle.Devin@ensic.inpl-nancy.fr
Durand, Alain .... (33) 383 17 52 92
  Alain.Durand@ensic.inpl-nancy.fr
Falk, Véronique .... (33) 383 17 52 75
  Veronique.Falk@ensic.inpl-nancy.fr
Ferrer, Monique .... (33) 383 17 50 58
  Monique.Ferrer@ensic.inpl-nancy.fr
Foucaut, Jean-françois .... (33) 383 17 50 60
  Jean-Francois.Foucaut@ensic.inpl-nancy.fr
Fournet, René .... (33) 383 17 51 01
  Rene.Fournet@ensic.inpl-nancy.fr
Gentric, Caroline .... (33) 383 17 53 38
  Caroline.Gentric@ensic.inpl-nancy.fr
Gigante, Alexandra .... (33) 383 17 51 11
  Alexandra.Gigante@ensic.inpl-nancy.fr
Gorner, Tatiana .... (33) 383 17 51 46
  Tatiana.Gorner@ensic.inpl-nancy.fr
Hubert, Nathalie .... (33) 383 17 50 64
  Nathalie.Hubert@ensic.inpl-nancy.fr
Ivanaj, Véra
  Vera.Ivanaj@ensic.inpl-nancy.fr
Jaubert, Jean-Noël .... (33) 383 17 50 81
  Jean-Noel.Jaubert@ensic.inpl-nancy.fr
Jaubert, Lucie .... (33) 383 17 50 25
  Lucie.Jaubert@ensic.inpl-nancy.fr
Lesage, François .... (33) 383 17 50 00
  Francois.Lesage@ensic.inpl-nancy.fr
Mauviel, Guillain .... (33) 383 17 52 07
  Guillain.Mauviel@ensic.inpl-nancy.fr
Muhr, Laurence .... (33) 383 17 53 11
  Laurence.Muhr@ensic.inpl-nancy.fr
Mutelet, Fabrice .... (33) 383 17 51 31
  Fabrice.Mutelet@ensic.inpl-nancy.fr
Nouvel, Cécile .... (33) 383 17 52 29
  Cecile.Nouvel@ensic.inpl-nancy.fr
Perrin, Laurent .... (33) 383 17 50 96
  Laurent.Perrin@ensic.inpl-nancy.fr
Petit, Alain .... (33) 383 17 50 95
  Alain.Petit@ensic.inpl-nancy.fr
Petitjean, Dominique .... (33) 383 17 50 31
  Dominique.Petitjean@ensic.inpl-nancy.fr
Poncin, Souhila .... (33) 383 17 52 23
  Souhila.poncin@ensic.inpl-nancy.fr
Roques-Carmes, Thibault .... (33) 383 17 51 35
  Thibault.Roques-Carmes@ensic.inpl-nancy.fr
Rouat-Rode, Sabine .... (33) 383 17 51 56
  Sabine.Rode@ensic.inpl-nancy.fr

Sadtler, Véronique .... (33) 383 17 50 79
  Veronique.Sadtler@ensic.inpl-nancy.fr
Schaer, Eric .... (33) 383 17 53 04
  Eric.Schaer@ensic.inpl-nancy.fr
Schrauwen, Cornélius .... (33) 383 17 50 98
  Cornelius.Schrauwen@ensic.inpl-nancy.fr
Simon, Yves .... (33) 383 17 51 22
  Yves.simon@ensic.inpl-nancy.fr
Zahraa, Orfan .... (33) 383 17 51 18
  Orfan.Zahraa@ensic.inpl-nancy.fr

Accredited by: IChemE (UK)

Degrees granted 2004-2005:
  B.S.: 115   M.S.: 32   Ph.D.: 25

Graduate advisor: M. Matlosz

Placement service: G. Scacchi

Germany

Aachen Technical University

Dept. of Chemical Engineering
Turmstrasse 46
D-52056 Aachen

University phone......................... 49-241-980-1
Department phone 49-241-8095470
Department FAX 49-241-8092252

Professors

Buechs, Joachim (Chair) 49-241-80-95546
  buechs@bio-vt.rwth-aachen.de
Hartmann, Hugo (Emeritus) 49-241-80-95490
Marquardt, Wolfgang (Chair) 49-241-80-96712
  marquardt@lpt.rwth-aachen.de
Melin, Thomas (Chair) 49-241-80-95470
  melin@ivt.rwth-aachen.de
Modigell, Michael (Associate Chair) 49-241-80-95159
  modigell@ivt.rwth-aachen.de
Pfennig, Andreas (Chair) 49-241-80-95490
  andreas.pfennig@tvt.rwth-aachen.de
Schummer, Paul (Emeritus) 49-241-80-95476

Accredited by:

Degrees granted 2004-2005:
            M.S.: 25   Ph.D.: 7

Graduate advisor: betreuung-vt@lpt.rwth-aachen.de

Technische Universitaet Braunschweig

Department of Mechanical Engineering (Maschinenbau)
Division of Chemical Engineering (Energie und Verfahrenstechnik)
Box 3329
D-38023 Braunschweig
Deliveries: Schleinitzstrasse 20, D-38106 Braunschweig

University phone......................... 49 531-391-0
Department phone 49 531-391-7683
Department FAX 49 531-391-5947

Professors

Bohnet, Matthias (Emeritus) 49 531-391-2790
  m.bohnet@tu-bs.de
Hempel, Dietmar-Christian (Dean) 49 531-391-7650
  d.hempel@tu-bs.de
Koehler, Juergen (Head) 49 531-391-2627
  juergen.koehler@tu-bs.de
Kosyna, Gunter (Head) 49 531-391-2918
  g.kosyna@tu-bs.de
Leithner, Reinhard (Head) 49 531-391-3030
  r.leithner@tu-bs.de
Scholl, Stephan (Head) 49 531-391-2780
  s.scholl@tu-braunschweig.de
Schwedes, Joerg (Head) 49 531-391-9610
  j.schwedes@tu-bs.de

Accredited by:

Technical University of Clausthal

Institut fuer Chemische Verfahrenstechnik
Leibnizstrasse 17
38678 Clausthal-Zellerfeld
Germany

University phone......................... 49 5323 72-0
University FAX 49 5323 72-3500
Department phone 49 5323 72-2184
Department FAX 49 5323 72-2182

Professors

Hoffmann, Ulrich (Emeritus) .... 72-2534 [†]
  hoffmann@icvt.tu-clausthal.de
Turek, Thomas (Head) 72-2184
  turek@icvt.tu-clausthal.de

Associate Professors

Kunz, Ulrich 72-2534
  kunz@icvt.tu-clausthal.de

[†] Faculty numbers have prefix 49 5323.

Accredited by:

Darmstadt University of Technology

Fachbereich Maschinenbau
Fachgebiet Thermische Verfahrenstechnik
Petersenstr. 30
64287 Darmstadt

University phone.........................49 (6151) 16-0
Department phone 49 (6151) 162164
Department FAX 49 (6151) 164516

Professors

Hampe, Manfred J. (Chair) 49 6151 162164
  hampe@tvt.tu-darmstadt.de
Kast, W. (Emeritus) 49 6151 163364

Associate Professors

Jaeschke, Lothar (Emeritus)
Korkhaus, Juergen (Visiting)
Schadler, Norbert (Visiting)
  schadler@axiva.com
Schneider, Klaus (Emeritus)

Accredited by: Deutscher Akkreditierungsrat

Graduate advisor: Barbara Seifert

University of Dortmund

Fachbereich Bio- und Chemieingenieurwesen
Universität Dortmund
D-44221 Dortmund
Deliveries: Emil-Figge-Str. 70, D-44227 Dortmund

University phone............................ 231-755-1
Department phone 231-755-2362
Department FAX 231-755-2251

Professors

Agar, D. (Associate Dean) 231-755-2697
  agar@bci.uni-dortmund.de
  MO,PR,RE
Behr, A. 231-755-2310
  behr@bci.uni-dortmund.de
  MO,PR,RE
Ehrhard, P. 231-755-3252
  p.ehrhard@bci.uni-dortmund.de
  EN,MO,RE
Engell, S. 231-755-5126
  s.engell@bci.uni-dortmund.de
  MO,PR
Fahlenkamp, H. 231-755-2322
  h.fahlenkamp@bci.uni-dortmund.de
  EV,MO,PR,RE,SE
Friedrich, C. 231-755-5115
  c.friedrich@bci.uni-dortmund.de
  AM,BT,MO
Giesekus, H. (Emeritus) 231-755-2301
Górak, A. 231-755-2323
  a.gorak@bci.uni-dortmund.de
  MO,RE,SE
Köster, U. 231-755-2678
  koester@bci.uni-dortmund.de
  AM,MO,NT,PO,SU
Onken, U. (Emeritus) 231-755-2696
Sadowski, G. (Associate Dean) 231-755-2635
  g.sadowski@bci.uni-dortmund.de
  MO,PO,RE,SU
Schecker, H.-G. (Emeritus) 231-755-2014
  schecker@bci.uni-dortmund.de
Schembecker, G. 231-755-2338
  schembecker@bci.uni-dortmund.de
  MO,PO,PR,RE,SU
Schmid, A. 231-755-7380
  A.Schmid@bci.uni-dortmund.de
  BM,BT,MO,PO,SE
Schmidt-Traub, H. (Emeritus) 231-755-2338
  schmtr@bci.uni-dortmund.de
Schulz, S. (Emeritus) 231-755-2635
Schwind, H. (Emeritus) 231-755-2577
Simmrock, K.H. (Emeritus) 231-755-2310
Strauss, K. (Emeritus) 231-755-3214
  strauss@bci.uni-dortmund.de
Walzel, P. (Dean) 231-755-6088
  p.walzel@bci.uni-dortmund.de
  EN,MO,NT,PR
Weinspach, P.-M. (Emeritus) 231-755-2323
Weiss, E. (Emeritus) 231-755-2523
  weiss@bci.uni-dortmund.de
Werner, U. (Emeritus) 231-755-2326
Wichmann, R. 231-755-3205
  wichmann@bci.uni-dortmund.de
  BT,MO,PO,PR,RE

Associate Professors

Steiff, A. 231-755-2356

Assistant Professors

Joerissen, J. 231-755-2317
  joerissen@bci.uni-dortmund.de
  EN,MO
Kenig, E. 231-755-2357
  kenig@bci.uni-dortmund.de
  MO,RE,SE

Accredited by:

Degrees granted 2005-2006:

Ph.D.: 20

Universitat Erlangen-Nurnberg

Department of Chemical and Bioengineering
Cauerstrasse 4
D-91058 Erlangen

University phone........................... 49 9131 85 0
Department phone 49 9131 85 29597
Department FAX 49 9131 85 29503

Professors

Arlt, Wolfgang .... 85-27440 [†]
  wolfgang.arlt@cbi.uni-erlangen.de
  EV,SE
Buchholz, Rainer 85-23000
  rainer.buchholz@bvt.cbi.uni-erlangen.de
  BM,BT
Delgado, Antonio 85-29500
  antonio.delgado@lstm.uni-erlangen.de
  EN,MO
Durst, Franz 85-29500
  durst@lstm.uni-erlangen.de
  EN,MO
Emig, Gerhard (Emeritus) 85-27424
  emig@tc.uni-erlangen.de
  RE
Leipertz, Alfred (Dean) 85-29900
  sek@ltt.uni-erlangen.de
  EN
Molerus, Otto (Emeritus) 85-29401
  o.molerus@lfg.uni-erlangen.de
  SU
Peter, Siegfried (Emeritus) 85-27443
  siegfried.peter@rzmail.uni-erlangen.de
  SE
Peukert, Wolfgang 85-29400
  w.peukert@lfg.uni-erlangen.de
  NT,SU
Schlücker, Eberhard 85-29450
  sl@ipat.uni-erlangen.de
  AM,PR
Steiner, Rudolf (Emeritus) 85-27957
  rudolf.steiner@rzmail.uni-erlangen.de
  SE
Vetter, Gerhard (Emeritus) 85-29457
  vetter@ipat.uni-erlangen.de
  PR
Wasserscheid, Peter (Director) 85-27420
  wasserscheid@crt.cbi.uni-erlangen.de
  AM,RE

Associate Professors

Brunn, Peter O. 85-29430
  pbrunn@lstm.uni-erlangen.de
  MO

Dörnenburg, Heike 85-23005
heike.doernenburg@bvt.cbi.uni-erlangen.de
BM,BT
König, Axel 85-27446
axel.koenig@rzmail.uni-erlangen.de
SE
Schwieger, Wilhelm 85-28910
schwieger@rzmail.uni-erlangen.de
RE
Wensing, Michael 85-29782
michael.wensing@ltt.uni-erlangen.de
EN
Wenzel, Herbert 85-27442
herbert.wenzel@rzmail.uni-erlangen.de
SE
Wirth, Karl E. 85-29403
k.e.wirth@lfg.uni-erlangen.de
SU

† Numbers above have prefix 49 9131.

Accredited by: ACQUIN

Degrees granted 2005-2006:
    B.S.: 2   M.S.: 53   Ph.D.: 27

Graduate advisor: Dr. Adrian Melling

Undergraduate advisor: Dr. Lüder Depmeier, Dr. Stefan Becker

Student organization: Studentenvertretung Chemie- und Bioingenieurwesen
Advisor: Sebastian Werner

Department reports to: Dean, Faculty of Engineering Sciences

University of Hannover

Institut fuer Mehrphasenprozesse / Biomedizintechnik
Institute of Multiphase Processes / Biomedical Engineering
Callinstr. 36
D-30167 Hannover
Germany

University phone 49 511 762-0
University FAX 49 511 762-3456
Department phone 49 511 762-3828
Department FAX 49 511 762-3031

Professors

Glasmacher, Birgit (Chair) 49 511 762 3860
glasmacher@ifv.uni-hannover.de
BM,SU
Hallensleben, M.L. (Emeritus) 49 761 430452
PO
Härtel, V. (Adjunct) 49 511 9763346
volker.haertel@conti.de
PO
Luke, Andrea (Chair) 49 511 762 2877
luke@ift.uni-hannover.de
PR
Mewes, Dieter (Emeritus) 49 511 762 3638
mewes@ifv.uni-hannover.de
BM,BT,ME,MO,NT,PO,PR,RE,SE
Obrecht, W. (Adjunct) 49 2133 5122731
werner.obrecht@lanxess.com
PO
Schuster, R.H. (Chair) 49 511 842010
PO
Seume, Joerg 49 511 762 2733
seume@ifs.uni-hannover.de
EN
Stark, R. (Adjunct) 49 5131 51130
starkgarbsen@aol.com
PO

Associate Professors

Alshuth, T. 49 511 842010
Bederna, Ch. (Adjunct) 49 511 97645993
christoph.bederna@contitech.de
PO
Bode, H. 49 6196 750018
Freund, B. (Adjunct) 49 2233 964890
burkhard.freund@degussa.com
PO
Giese, U. (Adjunct) 49 511 8420143
PO
Guth, Wolfgang (Adjunct) 49 6203 924874
wg@warranty-chain-management.de
PO
Herrmann, W. (Adjunct) 49 511 93859501
wolfram.herrmann@contitech.de
PO
Kammann, A. (Adjunct) 49 6201 806562
andreas.kammann@freudenberg.de
PO
Lechtenböhmer, A. (Adjunct) 352 81993626
annette.lechtenboehmer@goodyear.com
PO
Luther, Sabine (Adjunct) 49 511 8420121
Sabine.Luther@dikautschuk.de
PO
Morgenstern, Ute (Adjunct) 49 351 46334228
ute.morgenstern@mailbox.tu-dresden.de
MO
Peinemann, K.-V. (Adjunct) 49 4152 872420
klaus-viktor.peinemann@gkss.de
EV
von Brook, U. (Adjunct) 49 40 7667 1416
ulrich.vonbroock@vibracoustic.de
PO
Wahl, G. (Adjunct) 49 511 9763046
guenter.wahl@conti.de
PO
Weiß, R. (Adjunct) 49 7754 701207
rainer.weiss@freudenberg.de
MO,PO
Wrana, C. (Adjunct) 49 214 3081048
claus.wrana@lanxess.com
PO

Accredited by:

Degrees granted 2005-2006:
    M.S.: 5   Ph.D.: 3

Graduate advisor: Professor L. Overmeyer

Undergraduate advisor: Professor L. Overmeyer

Student organization: International Office

Placement service: internationaloffice@uni-hannover.de

Universitat Kaiserslautern

Fachbereich Maschinenbau und Verfahrenstechnik
Studienrichtung Vefahrenstechnik
Gottlieb Daimler Strasse
D-67663 Kaiserslautern

University phone 0049 631 205 0
University FAX 0049 631 205 3200
Department phone 0049 631 205 2417
Department FAX 0049 631 205 3600

Professors

Bart, Hans-Jorg (Chair) .... 2414 [†]
    bart@mv.uni-kl.de
    BT,EV,MO,PR,RE,TP
Ebert, Fritz (Emeritus) 2415
    ebert@mv.uni-kl.de
    EV,MO,NT,PR,TP
Maurer, Gerd (Chair) 2410
    gmaurer@rhrk.uni-kl.de
    BT,HT,MO,PR,RE,SE,TH,TP
Ripperger, Siegfried (Chair) 2121
    ripperger@mv.uni-kl.de
    BT,EV,MO,NT,PR,SE,SU,TP
Ulber, Roland (Chair) 4043
    ulber@mv.uni-kl.de
    BM,BT,RE,TP
Wuestenberg, Dieter (Emeritus) 3031
    wuestenberg@mv.uni-kl.de
    EV,MO,PR,TP

Assistant Professors

Dau, Guenter (Director) 2560
    dau@mv.uni-kl.de
    EV,PR,TP
Kraetz, Lorenz 2765
    kraetz@mv.uni-kl.de
    BM,EV,PR,RE
Otah Watanabe, Erika (Visiting) 3556
    watanabe@rhrk.uni-kl.de
    EV,HT,TH
Pérez-Salado, Alvaro 2761
    salado@rhrk.uni-kl.de
    HT,MO,TH

[†] Faculty numbers are extensions of Dept. phone.

Accredited by: ASII

Degrees granted 2005-2006:

                        Ph.D.: 36

Department reports to: Helmut J. Schmidt, President

Placement service: Harald Kuehn, x2215

Universitat Karlsruhe

Fakultaet fuer Chemieingenieurwesen und Verfahrenstechnik
Kaiserstrasse 12
D - 76128 Karlsruhe

Department phone .... ++49 - 721 - 608 6378
Department FAX .... (721) 608-7531

Professors

Bockhorn, Henning .... (721) 608-2570
Braun, Andre M. .... (721) 608-2557
Buggisch, Hans (Emeritus) .... (721) 608-2661
Frimmel, Fritz H. .... (721) 608-2580
Kasper, Gerhard .... (721) 608-6561
Kind, Matthias .... (721) 608-2390
Kraushaar-Czarnetzki, B. (Dean) .... (721) 608-4133
    kraushaar@cvt.uka.de
Kraushaar-Czarnetzki, Bettina (Dean) .... (721) 608-4133
    kraushaar@cvt.uka.de
Leuckel, W. (Emeritus) .... (721) 608-2570
Nirschl, Hermann .... (721) 608-2404
Reimert, Rainer .... (721) 608-2560
Schaber, Karlheinz .... (721) 608-2321
Schluender, E. (Emeritus) .... (721) 608-2390
Schubert, Helmar (Emeritus) .... (721) 608-2497
Schuchmann, Heike .... (721) 608 2497
    Heike.Schuchmann@lvt.uni-karlsruhe.de
Stahl, Werner (Emeritus) .... (721) 608-2401
Syldatk, Christoph .... (721) 608 2123
    Christoph.Syldatk@ciw.uni-karlsruhe.de
Willenbacher, N.
    norbert.willenbacher@mvm.uni-karlsruhe.de

Associate Professors

Lintz, Hans G. (Emeritus) .... (721) 608-3939
Martin, Holger .... (721) 608-2386
Oellrich, Lothar .... (721) 608-2332
Posten, Clemens .... (721) 608-2410
Schaub, Georg .... (721) 608-2572
Zarzalis, Nikolaos .... (721) 608-4231

Accredited by: EVALAG

Degrees granted 2004-2005:

        M.S.: 60   Ph.D.: 25

Undergraduate advisor: oellrich@ttk.ciw.uni-karlsruhe.de

Student organization: fachschaft@fmc.uni-karlsruhe.de

Georg-Simon-Ohm Fachhochschule Nuernberg

Fachbereich Angewandte Chemie
Kesslerplatz 12
Postfach
D-90121 Nuernberg

University phone .... (0911) 5880-0
Department phone .... (0911) 5880-1260
Department FAX .... (0911) 5880-5139

Professors

Aust, Eberhard
Bartsch, Stephan
Bauer, Hermann
Dorn, Alfred
Herold, Thomas
Jacob, Karl-Heinz (Head)
Kinkel, Hans-Joachim
Stark, Walter
Stephan, Rainer
Volgnandt, Peter
Wehnert, Gerd

Accredited by:

Graduate advisor: Prof. Dr. Peter Volgnandt

Department reports to: Prof. Dr. Karl-Heinz Jacob, Dekan

Fachhochschule Mannheim, University of Applied Sciences

Fachbereich Verfahrens- und Chemietechnik
Windeckstr 110
68163 Mannheim

Department phone 49 621 292-6424
Department FAX 49 621 292-6555

Professors

Adrian, Till .... 6305 [†]
    t.adrian@fh-mannheim.de
Diewald, Werner 6489
    w.diewald@fh-mannheim.de
Fosshag, Erich 6585
    e.fosshag@fh-mannheim.de
Fritz, Wolfgang 6306
    w.fritz@fh-mannheim.de
Hagen, Jens 6382
    j.hagen@fh-mannheim.de
Hassenpflug, Hans-Uwe 6484
    hassenpflug@fh-mannheim.de
Hoyningen-Huene v., Dietmar 6400
    hoy@zv.fh-mannheim.de
Kern, Heinz 6305
Kunz, Peter 6304
    kunz.p.m.prof@t-online.de
Landwehr, Brigitta 6486
    b.landwehr@fh-mannheim.de
Michel, Hartmut 6306
    dr.h.michel@t-online.de
Mueller, Herbert 6305
Peschges, Klaus-Juergen 6498
    k.peschges@fh-mannheim.de
Raedle, Matthias 6330
    m.raedle@fh-mannheim.de
Schinke, Bernd (Head) 6387
    b.schinke@fh-mannheim.de
Schmidt, Volkmar 6307
    v.m.schmidt@t-online.de
Schmitt, Wolfgang 6489
    w.ph.schmitt@fh-mannheim.de
Steinert, Heike 6306
    h.steinert@fh-mannheim.de
Traegner, Ulrich 6296
    u.traegner@fh-mannheim.de

[†] Numbers are extensions of Dept. phone.

Accredited by: IChemE(UK) in progress

Department reports to: Hoyningen-Huene v., Dietmar; Rector; ext. 6400/6401

Greece

Aristotle University

Chemical Engineering Department
54124 Thessaloniki

Department phone 302310996267
Department FAX 302310996168

Professors

Anastasiadis, S. .... 4245 [†]
    spiros@auth.gr
Assael, M. J. 6163
    assael@auth.gr
Kiparissides, C. 6211
    cypress@eng.auth.gr
Liakopoloulou-Kyriakidou, M. 6193
    markyr@eng.auth.gr
Nychas, S. G. 6231
    nychas@eng.auth.gr
Panayiotou, C. 6223
    cpanayio@auth.gr
Papageorgiou, V. P. (Chair) 6241
    vaspap@cheng.auth.gr
Sakellaropoulos, G. P. (Associate Chair) .. 6271
    sakel@eng.auth.gr
Stoukides, M. 6165
    mstoukid@auth.gr

Associate Professors

Bacola-Christianopoulou, M. N. 6213
    bakola@eng.auth.gr
Kastrinakis, E. G. 6183
    kastr@eng.auth.gr
Kotali, A. 6253
    kotali@eng.auth.gr
Lemonidou, A. 6273
    lemonido@auth.gr
Markopoulos, J. 6203
    jonimark@eng.auth.gr
Paras, S. V. 6174
    paras@cheng.auth.gr
Salifoglou, A. 6179
    salif@cheng.auth.gr
Stamatoudis, M. 6233
    stamatou@eng.auth.gr
Yiantsios, S. 1293
    yiantsios@cheng.auth.gr

Assistant Professors

Adamopoulos, C. 6205
    costadam@eng.auth.gr
Kabasakalis, V. 6243
    kabak@eng.auth.gr
Kyriakou, G. 6238
    kyriakou@eng.auth.gr
Mitrakas, M. 6248
    manasis@eng.auth.gr
Sikalides, K. 6185
    sikalidi@eng.auth.gr
Stoforos, N. 6450
    stoforos@cheng.auth.gr
Tzimou-Tsitouridou, R. 6194
    roxani@eng.auth.gr
Zabaniotou, A. 6274
    sonia@cheng.auth.gr
Zlatanos, S. N. 6173
    szlatano@eng.auth.gr

[†] Direct dial with prefix 3031 99.

Degrees granted 2005-2006:

    M.S.: 75   Ph.D.: 15

University of Patras

University Campus
GR-26504 Patras
Greece

University phone 302610-991822
Department phone 302610-997580
Department FAX 302610-997849

Professors

Dassios, George 30610997373
    dassios@iceht.forth.gr
Dondos, Anastasios (Emeritus) 30610997652
    dondos@iceht.forth.gr
Koutsoukos, Petros G. 30610997265
    pgk@iceht.forth.gr
Kravaris, Costas 30610996339
    kravaris@chemeng.upatras.gr
Ladas, Spyridon 30610997631
    ladas@iceht.forth.gr
Lyberatos, Gerasimos 30610997573
    lyberatos@chemeng.upatras.gr
Nikolopoulos, Panajotis S. (Associate Chair) . 30610997563
    nikolop@chemeng.upatras.gr
Pandis, Spyros
    pandis@chemeng.upatras.gr
Papatheodorou, George N. 30610997570
    gpap@iceht.forth.gr
Pavlou, Stavros 30610997640
    sp@chemeng.upatras.gr
Payatakes, Alkiviades C. 30610997574
    acp@iceht.forth.gr
Rapakoulias, Dimitrios E. (Chair) 30610993361
    rap@chemeng.upatras.gr
Tsahalis, Demosthenes T. 30610997577
    tsahalis@lfme.chemeng.upatras.gr
Tsamopoulos, John 30610997203
    tsamo@chemeng.upatras.gr
Vayenas, Constantinos G. 30610997576
    cat@chemeng.upatras.gr
Verykios, Xenophon E. 30610991527
    verykios@iceht.forth.gr

Associate Professors

Kennou, Styliani 30610993255
    kennou@iceht.forth.gr
Mavrantzas, Vlasis
    vlasis@chemeng.upatras.gr
Staikos, George 30610997501
    staikos@iceht.forth.gr
Tsitsilianis, Constantinos 30610997500
    tsic@iceht.forth.gr

Assistant Professors

Angelopoulos, George N. 30610997509
    angel@chemeng.upatras.gr
Bebelis, Symeon I. 30610997756
    simeon@iceht.forth.gr
Boghosian, Soghomon B. 30610997854
    bogosian@iceht.forth.gr
Mataras, Dimitrios 30610996340
    dim@chemeng.upatras.gr

Accredited by:

Degrees granted 2003-2004:
    M.S.: 70   Ph.D.: 12

Graduate advisor: Gerasimos Lyberatos

Undergraduate advisor: Gerasimos Lyberatos

Hong Kong, SAR of China

Hong Kong University of Science and Technology

Department of Chemical Engineering
Clear Water Bay
Kowloon, Hong Kong

Department phone 852 2358-7130
Department FAX 852 2358-0054

Professors

Chan, Chi-Ming . . . . 2358-7125 [†]
    kecmchan@ust.hk
Ng, Ka M. (Head) 2358-7238
    kekmng@ust.hk
Yue, Po-Lock 2358-8370
    keplyue@ust.hk

Associate Professors

Barford, John 2358-7237
    barford@ust.hk
Chan, Chak K. 2358-7124
    keckchan@ust.hk
Chen, Guohua 2358-7138
    kechengh@ust.hk
Gao, Furong 2358-7139
    kefgao@ust.hk
Gao, Ping 2358-7126
    kepgao@ust.hk
Hsing, I-Ming 2358-7131
    kehsing@ust.hk
Hu, Xijun 2358-7134
    kexhu@ust.hk
Hui, David 2358-7137
    kehui@ust.hk
Mckay, Gordon 2358-8412
    kemckayg@ust.hk
Mi, Yongli 2358-7127
    keymix@ust.hk
Yeung, King L. 2358-7123
    kekyeung@ust.hk

Assistant Professors

Porter, John F. 2358-7132
    kejep@ust.hk

[†] Dial counrty code 852 for faculty numbers

Accredited by: Institution of Chemical Engineers, Hong Kong Institution of Chemical Engineers

Graduate advisor: Dr. Xijun Hu

Undergraduate advisor: Dr. C.K. Chan

Department reports to: Prof. Philip Chan, Acting Dean of Engineering (852 2358-6952)

Hungary

Budapest University of Technology and Economics

Faculty of Chemical Engineering
H-1521, Budapest, Muegyetem rakpart 3.

Department phone ....(36-1) 463-3571
Department FAX ....(36-1) 463-3570

Professors

Bitter, Istvan .... 463-1379 [†]
    ibitter@mail.bme.hu
Borsa, Judit (Associate Dean) 463-1376
    jborsa@mail.bme.hu
Faigl, Ferenc (Associate Dean) 463-5889
    ffaigl@mail.bme.hu
Fekete, Jeno 463-1596
    fekete@mail.bme.hu
Fogassy, Elemer 463-1883
    efogassy@mail.bme.hu
Fonyo, Zsolt (Chair) 463-3196
    fonyo@mail.bme.hu
Gal, Sandor 463-1216
    gal@mail.bme.hu
Grofcsik, Andras 463-1484
    agrofcsik@mail.bme.hu
Hargittai, Istvan 463-1286
    hargittai@mail.bme.hu
Hencsei, Pal 463-2294
    hencsei@mail.bme.hu
Horvai, Gyorgy (Chair) 463-1480
    ghorvai@mail.bme.hu
Huszthy, Peter (Chair) 463-1071
    huszthy@mail.bme.hu
Kalaus, Gyorgy 463-1285
    kalaus@mail.bme.hu
Keglevich, Gyorgy (Chair) 463-5883
    gkeglevich@mail.bme.hu
Kemeny, Sandor 463-2209
    kemeny@mail.bme.hu
Kubinyi, Miklos 463-2137
    kubinyi@mail.bme.hu
Lasztity, Radomir (Emeritus) 463-1627
    lasztity@mail.bme.hu
Mihaltz, Pal 463-2574
    mihaltz@mail.bme.hu
Novak, Bela 463-1364
    bnovak@mail.bme.hu
Novak, Lajos 463-2207
    l-novak@mail.bme.hu
Nyitrai, Jozsef 463-2205
    nyitrai@mail.bme.hu
Nyulaszi, Laszlo (Chair) 463-1281
    nyulaszi@mail.bme.hu
Orsi, Ferenc 463-2283
    orsi@mail.bme.hu
Pokol, Gyorgy (Dean) 463-1593
    pokol@mail.bme.hu
Pukanszky, Bela (Chair) 463-2015
    bpukanszky@mail.bme.hu
Pungor, Erno (Emeritus) 463-4054
    pungor@mail.bme.hu
Reffy, Jozsef 463-3281
    jreffy@mail.bme.hu
Salgo, Andras (Chair) 463-3854
    salgo@mail.bme.hu
Sevella, Bela (Chair) 463-2595
    bsevella@mail.bme.hu
Szantay, Csaba (Emeritus) 463-1195
    szantay@mail.bme.hu
Toke, Laszlo 463-3653
    ltoke@mail.bme.hu
Toth, Klara 463-2273
    ktoth@mail.bme.hu
Tungler, Antal (Associate Dean) 463-1203
    atungler@mail.bme.hu
Vesztpremi, Tamas 463-1793
    tveszpremi@mail.bme.hu
Zrinyi, Miklos (Chair) 463-3229
    zrinyi@mail.bme.hu

[†] Faculty numbers have prefix 36-1.

Graduate advisor: Dr. Mariann Lengyelne

Undergraduate advisor: Marta Takacs

Department reports to: Gyorgy Pokol, Dean of Faculty

Placement service: Dr. Mariann Lengyelne

India

Annamalai University

Department of Technology
Annamalai University, Annamalai Nagar, Tamil Nadu 608 002

University phone......................... 914144-238259
Department phone 914144-239737
Department FAX 914144-238275

Professors

Karunanithi, T. 914144-221235
    tkarunanithi@hotmail.com
Viruthagiri, T. (Head) 914144-239737
    drtvgiri@rediffmail.com

Assistant Professors

Anand, G. M. F.
    mariafanand@yahoo.co.in
Bhaba, P. K. 04144-238691
    Pkbhaba@yahoo.com
Dhanasekaran, R. 914144-238022
    rdhanasekar28@rediffmail.com
Karthikeyan, C. 914144-224679
    drcktech@rediffmail.com
Manickam, N. 914144-224494
Meyyappan, R. M. 221654
    vijimani_cdm@yahoo.com
Vaithiyanathan, K. 914144223026
Vijayagopal, V. 914144249236
    vevevin@yahoo.co.in

Accredited by: AICTE

Department reports to: Dr.T.Viruthagiri

Placement service: Dr.T.Viruthagiri

Birla Institute of Technology & Science (BITS) - Pilani

Department of Chemical Engineering
Vidya Vihar Campus
Birla Institute of Technology and Science (BITS)
PILANI - 333 031
(Rajasthan) India
University phone 91-1596-245073
University FAX 91-1596-244183
Department phone 91-1596-245073x224
Department FAX 91-1596-244183

Professors
Babu, B V (Head) . . . . x205/224 [†]
    bvbabu@bits-pilani.ac.in
    BT,EN,EV,MO,NT,PR,RE,SE
Mathur, T N S (Visiting) x8404
    tns@bits-pilani.ac.in
    EN,RE,SE
Natarajan, B R x239
    brnt@bits-pilani.ac.in
    PR
Vaid, R P x201
    rpvaid@bits-pilani.ac.in
    MO,PR,RE

Assistant Professors
Angira, Rakesh x216
    angira@bits-pilani.ac.in
    MO,PR,RE
Gulyani, Bharat B. x215
    gulyanibb@bits-pilani.ac.in
    EN,MO,NT,PR,SU
Jana, Amiya K. x215
    akjana@bits-pilani.ac.in
    MO,PR
Kundu, Madhusree x215
    mkundu@bits-pilani.ac.in
    MO,RE,SE
Mohanta, H K x215
    hkm@bits-pilani.ac.in
    PR
Munshi, B D x262
    munshi@bits-pilani.ac.in
    MO,PR,SE
Sharma, Arvind K. x215
    arvinds@bits-pilani.ac.in
    BT,EV,PO,SE,SU

[†] Faculty numbers are extensions of 91-1596-245073.

Accredited by: NAAC

Degrees granted 2005-2006:
    B.S.: 90   M.S.: 20   Ph.D.: 5

Graduate advisor: Prof. Ravi Prakash

Undergraduate advisor: Prof. B. V. Babu

Student organization: Chemical Engineering Association
Advisor: Dr. Rakesh Angira

Department reports to: Prof. A.K. Sarkar, Dean, Faculty Division-I

Placement service: Prof. G. Raghurama

Calcutta University

Department of Chemical Engineering
University College of Tech.
92 Acharyya Prafulla Chandra Rd.
Kolkata, 700 009
University phone 91-33-2241-0071
Department phone . . . . 91-33-2350-8386 ext 201
Department FAX 91-33-2351-9755

Professors
Banik, Ajit K.
Bhattacharjee, Sekhar . . . . 91-33-23508386 x 203
    sekharbhatta@rediffmail.com
Das, Manas . . . . 91-33-23508386 x 296
De, Parameswar. (Head) . . . . 91-33-23508386 x 217
    parameswar_de@rediffmail.com
Dutta, Benay. K. 91-33-23508386 x208
    dutta_asc@vsnl.net
Pal, Sitangshu S. . . . . 91-33-23508386 x 290
Ray, P. . . . . 91-33-23508386 x 218
    praycuce@vsnl.com
Saha, Ahin C. . . . . 91-33-23508386 x 203

Associate Professors
Ash, Soumendra N. . . . . 91-33-23508386 x 289
Basu, Ranjan K. . . . . 91-33-23508386 x 289
Chaudhury, Basab . . . . 91-33-23508386 x 203
    basabc@vsnl.net
Das, Sudip K. . . . . 91-33-23508386 x 296
    drsudipkdas@vsnl.net
Saha, Bibhuti R. . . . . 91-33-23508386 x 206

Assistant Professors
Das, Bhaskar C. . . . . 91-33-23508386 x 203
De, Asim K. . . . . 91-33-23508386 x 203
Ganguly, Kausik. . . . . 91-33-23508386 x 203
    kausik_ganguly_cal@yahoo.com

Accredited by: Indian Institute of Chemical Engineering, Institute of Engineers, All India Council for Technical Education

Degrees granted 2003-2004:
    B.S.: 35   M.S.: 5   Ph.D.: 2

Graduate advisor: Head of Department

Department reports to: Head of Department

Indian Institute of Science

Department of Chemical Engineering
Bangalore 560 012
Karnataka
University phone 91 80 23600411
University FAX 91 80 3600085
Department phone 91 80 22932318
Department FAX 91 80 23608121

Professors
Gandhi, K. S. 91 80 22932320
    gandhi@chemeng.iisc.ernet.in
Modak, J. M. (Chair) 91 80 22933108
    modak@chemeng.iisc.ernet.in
Rao, K. K. 91 80 22932341
    kesava@chemeng.iisc.ernet.in

Associate Professors

Ayappa, K. G. 91 80 22932769
ayappa@chemeng.iisc.ernet.in
Gupta, S. K. 91 80 22933110
sanjeev@chemeng.iisc.ernet.in
Kumaran, V. 91 80 22933112
kumaran@chemeng.iisc.ernet.in
Madras, Giridhar 91 80 22932321
giridhar@chemeng.iisc.ernet.in
Nott, Prabhu R. 91 80 22932317
prnott@chemeng.iisc.ernet.in

Assistant Professors

Dixit, Narendra M. 91 80 22932768
narendra@chemeng.iisc.ernet.in
Santhanam, Venugopal 91 80 22933113
venu@chemeng.iisc.ernet.in

Accredited by: AICTE

Degrees granted 2004-2005:
M.S.: 19   Ph.D.: 2

Graduate advisor: Prof. Giridhar Madras

Student organization: Chemical Engineering Association
Advisor: Prof. Giridhar Madras

Department reports to: Prof. M. L. Munjal, Chairman, Division of Mechanical Sciences

Placement service: Prof. Vikram Jairam, Advisor, Placement Ctr.

Indian Institute of Technology, Kanpur

Chemical Engineering Department
Kanpur, U.P., 208 016, India
University phone 91-512-2597629
University FAX 91-512-2590104
Department phone 91-512-2597406
Department FAX 91-512-2590104

Professors

Bhattacharya, P. K. 2597093
pkbhatta@iitk.ac.in
Chhabra, R. P. 2597393
chhabra@iitk.ac.in
Gupta, J. P. 2597175
jpg@iitk.ac.in
Gupta, S. K. 2597031
skg@iitk.ac.in
Khanna, Ashok 2597117
akhanna@iitk.ac.in
Kumar, Anil 2597195
anilk@iitk.ac.in
Kunzru, D.K. (Dean) 2597193
dkunzru@iitk.ac.in
Rao, D. P. 2597873
dprao@iitk.ac.in
Sharma, Ashutosh (Head) 2597026
ashutos@iitk.ac.in

Associate Professors

Deo, Goutom 2597363
goutam@iitk.ac.in
Verma, Nishith 2597704
nishith@iitk.ac.in

Assistant Professors

Bandyopadhyay, Rajdip 2597697
rajdip@iitk.ac.in
Garg, Sanjeev 2597736
sgarg@iitk.ac.in
Ghatak, Animangsu 2597146
aghatak@iitk.ac.in
Joshi, Yogesh M 2597629
ymjoshi@iitk.ac.in
Kaistha, Nitin 2597513
nkaistha@iitk.ac.in
Shankar, V 2597377
vshankar@iitk.ac.in

Accredited by:

Degrees granted 2003-2004:
B.S.: 45   M.S.: 36   Ph.D.: 4

Graduate advisor: Dr. Sanjeev Garg

Undergraduate advisor: Dr. Nitin Kaistha

Student organization: Student Gymkhana
Advisor: Dean of Students Affairs

Placement service: Dr. Ashok Khanna

Indian Institute of Technology, Madras

Chemical Engineering Department
IIT-Madras,
Chennai 600 036
University phone 91-44- 2257 8001
University FAX 91-44-2257 0509
Department phone 91-44-2257 4150
Department FAX 91-44-2257 0509

Professors

Ananth, M.S. (Director) .... +91-44-2257 8001
ananth@iitm.ac.in
AM,EN,MO
Balakrishnan, A.R. (Head) .... +91-44-2257 4151
arbala@iitm.ac.in
EN,MO
Chidambaram, M. .... +91-44-2257 4155
chidam@iitm.ac.in
PR
Jayanti, Sreenivas .... +91-44-2257 4168
sjayanti@iitm.ac.in
MO
Krishnaiah, K. .... +91-44-2257 4156
krishnak@iitm.ac.in
RE,SU
Murthy, D.V.S. .... +91-44-2257 4157
dvs@iitm.ac.in
EV
Panda, T. .... +91-44-2257 4160
tpanda@che.iitm.ac.in
BM,BT
Pushpavanam, S. .... +91-44-2257 4161
spush@iitm.ac.in
EV,MO,RE

R., Nagarajan ....+91-44-2257 4158
nag@iitm.ac.in
AM,NT,SU
Rao, V. S. R. ....+91-44-2257 4162
vsrr@iitm.ac.in
PR
Ravi, R. ....+91-44-2257 4167
rravi@iitm.ac.in
MO,SE
Sai, P. S. T. ....+91-44-2257 4163
psts@iitm.ac.in
RE,SE
Shankar, Narasimhan ....+91-44-2257 4165
naras@iitm.ac.in
PR
Swaminathan, T. ....+91-44-2257 4166
tswami@iitm.ac.in
EV

Associate Professors

Deshpande, Abhijit P. ....+91-44-2257 4169
abhijit@iitm.ac.in
EV,PO
Kannan, A. ....+91-44-2257 4170
kannan@iitm.ac.in
EN,MO,SE

Assistant Professors

Basak, Tanmay ....+91-44-2257 4173
tanmay@iitm.ac.in
EN,MO
R., Ramnarayanan ....+91-44-2257 4174
ramna@iitm.ac.in
AM,MO
Srinivasan, Ramanathan ....+91-44-2257 4171
srinivar@iitm.ac.in
EV,ME,NT
Tangiralla, Arun ....+91-44-2257 4181
arunkt@iitm.ac.in
EN,PR
Varughese, Susy ....+91-44-2257 4172
susy@iitm.ac.in
EV,PO

Accredited by:

Degrees granted 2005-2006:
B.S.: 50   M.S.: 30   Ph.D.: 5

Indian Institute of Technology, Delhi

Department of Chemical Engineering,
Indian Institute of Technology - Delhi
Hauz Khas, New Delhi 110016
INDIA
University phone 91 11 2658 2222
University FAX 91 11 2658 2037
Department phone 91 11 2659 1021
Department FAX 91 11 2658 1120

Professors

Guha, B. K. (Head) 91 11 2659 1038
bkguha_iitd@rediffmail.com
Gupta, A. K. 91 11 2659 1019
guptaak@chemical.iitd.ac.in
Gupta, S. K. (Associate Dean) 91 11 2659 1023
sgupta@chemical.iitd.ac.in

Nigam, K. D. P. 91 11 2659 1020
drkdpn@gmail.com
Pitchumani, B. 91 11 2659 1022
bpmani@chemical.iitd.ac.in
Rao, D. P. 91 11 2659 1016
dprao@chemical.iitd.ernet.in
Rao, D. S. 91 11 2659 1018
subbarao@chemical.iitd.ac.in
Rao, T. R. 91 11 2659 1026
trrao@chemical.iitd.ac.in
Walia, D. S. 91 11 2659 1017
dswalia@chemical.iitd.ac.in

Associate Professors

Basu, S. 91 11 2659 1035
sbasu@chemical.iitd.ac.in
Bhaskarwar, A. N. 91 11 2659 1028
ashoknb@chemical.iitd.ac.in

Assistant Professors

Amar, O. P. 91 11 2659 1030
Khanna, Rajesh 91 11 2659 1031
rajesh@chemical.iitd.ac.in
Krishnan, V. V. 91 11 2659 1024
vvkrish@chemical.iitd.ac.in
Mohan, Ratan 91 11 2659 1033
ratan@chemical.iitd.ac.in
Pant, K. K. 91 11 2659 6172
kkpant@chemical.iitd.ac.in
Roy, Shantanu 91 11 2659 6021
roys@chemical.iitd.ac.in
Saroha, A. K. 91 11 2659 1032
aksaroha@chemical.iitd.ac.in

Accredited by:

Degrees granted 2004-2005:
B.S.: 45   M.S.: 40   Ph.D.: 6

Student organization: CHES
Advisor: Dr. Rajesh Khanna

Department reports to: Director, IITD

Placement service: Y. S. Goel, Placement Cell

Laxminarayan Institute of Technology, Nagpur University

Chemical Engineering Department
Nagpur, 440 010
Department phone 531659 (0712)

Professors

Nageshwar, G. D. (Director)

Associate Professors

Chandak, B. S.
Sonolikar, R. L.
Thorat, R. T.

Assistant Professors

Bhagade, S. S.
Dawande, S. D.
Koranne, K. V.
Pandharipande, S. L.
Thorat, R. T.

Vyas, R. P.

Accredited by: Indian Institute of Chemical Engineers

Department reports to: G. D. Nageshwar, Director

Placement service: S. P. Ghisad

Manipal Institute of Technology

Department of Chemical Engineering
University of MAHE
Manipal (S.K.)
Postal Pin Code 576 119

Department phone .... (08252)71061 ext 24311
Department FAX .... (08252) 71071

Professors
Bhat, Jayadev (Associate Dean) .... x24311 [†]
    soodajbhat@eudoramail.com
Muniswaran, P. K. A. (Head) x24311
    munishpka@yahoo.com
Murthy, P. S. x24316
    parimimurthy@rediffmail.com

Associate Professors
Prabhu, Balakrishna x24312
    balakrishnaprabhu@yahoo.com
Reddy, B. N. x24316
    reddy_botta@hotmail.com
Sivasankaran, S. x24314
    sivasankaraniyer@yahoo.com

Assistant Professors
Guru, Bharath R. x24312
    bharathrajag@yahoo.co.in
Kini, Srinivasa x24315
    srinivas_kini@rediffmail.com
Kumar, S. H. x24314
    harishmanipal@rediffmail.com
Murthy, V. R. x24314
    vytlarama@yahoo.com

[†] Faculty phone numbers are extensions of 2571060.

Accredited by: AICTE, National Board of Accreditation (Technical Education)

Department reports to: Dr. B.S. Prabhu, Director Office : 71072

Placement service: K. J. Kamath

Malaviya National Institute of Technology, Jaipur

Chemical Engineering Department
Malaviya National Institute of Technology
Jaipur 302 017
Rajasthan

University phone....................... 91 141 2702954
Department phone 91 141 2702591

Professors
Gupta, Alok 91 141 2702591
    agch1@rediffmail.com

Associate Professors
Chaurasia, S. P. (Head) 91-141-2702591
    chch3@mnit.ac.in
Vyas, R.K. (Head) 91-141-2702591
    rkvyas2@rediffmail.com

Assistant Professors
Agarwal, Madhu
George, Suja
Jana, Susanta K.
Pandit, Prabhat
Singh, Kailash 91-141-2702591
    ksch@mnit.ac.in
Vasistha, Manish

Accredited by: AICTE

Undergraduate advisor: Dr. R. K. Vyas

Student organization: Chemical Engineering Students' Society (ChESS)
Advisor: Dr. S.P. Chaurasia

Department reports to: Director

Placement service: Professor, Training and Placement.

University of Mumbai

Chemical Engineering Division
Institute of Chemical Technology
Matunga, Mumbai
400 019

University phone........................ 91-22-2703248
Department phone 91-22-4145616
Department FAX 91-22-4145614

Professors
Gaikar, V. G. .... x286 [†]
    v.g.gaikar@udct.org
Joshi, J. B. (Director) x282
    jbj@udct.org
Mahajani, V. V. x287
    vvm@udct.org
Mhaskar, R. D. x381
    rdm@udct.org
Pandit, A. B. x278
    abp@udct.org
Pangarkar, V. G. (Head) x284
    vgp@udct.org
Sawant, S. B. x281
    sbs@udct.org
Yadav, G. D. (Chair) x291
    gdy@udct.org

Associate Professors
Bhagwat, S. S. x285
    ssb@udct.org
Lali, A. M. x292
    aml@udct.org
Marathe, K. V. x288
    kvm@udct.org
Thorat, B. N. x289
    bnt@udct.org

Assistant Professors

Patwardhan, A. W. x283
awp@udct.org
Rathod, V. K. x293
vkr@udct.org

† Faculty numbers are extensions of 91-22-4145616.

Accredited by: All India Council for Technical Education

Degrees granted 2004-2005:

    B.S.: 75   M.S.: 38   Ph.D.: 18

Graduate advisor: Head Chemical Engineering

Undergraduate advisor: V. G. Gaikar

Student organization: Technological Association
Advisor: V.G. Gaikar

Department reports to: J.B.Joshi, Director, Institute of Chemical Technology, 91-22-4140865

Placement service: B. N. Thorat

National Institute of Technology Warangal

Department of Chemical Engineering
National Institute of Technology
Warangal - 506 004 (A.P.)INDIA

University phone 091-870-2459191
University FAX 091-870-2459547
Department phone 091-870-2462601
Department FAX 091-870-2459547

Professors

G., Venkat R. (Dean) .... 2462602 †
BT,MO
R.C., Sastry 2462603
sastry@nitw.ac.in
PR,SE
Y., Pydisetty (Head) 2462600, 2462611
psetty@nitw.ac.in
RE,SE

Associate Professors

A., Sarat B. 2462610
sarat@nitw.ac.in
MO,RE

Assistant Professors

A., Venu V. 2462621
avv@nitw.ac.in
BT,TP
K. (Mrs.), Srivani 2462620
vani@nitw.ac.in
HT,PO
K., Anand K. 2462623
kola@nitw.ac.in
BT,MO
L. (Mrs.), Prasanna L. 2462622
prasan@nitw.ac.in
BT,EV
S., Srinath 2462624
srinath@nitw.ac.in
MO,SE

† Faculty numbers are extensions of 091-870.

Accredited by: NBA

Degrees granted 2005-2006:

    B.S.: 28   M.S.: 13   Ph.D.: 1

Student organization: Chemical Engineering Association
Advisor: Dr. A. Venu Vinod

Placement service: Training & Placement Section

Regional Engineering College, Affiliated with Bharathidasan University

Chemical Engineering Department
National Institute of Technology
Tiruchirappalli-620 015
(Tamil Nadu)

University phone 91-431-2501801
University FAX 91-431-2500133
Department phone .... 91-431-2501811 EXT. 2901
Department FAX 91-431-2500133

Professors

Pandey, S. K.
Sundaram, S. (Head) 91 431 500 295
Venkataramani, V.

Assistant Professors

Anantharaman, N
SE
Jakka, Sarat C. B. 91 431 501 280
sarat@rect.ernet.in
NT
Prabhu, H. J.
Radhakrishnan, T. K.
radha@nitt.edu
PR
Shanmugasundaram, P.
Sivashanmugam, P.

Accredited by: NBA

Degrees granted 2005-2006:

    B.S.: 45   M.S.: 40   Ph.D.: 1

Graduate advisor: S. Sundaram

Undergraduate advisor: S. Sundaram

Student organization: Chemical Engineering Association
Advisor: Dr. Sarat Chandra Babu

Department reports to: Dr. S. Sundaram

Placement service: Dr. T. Srinivasa Rao, Head-in-Charge, Training and Placement Cell

Indian Institute of Technology

Chemical Engineering Department
Roorkee (Uttranchal)
PIN-247667

University phone 91-1332-272349
Department phone 91-1332-276534
Department FAX 91-1332-276535

Professors

Agarwal, C. P. .... 285712 [†]
    cpaurfch@iitr.ernet.in
Bhattacharya, S. D. 285708
    sdbiofch@iitr.ernet.in
Gupta, S. C. 285709
    satisfch@iitr.ernet.in
Kumar, Surendra 285714
    skumar@iitr.ernet.in
Mishra, I. M. 285715
    imishfch@iitr.ernet.in
Mohanty, B. (Head) 285710
    bmohanty@iitr.ernet.in

Associate Professors

Chand, Shri 285383
    schanfch@iitr.ernet.in
Mall, I. D. 285319
    invigfch@iitr.ernet.in

Assistant Professors

Agarwal, V. K. 285718
    vijayfch@iitr.ernet.in
Bhargava, R. 285382
    ravibfch@mail.iitr.ernet.in
Majumder, C. 285321
    chandfch@iitr.ernet.in
Prasad, B. 285323
    bashefch@iitr.ernet.in
Shashi 285672
    shashifch@iitr.ernet.in
Sinha, S. N. 285384
    sudhifch@iitr.ernet.in

[†] All faculty numbers are preceded by 91-1332

Accredited by: AICTE, Indian Institute of Chemical Engineers, Institution of Engineers (India)

Graduate advisor: Dr. B. Prasad

Undergraduate advisor: Dr. Shri Chand

Student organization: Institution of Engineers(India)
Advisor: Dr. V. K. Agarwal

Department reports to: D. V. Singh (72742), Vice Chancellor

Placement service: H. K. Verma (85245), Chief, Employment Bureau

SDM College of Engineering and Technology

Department of Chemical Engineering
SDM College of Engineering and Technology
Dhavalagiri
Dharwad-580002, INDIA

University phone 91-836-2447465
University FAX 91-836-2464638
Department phone 91-836-2447465x364
Department FAX 91-836-2464638

Assistant Professors

Bhat, Ramanand 91-836-2447465
    bhatchem@rediffmail.com
Desai, Sudhanva 91-836-2447465
    nivu_sudhu@rediffmail.com
Shivanand, Adaganti 91-836-2447465
    shivanand@sify.com

Accredited by: NBA(National Board of Accreditation)

Graduate advisor:

Department reports to: Principal,Sudhaker Nayak

Placement service: Mr.Nitin Kulkarni

Siddaganga Institute of Technology, Affiliated with Viveswaraiah University

Chemical Engineering Department
Tumkur (Karnataka)
Pin 572103

Department phone 91- 816-2092070
Department FAX 91-816-2282994

Professors

Nirgunababu, P. 91 9880031145

Associate Professors

B.S., Gowrishankar 91 9444190002
    bsgowri@rediffmail.com
K.R.S., Murthy 91 9844327238
Nagaraj, Naveen 91 9880385009
    navee74@yahoo.co.in
Rajaguru, S. 91 8162274167
Shivabasappa, K. L. 91 9845768784
    shivukl@yahoo.com

Assistant Professors

Hegde, Vinayak. M. 91 8162290818
    vinumh@rediffmail.com
Suma, G R

Accredited by: NBA

Degrees granted 2004-2005:
    B.S.: 66   M.S.: 5

Undergraduate advisor: P Nirguna Babu

Student organization: IIChE
Advisor: Aayippa

Department reports to: Dr. M. N. Channabasappa, Principal

Placement service: Naveen Nagaraj, Placement Officer

S. V. National Institute of Technology

Department of Chemical Engineering
Ichchhanath
Surat 395 007

University phone .... 91 261 2223371 - 74
University FAX 91 261 2227334, 2229394
Department phone .... 91 261 2223371 Ext. 332
Department FAX 91 261 2227334

Associate Professors

Murthy, Z. V. P. .... 334 [†]
    zvpm2000@yahoo.com
Parikh, P.A (Head) 331
    parimal_svr@yahoo.co.uk

Assistant Professors

Chakraborty, M. 332
      mousumi_chakra@yahoo.com
Mukhopadhyay, M. 333
      mausumi_mukhopadhyay@yahoo.com

† Faculty numbers are extensions of university main number.

Accredited by: AICTE

Degrees granted 2003-2004:
  B.S.: 60

Undergraduate advisor: Dr. Z.V.P. Murthy

Placement service: Prof. D. B. Naik

Thapar Institute of Engineering and Technology

Department of Chemical Engineering
Thapar Technology Campus
Patiala-147004

University phone 91 175 2393001
University FAX 91 175 2364498
Department phone 91 175 2393063
Department FAX 91 175 2364498

Professors

Bajpai, Pramod K. (Adjunct) 91 175 2393551
      pkbajpai@tiet.ac.in
Jain, S. C. (Visiting) 91 175 2393442
Kumar, Vineet (Head) 91 175 2393063
      vkumar@tiet.ac.in

Associate Professors

Gangacharyulu, D. 91 175 2393305
      dgangacharyulu@tiet.ac.in
Mehta, Rajeev 91 175 2393440
      rmehta@tiet.ac.in
Sinha, Shishir 91 175 2393308
      ssinha@tiet.ac.in

Assistant Professors

Ahuja, Sanjeev 91 175 2393388
      skahuja@tiet.ac.in
Bhunia, H. 91 175 2393120
      hbhunia@tiet.ac.in
Gupta, Raj K. 91 175 2393310
      rkgupta@tiet.ac.in
Yadav, M. K. 91 175 2393309

Accredited by: All India Council For Technical Education, University Grants Commission

Degrees granted 2004-2005:
  B.S.: 43   Ph.D.: 1

Graduate advisor: Dr. D. Gangacharyulu

Undergraduate advisor: Mr. Rajeev Mehta

Student organization: I.I.Ch.E.
Advisor: Dr. Shishir Sinha

Placement service: 91 175 2393005

Vellore Institute of Technology

Room 114, Hexagon Building
Katpadi-Thiruvalam Road
Deemed University
Vellore, Tamil Nadu, 632014

University phone 91-0416-2243091
University FAX 91-0416-2243092
Department phone 91-416-2202523
Department FAX 91-416-2243092

Professors

Pullabhotla, S. R. (Dean) 0416-2202144
      dean.placement@vit.ac.in
Rao, Y.V.C. 0416-2202537
      yvchalapatirao@vit.ac.in

Assistant Professors

David, K.Daniel 0416-2202515
      davidkdaniel@vit.ac.in
Kumar, P. 0416-2202557
      drkumarperumal_vit@yahoo.com
Murthy Shekhar, S. (Head) 0416-2202523
      murthyshekhar@yahoo.co.in

Accredited by: NAAC/New Delhi, NBA, ISO

Graduate advisor: Head/Chemical

Undergraduate advisor: Head/chemical

Student organization: Society of Chemical Engineers

Department reports to: Dr. S. Murthy Shekhar

Placement service: Prof. S. R. Pullabhotla

Indonesia

Institut Teknologi Bandung

Department of Chemical Engineering
Faculty of Industrial Technology
Jalan Ganesha 10
Bandung 40132, Indonesia

University phone 62 22 250 3147
University FAX 62 22 250 0935
Department phone 62 22 250 0989
Department FAX 62 22 250 1438

Professors

Susanto, Herri .... 250 0989 ext 402
      herri@che.itb.ac.id
      EN,MO

Associate Professors

Achmad, Hilman .... 250 0989 ext 422
      ha@che.itb.ac.id
      AM
Adisasmito, Sanggono (Chair) .... 250 0989 ext 237
      gold@bdg.centrin.net.id
      BT,EN,EV
Ariono, Danu .... 250 0989 ext 448
      danu@che.itb.ac.id
      MO,SE

Noezar, Irwan .... 250 0989 ext 310
inoezar@che.itb.ac.id
AM,PO,SE
Purwasasmita, Mubiar .... 250 0989 ext 449
mp@che.itb.ac.id
SE
Sasongko, Dwiwahju (Dean) .... 250 0989 ext 409
sasongko@che.itb.ac.id
EN,MO
Setiadi, Tjandra .... 250 0989 ext 331
tjandra@che.itb.ac.id
BT,EV
Sitompul, Johnner P. .... 250 0989 ext 410
sitompul@che.itb.ac.id
MO,PR
Sjamsuriputra, Achmad A. .... 250 0989 ext 333
aas@che.itb.ac.id
BT
Soerawidjaja, Tatang H. .... 250 0989 ext 411
tatanghs@che.itb.ac.id
EN,PR
Subagjo, ... 250 0989 ext 431
subagjo@che.itb.ac.id
RE
Sukandar, Ukan .... 250 0989 ext 323
usukandar@che.itb.ac.id
BT

Assistant Professors

Adhi, Tri P. .... 250 0989 ext 322
tpadhi@che.itb.ac.id
MO,PR
Argasetya , G. H. .... 250 0989 ext 304
handi@che.itb.ac.id
AM,PR
Bindar, Yazid .... 250 0989 ext 403
yazid@che.itb.ac.id
EN,MO
Laniwati, Melia .... 250 0989 ext 429
melia@che.itb.ac.id
MO,PR,RE
Makertihartha, IGBN .... 250 0989 ext 430
makertia@che.itb.ac.id
MO,RE
Nurdin, Isdiriayani .... 250 0989 ext 437
isdi@che.itb.ac.id
AM,EN
Prakoso, Tirto .... 250 0989 ext 404
tirto@che.itb.ac.id
EN
Trianto, Azis .... 250 0989 ext 436
trianto@che.itb.ac.id
EN,MO,RE

Accredited by: National Accreditation Board of Indonesia

Degrees granted 2005-2006:
    B.S.: 95   M.S.: 17   Ph.D.: 2

Graduate advisor: Susanto, Herri

Undergraduate advisor: Adisasmito, Sanggono

Student organization: HIMATEK ITB
Advisor: Prakoso, Tirto

Department reports to: Faculty of Industrial Technology - Institut Teknologi Bandung

Placement service: Chemical Engineering Department Chairman

Iran

Amirkabir University of Technology(Tehran Polytechnic)

Chemical Engineering Department
No. 424, Hafez Avenue
Tehran, Iran
P.O.Box 15875-4413

Department phone .... (9821) 6499066
Department FAX .... (9821) 6405847

Professors

Dabir, Bahran
dabir@aut.ac.ir
Edrisi, Mohammad
edrisi@aut.ac.ir
Modarress, Hamid M.
hmodares@aut.ac.ir
Rashidi, Fariborz
rashidi@aut.ac.ir
Sohrabi, Morteza
sohrabi@aut.ac.ir
Jamshidi, Esmail
jamshidi@aut.ac.ir
Kaghazchi, Tahereh
kaghazch@aut.ac.ir

Associate Professors

Kalbasi, Mansour
mkabasi@aut.ac.ir

Accredited by:

Graduate advisor:

Undergraduate advisor: Eng. Mohammadhossein Namazi Ghadim

Department reports to:
http://www.aut.ac.ir/departments/chem/1024/index.htm

Placement service: Chem.Eng@aut.ac.ir

Isfahan University of Technology

Department of Chemical Engineering
Isfahan, Iran
Post Code: 8415683111

University phone .... (98) (311) 3912737
University FAX .... (98) (311) 3912862
Department phone .... (98) 311 3915618
Department FAX .... (98) 311 3912677

Associate Professors

Shams, Kayghobad
k_shams@cc.iut.ac.ir
Ehsani, Mohammad R. .... (98) 311 3915607
ehsanimr@cc.iut.ac.ir
Etemad, S.Gholamreza
etemad@cc.iut.ac.ir
Ghoreishi, Seyyed M. 98 311 3915604
ghoreshi@cc.iut.ac.ir
Mehrabani Z., Arjomand
Arjomand@cc.iut.ac.ir

Assistant Professors

Roodpeyma, Shapoor
                          shapoor@cc.iut.ac.ir

Accredited by:

Graduate advisor: Ahmad Moheb

Sharif University of Technology

Chemical and Petroleum Engineering Department
P. O. Box 11365-9465
Azadi Avenue, Tehran

University phone	98-21-6600 5210
Department phone	98-21-6600 5819
Department FAX	98-21-6602 2853

Professors

Alemzadeh, Iran .... 98 21 6616 5486 [†]
    alemzadeh@sharif.edu
Badakhshan, Amir (Visiting) 98 21 66165461
    abadakhs@sharif.edu
Ghotbi, Cyrus (Chair) 98 21 6600 5819
    ghotbi@sharif.edu
Goodarznia, Iraj 98 21 6616 5456
    goodarz@sharif.edu
Kazemeini, Mohammad 98 21 66165425
    kazemeini@sharif.edu
Razavi, Jalil 98 21 66165414
    razavi@sharif.edu
Safe Kordi, Ali A. 98 21 6616 5418
    safekordi@sharif.edu
Shahrokhi, Mohammad 98 21 66165427
    shahrokhi@sharif.edu
Shayegan, Jaleleddin 98 21 66165420
    shayegan@sharif.edu
Soltanieh, Mohammad 98 21 66165417
    soltani@sharif.edu
Vosoughi, Manouchehr 98 21 66165478
    vosoughi@sharif.edu

Associate Professors

Abdekhodaie, Mohammad J. 98 21 6616 5426
    abdmj@sharif.edu
Bastani, Daryoush 98 21 66165409
    bastani@sharif.edu
Borghei, Mehdi 98 21 66165483
    mborghei@sharif.edu
Bozorgmehri, Ramin (Associate Chair) 98 21 6600 5819
    Bozorgmehri@sharif.edu
Farhadi, Fathollah (Associate Chair) 98 21 6600 5819
    farhadi@ sharif.edu
Frounchi, Masoud 98 21 6616 5422
    Frounchi@sharif.edu
Khorasheh, Farhad 98 21 66165411
    khorashe@sharif.edu
Ramazani, Ahmad 98 21 66165431
    ramazani@sharif.edu
Rashtchian, Davood (Associate Chair) 98 21 6600 5819
    rashtchian@sharif.edu
Taghikhani, Vahid 98 21 66165458
    Taghikhani@sharif.edu
Yaghmaei, Soheila (Associate Chair) 98 21 66165428
    yaghmaei@sharif.edu

Assistant Professors

Aghajan, Abdol H. (Emeritus) 98 21 6616 5408
    aghajan@sharif.edu
baghalha, Morteza 98 21 6616 5469
Kazemi, Akhtarelmolouk 98 21 66165484
    a_kazemi@sharif.edu
Molaee dehkordi, Asghar 98 21 6616 5411
Pishvaie, Mohammad R. 98 21 66165429
    Pishvaie@sharif.edu
Roosta Azad, Reza (Associate Chair) 98 21 6600 5417
    roosta@sharif.edu
Shojaee, Akbar 98 21 6616 5429
    shojaee@sharif.edu

[†] Dial department number and ask for extension.

Accredited by:

Degrees granted 2004-2005:
    B.S.: 140   M.S.: 80   Ph.D.: 10

Graduate advisor: Davood Rashtchian

Undergraduate advisor: R. Bozorgmehri

Student organization: Soheila Yaghmaei
Advisor: Soheila Yaghmaei

Placement service: Cyrus Ghotbi

Ireland

Cork Institute of Technology

Department of Chemical Engineering
Rossa Avenue
Cork, Ireland
Deliveries: Mrs. P. Copley

University phone	353-21-4326100
Department phone	353-21-4326885
Department FAX	353-21-4326851

Professors

Cunningham, J. D. 353 21 344864
    cleantechnology@cit.ie
Duffy, N. B. 353 21 324407
    nduffy@cit.ie
O'Shea, J. T. (Head) 353 21 326404
    joshea@cit.ie

Associate Professors

Moroney, P. J.
O'Gorman, A. 353 21 326411
    aogorman@cit.ie
Petrie, G. R. 353 21 326430
    gpetrie@cit.ie
Power, P. C.
Raissian, S. M. 353 21 326430
    mraissian@cit.ie
Ryan, L. B. 353 21 326411
    bryan@cit.ie

Assistant Professors

Doyle, B.
Kennedy, P 353 21 326416
    pkennedy@cit.ie
O'Suilleabhain, C 353 21 326416
    cosuilleabhain@cit.ie

184 Israel

Accredited by: National Council for Educational Awards, Institution of Chemical Engineers (UK), Institution of Engineers of Ireland

Degrees granted 2003-2004:
    B.S.: 29   M.S.: 2   Ph.D.: 1

Graduate advisor: A.O'Gorman

Undergraduate advisor: J. T. O'Shea

Student organization: A.O'Gorman
Advisor: A.O'Gorman

Department reports to: Dr B. J. Murphy, Director

Placement service: J. T. O'Shea

University College Cork

Department of Process & Chemical Engineering
University College Cork
Ireland

University phone 353-21-490 3000
Department phone .... +353-21-490 2389
Department FAX .... +353-21-427 0249

Professors
Oliveira, Fernanda A.R. (Chair) .... +353 21 490 2383
    f.oliveira@ucc.ie

Associate Professors
Oliveira, Jorge .... +353 21 490 2006
    j.oliveira@ucc.ie

Assistant Professors
Byrne, Edmond P. .... +353 21 490 3094
    e.byrne@ucc.ie
Cronin, Kevin 353 21 490 2644
    k.cronin@ucc.ie
de Sousa Gallagher, Maria J. C.F. 353 21 490 3594
    m.desousagallagher@ucc.ie
Fitzpatrick , John J. 353 21 490 3089
    j.fitzpatrick@ucc.ie

Accredited by: Engineers Ireland, Institution of Chemical Engineers (UK)

Degrees granted 2005-2006:
    B.S.: 24   Ph.D.: 3

Graduate advisor:

Student organization: Process & Chemical Engineering Society
Advisor: Dr Edmond Byrne

Department reports to: College of Science, Engineering & Food Science

University College Dublin

School of Chemical & Bioprocess Engineering
College of Engineering, Mathematical and Physical Sciences
University College Dublin
Belfield, Dublin 4, Ireland

University phone 353-1-716 7777
Department phone 353-1-716 1825
Department FAX 353-1-716 1177

Professors
Al-Rubeai, Mohamed 353-1-716 1862
    m.al-rubeai@ucd.ie
MacElroy, J. M. D. (Head) 353-1-716-1824
    don.macelroy@ucd.ie

Associate Professors
Glennon, Brian A. 353-1-716-1954
    brian.glennon@ucd.ie
Kieran, Patricia M. 353-1-716-1956
    patricia.kieran@ucd.ie
MacLoughlin, P. F. 353-1-716-1894
    frank.macloughlin@ucd.ie
Malone, Dermot M. 353-1-716-1895
    dermot.malone@ucd.ie

Assistant Professors
Casey, Eoin 353-1-716 1877
    eoin.casey@ucd.ie
McDonnell, Susan 353-1-716 1893
    susan.mcdonnell@ucd.ie
Mooney, Damian 353-1-716 1827
    damian.mooney@ucd.ie
Stubenrauch, Cosima 353-1-716 1923
    cosima.stubenrauch@ucd.ie

Accredited by: IChemE (UK), IEI (Ireland)

Degrees granted 2004-2005:
    B.S.: 32   Ph.D.: 4

Undergraduate advisor: Colleen Blaney

Student organization: Students' Union

Department reports to: The President

Placement service: Colm Tobin

Israel

Ben-Gurion University of the Negev

Chemical Engineering Department
Beer Sheva 84105
P.O.B. 653

University phone 972-8-6461-111
Department phone 972-8-6461-480
Department FAX 972-8-6472-916

Professors
Apelblat, Alexander .... 6461-487 [†]
    apelblat@bgumail.bgu.ac.il
Gottlieb, Moshe 6461-486
    mosheg@inca.bgu.ac.il
Herskowitz, Mordechai (Dean) 6461-482
    herskow@bgumail.bgu.ac.il
Kost, Joseph 6461-766
    kost@bgumail.bgu.ac.il
Lang, Sidney B. 6461-490
    lang@bgumail.bgu.ac.il
Loeb, Sidney (Emeritus)
    sidloeb@bgumail.bgu.ac.il
Merchuk, Jose C. 6461-768
    jcm@bgumail.bgu.ac.il
Shacham, Mordechai 6461-481
    shacham@bgumail.bgu.ac.il

Tamir, Abraham 6461-765
    atamir@bgumail.bgu.ac.il
Wisniak, Jaime 6461-479
    wisniak@bgumail.bgu.ac.il
Wolf, David (Emeritus) 6461-489
    dwolf@bgumail.bgu.ac.il

Associate Professors

Korin, Eli (Head) 6461-820
    ekorin@bgumail.bgu.ac.il
Landau, Miron 6472-141
    mlandau@bgumail.bgu.ac.il
Regev, Oren 6472-145
    oregev@bgumail.bgu.ac.il
Roy, Ahron (Emeritus) 6461-484
    royaaron@bgumail.bgu.ac.il

Assistant Professors

Barenheim-Groswasser, Anne 6472-129
    bernheim@bgumail.bgu.ac.il
Yerushalmi-Rosen, Rachel 6472-140
    rachely@bgumail.bgu.ac.il

† Direct dial with prefix 972-7.

Accredited by: Israel Council for Higher Education

Graduate advisor: J. Kost

Undergraduate advisor: R. Yerushalmi-Rozen

Department reports to: Yigal Ronen, Dean of Engineering

Technion-Israel Institute of Technology

Chemical Engineering Department
Technion City
Haifa 32000
University phone 972-4-829-2111
Department phone 972-4-829-2820
Department FAX 972-4-829-5672

Professors

Cohen, Yachin (Chair) 972-4-8292010
    yachinc@tx.technion.ac.il
    AM,NT,PO
Grader, Gideon 972-4-8292008
    grader@tx.technion.ac.il
Hasson, David (Emeritus) 972-4-8292936
    hasson@tx.technion.ac.il
    EN,EV,SE,SU
Kehat, Ephraim (Emeritus) 972-4-8292935
    cerekek@tx.technion.ac.il
    EV,MO,NT,SE
Lavie, Ram (Emeritus) 972-4-8292934
    lavie@tx.technion.ac.il
    EN,EV,MO,PR,RE,SE
Lewin, Daniel 972-4-8292006
    dlewin@tx.technion.ac.il
    EN,ME,MO,PR
Marmur, Abraham 972-4-8293088
    marmur@tx.technion.ac.il
    BM,MO,NT,SU
Narkis, Moshe (Emeritus) 972-4-8292937
    narkis@tx.technion.ac.il
    AM,NT,PO
Nir, Avinoam 972-4-8292119
    avinir@tx.technion.ac.il
Pismen, Leonid (Emeritus) 972-4-8293086
    pismen@tx.technion.ac.il
    MO,RE,SU
Ram, Arie (Emeritus) 972-4-8292933
    cerarar@tx.technion.ac.il
Rigbi, Zvi (Emeritus) 972-4-8293089
Semiat, Raphael 972-4-8292009
    cesemiat@tx.technion.ac.il
Sheintuch, Moshe 972-4-8292823
    cermsll@tx.technion.ac.il
    EV,MO,RE
Tadmor, Zehev (Emeritus) 972-4-8293087
    tadmorz@tx.technion.ac.il
Talmon, Yeshayahu 972-4-8292007
    ishi@tx.technion.ac.il
    AM,NT,PO,SU

Associate Professors

Brandon, Simon 972-4-8292822
    cersbsb@tx.technion.ac.il
Orell, Aluf (Emeritus) 972-4-8292020
    orell@tx.technion.ac.il
Paz, Yaron 972-4-8292486
    paz@tx.technion.ac.il
    AM,EN,EV,ME,NT,PO,RE,SU
Tannenbaum, Rina 972-4-8292933
    rinatan@tx.technion.ac.il

Assistant Professors

Bianco-Peled, Havazelet 972-4-8293588
    bianco@tx.technion.ac.il
    AM,BM,NT,PO
Brenner, Naama 972-4-8293578
    nbrenner@tx.technion.ac.il
Haick, Hossam 972-4-8293087
    hhossam@tx.technion.ac.il
    AM,ME,NT,SU
Leshansky, Alexander 972-4-8292119
    lisha@tx.technion.ac.il
    BT,MO,SU
Srebnik, Simcha 972-4-8293584
    simchas@tx.technion.ac.il
    AM,BT,MO,NT,PO,SU
Tsur, Yoed 972-4-8293586
    tsur@tx.technion.ac.il
    AM,EV,ME,NT,SU

Accredited by: Israel Council for Higher Education

Degrees granted 2005-2006:
    B.S.: 81   M.S.: 9   Ph.D.: 3

Undergraduate advisor: Daniel R. Lewin

Department reports to: Prof. Aviv Rosen, Senior Vice President

Italy

Universita di Bologna

DICMA - Dipartimento di Ingegneria Chimica,
Mineraria e delle Tecnologie Ambientali
viale Risorgimento, 2
I-40136 Bologna

University phone	(39) 051-2099349
Department phone	(39) 051-2093135
Department FAX	(39) 051-581200

Professors

Doghieri, Ferruccio .... (39) 051-2090426
ferruccio.doghieri@mail.ing.unibo.it
Gostoli, Carlo .... (39) 051-2093144
carlo.gostoli@mail.ing.unibo.it
Magelli, Franco (Chair) .... (39) 051-2093147
franco.magelli@mail.ing.unibo.it
Santarelli, Francesco .... (39) 051-2093148
francesco.santarelli@mail.ing.unibo.it
Sarti, Giulio C. .... (39) 051-2093142
giulio.sarti@mail.ing.unibo.it
Spadoni, Gigliola .... (39) 051-2093146
gigliola.spadoni@mail.ing.unibo.it

Associate Professors

Bandini, Serena .... (39) 051-2093138
serena.bandini@mail.ing.unibo.it
Camera-Roda, Giovanni .... (39) 051-2093137
giovanni.camera.roda@mail.ing.unibo.it
Cozzani, Valerio .... (39) 051-2093140
valerio.cozzani@mail.ing.unibo.it
Gatta, Alceo .... (39) 051-2093154
alceo.gatta@mail.ing.unibo.it
Nocentini, Massimo .... (39) 051-2090424
massimo.nocentini@mail.ing.unibo.it
Pasquali, Gabriele .... (39) 051-2093139
gabriele.pasquali@mail.ing.unibo.it
Pinelli, Davide .... (39) 051-2090427
davide.pinelli@mail.ing.unibo.it
Stramigioli, Carlo .... (39) 051-2093153
carlo.stramigioli@mail.ing.unibo.it

Assistant Professors

Bonvicini, Sarah .... (39) 051-2093141
sarah.bonvicini@mail.ing.unibo.it
Giacinti Baschetti, Marco .... (39) 051-2090408
marco.giacinti@mail.ing.unibo.it
Leonelli, Paolo .... (39) 051-2093150
paolo.leonelli@mail.ing.unibo.it
Montante, Giusi .... (39) 051-2090406
giusi.montante@mail.ing.unibo.it

Department reports to: F. Magelli, Chairman

Universita di Cagliari

Dipartimento di Ingegneria Chimica e Materiali
Piazza d'Armi
I-09123 Cagliari

University phone	390706751
Department phone	390706755064
Department FAX	390706755067

Professors

Baratti, Roberto (Chair) 390706755056
baratti@dicm.unica.it
MO,PR,RE
Cao, Giacomo 390706755058
cao@dicm.unica.it
AM,BT,MO
Massidda, Luigi 390706755080
massidda@dicm.unica.it
AM
Polcaro, Anna M. 390706755059
polcaro@dicm.unica.it
RE
Sanna, Ulrico 390706755063
sanna@dicm.unica.it
AM

Associate Professors

Carta, Renzo 390706755068
carta@dicm.unica.it
RE
Dernini, Stella 390706755077
dernini@dicm.unica.it
Lallai, Antonio 390706755060
lallai@dicm.unica.it
EV
Mura, Giampaolo 390706755051
mura@dicm.unica.it
EV,MO
Orru', Roberto 390706755076
orru@dicm.unica.it
AM
Tola, Giuseppe 390706755074
tola@dicm.unica.it
PR
Usai, Giorgio 390706755061
usai@dicm.unica.it

Assistant Professors

Cincotti, Alberto 390706755066
cincotti@dicm.unica.it
AM,MO
Grosso, Massimiliano 390706755075
grosso@dicm.unica.it
MO,PR
Locci, Antonio 390706755066
locci@dicm.unica.it
AM,MO
Mascia, Michele 390706755066
mmascia@dicm.unica.it
PR,RE
Meloni, Paola 390706755077
meloni@dicm.unica.it
Palmas, Simona 390706755069
spalmas@dicm.unica.it
RE
Pilloni, Giovanni 390706755072
pilloni@dicm.unica.it
Tronci, Stefania 390706755073
tronci@dicm.unica.it
MO

Accredited by:

Degrees granted 2005-2006:
  B.S.: 11   M.S.: 4   Ph.D.: 3

Universita de L'Aquila

Dipartimento di Chimica
Ingegneria Chimica e Materiali
I-67040 Monteluco Di Roio - L'Aquila
University phone .... (39) 0862-43-4002
Department phone .... (39) 862-43-4214
Department FAX .... (39) 862-43-4203

Professors

Andruzzi, Romano .... (39) 0862-434216
    nike@ing.univaq.it
Barba, Diego .... (39) 0862-434238
    moscalia@ing.univaq.it
Brandani, Vincenzo .... (39) 0862-434229
    brandani@ing.univaq.it
Cantarella, Maria .... (39) 0862-434215
    cantarel@ing.univaq.it
Del Re, Giovanni .... (39) 0862-434219
    delre@ing.univaq.it
Di Giacomo, Gabriele .... (39) 0862-434225
    digiacom@ing.univaq.it
Foscolo, Pier U. (Associate Dean) .... (39) 0862-434214
    foscolo@ing.univaq.it
Gibilaro, Giovanni L. .... (39) 0862-434214
    gibilaro@ing.univaq.it
Inesi, Achille .... (39) 0862-434246
    inesi@ing.univaq.it
Pelino, Mario .... (39) 0862-434224
    pelino@ing.univaq.it
Scoccia, Giancarlo .... (39) 0862-434222
    scoccia@ing.univaq.it
Villa, Pierluigi .... (39) 0862-434245
    villa@ing.univaq.it
Volpe, Roberto (Head) .... (39) 0862-434222
    volpe@ing.univaq.it

Associate Professors

Cantalini, Carlo .... (39) 0862-434233
    canta@ing.univaq.it
Evangelista, Franco .... (39) 0862-434243
    frev@ing.univaq.it
Fantauzzi, Felice .... (39) 0862-434216
    felice@ing.univaq.it
Fumarola, Giuseppe .... (39) 0862-434242
    fumarola@ing.univaq.it
Gallifuoco, Alberto .... (39) 0862-434232
    gallifuo@ing.univaq.it
Germanà, Antonino .... (39) 0862-434214
    germana@ing.univaq.it
Pajewsky, Leonardo .... (39) 0862-434227
    pajewsky@ing.univaq.it
Quaresima, Raimondo .... (39) 0862-434226
    raimondo@ing.univaq.it
Rossi, Leucio .... (39) 0862-434246
    rossil@ing.univaq.it
Vegliò, Francesco .... (39) 0862-434223
    veglio@ing.univaq.it

Assistant Professors

Beolchini, Francesca .... (39) 0862-434236
    beolffra@ing.univaq.it
Jand, Nader .... (39) 0862-434237
    nader@ing.univaq.it
Mucciante, Vittoria .... (39) 0862-434216
    nike@ing.univaq.it
Prisciandaro, Marina .... (39) 0862-434241
    maripri@ing.univaq.it
Taglieri, Giuliana .... (39) 0862-434227
    taglieri@ing.univaq.it

Graduate advisor: Prof. Giovanni Del Re

Undergraduate advisor: Prof. Giovanni Del Re

Student organization: Dr. Katia Gallucci
Advisor: Dr. Nader Jand

Department reports to: Prof. Pier Ugo Foscolo, Director of Studies

Placement service: Prof. Alberto Gallifuoco

Universita di Pisa

Dipartimento di Ingegneria Chimica,
Chimica Industriale e Scienza dei Materiali
Via Diotisalvi, 2
56126 Pisa
Italy
University phone 39-050-2212-111
Department phone 39-050-511-111
Department FAX 39-050-511-266

Professors

Brambilla, Alessandro .... 39-050-511-243 [†]
    a.brambilla@ing.unipi.it
Butta, Enzo (Emeritus) 39-050-511-203
Giusti, Paolo 39-050-511-287
    p.giusti@ing.unipi.it
Levita, Giovanni (Associate Chair) 39-050-511-201
    levita@ing.unipi.it
Magagnini, Pierluigi (Chair) 39-050-511-222
    Magagnini@ing.unipi.it
Marconi, Pierfilippo 39-050-511-211
    marconi@ing.unipi.it
Nencetti, Gianfranco 39-050-511-215
    g.nencetti@ing.unipi.it
Scali, Claudio 39-050-511-241
    scali@ing.unipi.it
Tartarelli, Roberto 39-050-511-256
    r.tartarelli@ing.unipi.it
Tognotti, Leonardo 39-050-511-240
    tognotti@ing.unipi.it
Zanelli, Severino 39-050-511-212
    zanelli@ing.unipi.it

Associate Professors

Bartolozzi, Mauro 39-050-511209
    m.bartolozzi@ing.unipi.it
Davini, Paolo 39-050-511-216
    davini@ing.unipi.it
De Sanctis, Massimo 39-050-511-227
    desanctis@ing.unipi.it
Di Maina, Marcello 39-050-511-111
Lazzeri, Andrea 39-050-511-207
    a.lazzeri@ing.unipi.it
Lazzeri, Luigi 39-050-511-277
    lu.lazzeri@ing.unipi.it
Lupinacci, Domenico 39-050-511-231
    d.lupinacci@ing.unipi.it
Marchetti, Augusto 39-050-511-207
    marchetti@ing.unipi.it
Mauri, Roberto 39-050-511-248
    r.mauri@ing.unipi.it
Nicolella, Cristiano 39-050-511294
    c.nicolella@ing.unipi.it

Paci, Massimo	39-050-511-230
	m.paci@ing.unipi.it
Petarca, Luigi	39-050-511-224
	petarca@ing.unipi.it
Solina, Adriano	39-050-511-257
	solina@ing.unipi.it
Vitolo, Sandra	39-050-511-278
	vitolo@ing.unipi.it

Assistant Professors

Bresci, Bruno	39-050-511-278
Brunazzi, Elisabetta	39-050-511-213
	e.brunazzi@ing.unipi.it
Cascone, Maria G.	39-050-511-221
	mg.cascone@ing.unipi.it
Polacco, Giovanni	39-050-511-220
	g.polacco@ing.unipi.it
Rizzo, Cosimo	39-050-511-204
	c.rizzo@ing.unipi.it
Seggiani, Maurizia	39-050-511-281
	m.seggiani@ing.unipi.it
Tricoli, Vincenzo	39-050-511-246
	v.tricoli@ing.unipi.it
Valentini, Renzo	39-050-511-259
	r.valentini@ing.unipi.it
Vatistas, Nicolaos	39-050-511-239
	vatistas@ing.unipi.it

† Direct dial with prefix 39-50.

Accredited by:

Degrees granted 2003-2004:

M.S.: 112   Ph.D.: 2

Graduate advisor: Prof. L. Tognotti

Undergraduate advisor: Prof. R. Mauri

University of Salerno

Dipartimento di Ingegneria Chimica e Alimentare
Via Ponte Don Melillo
I-84084 Fisciano (SA)
Deliveries: Prof. Ernesto Reverchon

University phone	39 089 964185
Department phone	39 089 964185
Department FAX	39 089 964 057

Professors

Ciambelli, Paolo	39 089 96 4151
	pciambelli@unisa.it
D'Amore, Matteo	39 089 96 4136
	damore@dica.unisa.it
Di Matteo, Marisa	39 089 96 4131
	mdimatteo@unisa.it
Donsi', Giorgio	39 089 69 4150
	donsi@unisa.it
Ferrari, Giovanna	39 089 96 4134
	gferrari@unisa.it
Iannelli, Pio	39 089 96 4103
	piannelli@unisa.it
Reverchon, Ernesto (Director)	39 089 96 4116
	ereverchon@unisa.it
Titomanlio, Giuseppe	39 089 96 4152
	gtitomanlio@unisa.it
Vittoria, Vittoria	39 089 96 4114
	vittoria@dica.unisa.it

Associate Professors

Dovinola, Vincenzo	39 089 96 4146
	vdovinola@unisa.it
Incarnato, Loredana	39 089 96 4144
	lincarnato@unisa.it
Miccio, Michele	39 089 96 4148
	mmiccio@unisa.it
Nobile, Maria R.	39 089 96 4143
	mrnobile@unisa.it
Palma, Vincenzo	39 089 96 4147
	vpalma@unisa.it
Parascandola, Palma	39 089 96 4078
	pparascandola@unisa.it
Poletto, Massimo	39 089 96 4132
	mpoletto@unisa.it
Romano, Vittorio	39 089 96 4079
	vromano@unisa.it
Sesti Osseo, Libero	39 089 96 4133
	lsestiosseo@unisa.it
Vaccaro, Salvatore	39 089 96 4153
	svaccaro@unisa.it

Assistant Professors

Albanese, Donatella	39 089 96 4129
	dalbanese@unisa.it
Barletta, Diego	39 089 96 4014
	dbarletta@unisa.it
Caputo, Giuseppe	39 089 96 4091
	gcaputo@unisa.it
Concilio, Simona	39 089 96 4019
	sconcilio@unisa.it
Di Maio, Luciano	39 089 96 4102
	ldimaio@unisa.it
Gorrasi, Giuliana	39 089 96 4115
	ggorrasi@unisa.it
Guadagno, Liberata	39 089 96 4115
	lguadagno@unisa.it
Lamberti, Gaetano	39 089 96 4026
	glamberti@unisa.it
Marra, Francesco	39 089 96 4064
	fmarra@unisa.it
Pantani, Roberto	39 089 96 4141
	rpantani@unisa.it
Russo, Paola	39 089 96 4027
	parusso@unisa.it
Sannino, Diana	39 089 96 4027
	sannino@dica.unisa.it
Scarfato, Paola	39 089 96 3404
	pscarfato@unisa.it

Accredited by:

Degrees granted 2004-2005:

B.S.: 30   M.S.: 37   Ph.D.: 4

Undergraduate advisor: Prof. G. Titomanlio

Japan

Fukuoka University

Department of Chemical Engineering
Fukuoka University
8-19-1 Nanakuma, Jonan-ku,
Fukuoka 814-0180

University phone 81-92-871-6631
University FAX 81-92-865-6031
Department phone 81-92-871-6631
Department FAX 81-92-865-6031

Professors
Ide, Mitsuharu
Ishikura, Toshifumi 8-92-871-6631
    ishikura@fukuoka-u.ac.jp
Kariyasaki, Akira
Morooka, Shigeharu (Chair) .... 8-92-871-6631 ex 6423
    smorooka@fukuoka-u.ac.jp
Nakahara, Shunsuke
Nakano, Katsuyuki
Tanaka, Ryuichi

Associate Professors
Kato, Takafumi
Mishima, Kenji
Shibata, Hiromichi 8-92-871-6631
    hshibata@fukuoka-u.ac.jp

Accredited by: JABEE

Degrees granted 2004-2005:
    B.S.: 138   M.S.: 1

Department reports to: S. Morooka

Placement service: T. Kato

Hiroshima University

Department of Chemical Engineering
Graduate School of Engineering
Kagamiyama 1-4-1
Higashi-Hiroshima, 739-8527

Department phone .... (81)82-424-7718
Department FAX .... (81)82-424-5494

Professors
Asaeda, Masashi .... (81)82-424-7719
    asaeda@hiroshima-u.ac.jp
Okada, Mitsumasa 24-5526
    mokada@hiroshima-u.ac.jp
Okuyama, Kikuo (Chair) .... (81)82-424-7716
    okuyama@hiroshima-u.ac.jp
Sakohara, Shuji .... (81)82-424-7720
    sakohara@hiroshima-u.ac.jp
Takishima, Shigeki .... (81)82-424-7713
    r736735@hiroshima-u.ac.jp
Yoshida, Hideto .... (81)82-424-7853
    r736619@hiroshima-u.ac.jp

Associate Professors
Iizawa, Takashi .... (81)82-424-7711
    tiizawa@hiroshima-u.ac.jp
Isomoto, Yoshinori .... (81)82-424-7845
    iyoshi@hiroshima-u.ac.jp
Kitamura, Mitsutaka .... (81)82-424-7715
    mkitamu@hiroshima-u.ac.jp
Nishijima, Wataru 24-6199
    wataru@hiroshima-u.ac.jp
Shimada, Manabu .... (81)82-424-7717
    smd@hiroshima-u.ac.jp
Takimoto, Kazuto .... (81)82-424-7621
    takimoto@hiroshima-u.ac.jp
Tsuru, Toshinori .... (81)82-424-7714
    tsuru@hiroshima-u.ac.jp

Assistant Professors
Fukui, Kunihiro .... (81)82-424-7853
    kfukui@hiroshima-u.ac.jp
Gotoh, Takehiko .... (81)82-424-7720
    tgoto@hiroshima-u.ac.jp
Lenggoro, Wuled .... (81)82-424-7850
    wld@hiroshima-u.ac.jp
Mukai, Tetsuo .... (81)82-424-7626
    tmukai@hiroshima-u.ac.jp
Nakano, Yoichi 24-6195
    ynakano@hiroshima-u.ac.jp
Okuda, Tetsuji 24-6197
    aqua@hiroshima-u.ac.jp
Tokuyama, Hideaki .... (81)82-424-7720
    tokuyama@hiroshima-u.ac.jp
Tsumura, Toshinori .... (81)82-424-7723
    ttsumura@hiroshima-u.ac.jp
Yabuki, Akihiro .... (81)82-424-7852
    ayabuki@hiroshima-u.ac.jp
Yang, Kwan-sik .... (81)82-424-7853
    yangks@hiroshima-u.ac.jp
Yoshioka, Tomohisa .... (81)82-424-7719
    tom@hiroshima-u.ac.jp

Accredited by:

Degrees granted 2003-2004:
    B.S.: 34   M.S.: 31   Ph.D.: 11

Kanazawa University

Chemistry and Chemical Engineering Department
Kakuma-machi
Kanazawa, 920-1192

University phone 076-264-5111
Department phone 076-234-4830
Department FAX 076-234-4829

Professors
Hayashi, Yoshishige 234-4806
    yohayasi@t.kanazawa-u.ac.jp
Kanoh, Shigeyoshi 234-4782
    kanoh@t.kanazawa-u.ac.jp
Komura, Teruhisa (Emeritus) 234-4770
    komura@t.kanazawa-u.ac.jp
Kunimoto, Kouki 264-5805
    lee@t.kanazawa-u.ac.jp
Miyagishi, Shigeyoshi (Head) 234-4764
    miyagisi@t.kanazawa-u.ac.jp
Mori, Shigeru 234-4825
    smori@t.kanazawa-u.ac.jp
Motoi, Masatoshi 234-4781
    motoi@t.kanazawa-u.ac.jp
Nakamoto, Yoshiaki 234-4774
    nakamoto@t.kanazawa-u.ac.jp

Nitta, Koh-hei 234-4818
  nitta@t.kanazawa-u.ac.jp
Ohta, Tatsuhiko 234-4803
  tohta@t.kanazawa-u.ac.jp
Otani, Yoshio 234-4813
  otani@t.kanazawa-u.ac.jp
Senda, Hitoshi 264-5802
  senda@t.kanazawa-u.ac.jp
Shimizu, Nobuaki 234-4807
  nshimizu@t.kanazawa-u.ac.jp
Takahashi, Koushin 234-4771
  ktakaha@t.kanazawa-u.ac.jp
Tamura, Kazuhiro (Director) 234-4804
  tamura@t.kanazawa-u.ac.jp
Ueda, Kazumasa 234-4791
  kueda@t.kanazawa-u.ac.jp
Yamada, Toshiro 234-4802
  tyamada@t.kanazawa-u.ac.jp

Associate Professors
Asakawa, Tsuyoshi 234-4765
  asakawa@t.kanazawa-u.ac.jp
Hasegawa, Hiroshi 234-4792
  hhiroshi@t.kanazawa-u.ac.jp
Kawanishi, Takuya 234-4809
  kawanisi@t.kanazawa-u.ac.jp
Kumita, Mikio 234-4827
  kumita@t.kanazawa-u.ac.jp
Nakamura, Yoshitoshi 234-4819
  ynakamu@t.kanazawa-u.ac.jp
Segi, Masahito 234-4787
  segi@t.kanazawa-u.ac.jp
Takahashi, Kenji 234-4828
  ktkenji@t.kanazawa-u.ac.jp
Yamagishi, Tadaaki 234-4776
  yamagisi@t.kanazawa-u.ac.jp

Accredited by: JABEE

Degrees granted 2005-2006:
    B.S.: 52   M.S.: 25   Ph.D.: 5

Department reports to: Chairman of Department

Placement service: Chairman of Department

Kansai University

Faculty of Engineering
Chemical Engineering Department
Suita Osaka 564-8680
University phone 81-6-6368-1121
Department phone 81-6-6368-1121
Department FAX 81-6-6388-8869

Professors
Miyake, Takanori .... 81-6-6368-0918 [†]
  tmiyake@ipcku.kansai-u.ac.jp
Miyake, Yoshikazu 81-6-6368-0950
  ymiyake@kansai-u.ac.jp
Muroyama, Katsuhiko 81-6-6368-0945
  muroyama@kansai-u.ac.jp
Oda, Hirokazu 81-6-6368-0808
  oda@ipcku.kansai-u.ac.jp
Okada, Yoshiki 81-6-6368-0868
  yokada@ipcku.kansai-u.ac.jp
Shibata, Junji (Chair) 81-6-6368-0856
  shibata@kansai-u.ac.jp
Suzuki, Toshimitsu 81-6-6368-0865
  tsuzuki@ipcku.kansai-u.ac.jp
Yamamoto, Hideki 81-6-6368-0972
  yhideki@kansai-u.ac.jp

Associate Professors
Hayashi, Junichi 81-6-6368-0913
  hayashi7@ipcku.kansai-u.ac.jp
Ikenaga, Naiki 81-6-6368-0792
  ikenaga@ipcku.kansai-u.ac.jp
Iyoki, Shigeki 81-6-6368-0786
  iyoki@ipcku.kansai-u.ac.jp

[†] Dial 81-6-368-1121 and ask for extension.

Accredited by:

Graduate advisor: Junji Shibata

Undergraduate advisor: Junji Shibata

Department reports to: Junji Shibata

Kyoto University

Department of Chemical Engineering
Faculty of Engineering
Kyoto University Katsura, Nishikyo-ku
Kyoto, 615-8510
University phone 81-75-753-7531
Department phone 81-75-383-2071
Department FAX 81-75-383-2078

Professors
Adachi, Motonari (Joint) 81-774-38-3518
  adachi@iae.kyoto-u.ac.jp
Harada, Makoto (Emeritus)
Hasebe, Shinji 81-75-383-2667
  hasebe@cheme.kyoto-u.ac.jp
Hashimoto, Iori (Emeritus)
  iori@cheme.kyoto-u.ac.jp
Hashimoto, Kenji (Emeritus)
Higashitani, Ko (Head) 81-75-383-2662
  k_higa@cheme.kyoto-u.ac.jp
Mae, Kazuhiro 81-75-383-2668
  kaz@cheme.kyoto-u.ac.jp
Masuda, Hiroaki 81-75-383-2665
  masuda@cheme.kyoto-u.ac.jp
Miura, Kouichi 81-75-383-2663
  miura@cheme.kyoto-u.ac.jp
Ogino, Fumimaru (Emeritus)
Ohshima, Masahiro 81-75-383-2666
  oshima@cheme.kyoto-u.ac.jp
Okazaki, Morio (Emeritus)
Sada, Eizo (Emeritus)
Takamatsu, Takeichiro (Emeritus)
Tamon, Hajime 81-75-383-2664
  tamon@cheme.kyoto-u.ac.jp
Tanigaki, Masataka (Joint) 81-75-753-5566
  tanigaki@cheme.kyoto-u.ac.jp
Yoshida, Fumitake (Emeritus)

Associate Professors
Kano, Manabu 81-75-383-2687
  kano@cheme.kyoto-u.ac.jp
Kawase, Motoaki 81-75-383-2683
  kawase@cheme.kyoto-u.ac.jp
Kinoshita, Masahiro (Joint) 81-774-38-3503
  kinoshit@iae.kyoto-u.ac.jp

Maruyama, Toshiro 81-75-383-2631
maruyama@cheme.kyoto-u.ac.jp
Matsusaka, Shuji 81-75-383-2685
matsu@cheme.kyoto-u.ac.jp
Miyahara, Minoru 81-75-383-2682
miyahara@cheme.kyoto-u.ac.jp
Mukai, Shin 81-75-383-2684
mukai@cheme.kyoto-u.ac.jp

Accredited by:

Degrees granted 2003-2004:
B.S.: 51   M.S.: 25   Ph.D.: 4

Kyushu University

Department of Chemical Engineering
Faculty of Engineering
744 Motooka, Nishi-ku
Fukuoka, 819-0395

University phone 81-92-642-2111
Department phone 81-92-802-2801
Department FAX 81-92-802-2800

Professors

Arai, Yasuhiko 81-92-802-2741
arai@chem-eng.kyushu-u.ac.jp
Fukai, Jun 81-92-802-2744
jfukai@chem-eng.kyushu-u.ac.jp
Funatsu, Kazumori (Emeritus)
Goto, Masahiro 81-92-802-2806
mgototcm@mbox.nc.kyushu-u.ac.jp
Kajiwara, Toshihisa 81-92-802-2746
kajiwara@chem-eng.kyushu-u.ac.jp
Kamihira, Masamichi 81-92-802-2743
kamihira@chem-eng.kyushu-u.ac.jp
Kawakami, Koei 81-92-802-2748
kawakami@chem-eng.kyushu-u.ac.jp
Kishida, Masahiro 81-92-802-2742
kishida@chem-eng.kyushu-u.ac.jp
Kusunoki, Koichiro (Emeritus)
Matsuyama, Hisayoshi (Emeritus)
Minemoto, Masaki 81-92-802-2745
minemoto@chem-eng.kyushu-u.ac.jp
Miyatake, Osamu (Emeritus)
Miyazaki, Noriyuki (Emeritus)
Morooka, Shigeharu (Emeritus)
Nakashio, Fumiyuki (Emeritus)
Tsuge, Yoshifumi 81-92-802-2747
tsuge@chem-eng.kyushu-u.ac.jp
Wakabayashi, Katsuhiko (Emeritus)

Associate Professors

Ijima, Hiroyuki 81-92-802-2758
ijima@chem-eng.kyushu-u.ac.jp
Ito, Akira 81-92-802-2753
akira@chem-eng.kyushu-u.ac.jp
Iwai, Yoshio 81-92-802-2751
iwai@chem-eng.kyushu-u.ac.jp
Kamiya, Noriho 81-92-802-2807
noritcm@mbox.nc.kyushu-u.ac.jp
Matsukuma, Yosuke 81-92-802-2755
ymatsu@chem-eng.kyushu-u.ac.jp
Mizumoto, Hiroshi 81-92-802-2759
hiroshi@chem-eng.kyushu-u.ac.jp
Takenaka, Sakae 81-92-802-2752
takenaka@chem-eng.kyushu-u.ac.jp
Yamamoto, Tsuyoshi 81-92-802-2754
yamamoto@chem-eng.kyushu-u.ac.jp

Accredited by:

Degrees granted 2005-2006:
B.S.: 44   M.S.: 35   Ph.D.: 3

Nagoya University

Department of Chemical Engineering
Furo-cho, Chikusa-ku
Nagoya, 464-8603

University phone 052-781-5111
Department phone 052-789-4822
Department FAX 052-789-3180

Professors

Goto, Shigeo (Emeritus)
Hasatani, Masanobu (Emeritus)
Horizoe, Hirotoshi 052-789-3618
horizoe@nuce.nagoya-u.ac.jp
Iritani, Eiji (Chair) 052-789-3374
iritani@nuce.nagoya-u.ac.jp
SE
Koda, Shinobu 052-789-3275
koda@nuce.nagoya-u.ac.jp
Kurimoto, Hidekazu 052-789-4757
kurimoto@nuce.nagoya-u.ac.jp
Matsubara, Masakazu (Emeritus)
Matsuda, Hitoki 052-789-3382
matsuda@nuce.nagoya-u.ac.jp
Miyahara, Yutaka (Emeritus)
Mori, Shigekatsu (Emeritus)
Morisue, Toshiya (Emeritus)
Murase, Toshiro (Emeritus)
Nakamura, Masaaki (Emeritus)
Nomura, Hiroyasu (Emeritus)
Onogi, Katsuaki 052-789-3263
onogi@nuce.nagoya-u.ac.jp
Shirato, Mompei (Emeritus)
Sugiyama, Sachio (Emeritus)
Suzuki, Kenji 052-789-5537
k-suzuki@esi.nagoya-u.ac.jp
Tagawa, Tomohiko 052-789-3388
tagawa@park.nuce.nagoya-u.ac.jp
Takahashi, Katsuroku (Emeritus)
Takeuchi, Hiroshi (Emeritus)
Tanabe, Yasuhiro 052-789-3377
y.tanabe@nuce.nagoya-u.ac.jp
Toyama, Shigeki (Emeritus)
Tsubaki, Junichiro 052-789-3096
tsubaki@nuce.nagoya-u.ac.jp

Associate Professors

Bando, Yoshiyuki 052-789-3622
bando@nuce.nagoya-u.ac.jp
Deguchi, Seiichi 052-789-2733
deguchi@nuce.nagoya-u.ac.jp
Hashizume, Susumu 052-789-3594
hashi@nuce.nagoya-u.ac.jp
Itaya, Yoshinori 052-789-3378
yitaya@nuce.nagoya-u.ac.jp
Kobayashi, Noriyuki 052-789-3383
koba@nuce.nagoya-u.ac.jp
Kojima , Yoshihiro 052-789-3912
ykojima@esi.nagoya-u.ac.jp

Matsuoka, Tatsuro 052-789-3274
　　　matsuoka@nuce.nagoya-u.ac.jp
Mukai, Yasuhito 052-789-3375
　　　mukai@nuce.nagoya-u.ac.jp
　　　SE
Nii, Susumu 052-789-3390
　　　nii@nuce.nagoya-u.ac.jp
Yasuda, Keiji 052-789-3623
　　　yasuda@nuce.nagoya-u.ac.jp

Accredited by:

Niigata University

Department of Chemistry and Chemical Engineering
Ikarashi 2-8050
Niigata ZC 950-2181

University phone..................... 81 25 262 7000
University FAX....................... 81 25 262 6539
Department phone 81 25 262 6781
Department FAX 81 25 262 7010

Professors
Ohkawa, Akira .... x6786 †
　　　biochem@eng.niigata-u.ac.jp
Taguchi, Yoji (Head) x6787
　　　taguchi1@eng.niigata-u.ac.jp
Tanaka, Masato x6791
　　　tanaka@eng.niigata-u.ac.jp
Watanabe, Atsuo x7958
　　　watanabe@gs.niigata-u.ac.jp
Yamagiwa, Kazuaki x6785
　　　yamagiwa@eng.niigata-u.ac.jp

Associate Professors
Hotta, Noriyasu x6782
　　　nhotta@eng.niigata-u.ac.jp
Ito, Akira x6789
　　　aito@eng.niigata-u.ac.jp
Kimura, Isao x7194
　　　ikim@eng.niigata-u.ac.jp
Shimizu, Tadaaki x6783
　　　tshimizu@eng.niigata-u.ac.jp

Assistant Professors
Taguchi, Yoshinari x6784
　　　puchi@eng.niigata-u.ac.jp
Yoshida, Masanori x7343
　　　myoshida@eng.niigata-u.ac.jp

† Dial 025-262-6098 and ask for extension.

Accredited by: JABEE

Osaka University

Division of Chemical Engineering
Department of Materials Engineering Science
Graduate School of Engineering Science
Toyonaka
Osaka 560-8531

University phone......................06-6877-5111
University FAX........................06-6879-7106
Department phone 06-6850-6295
Department FAX 06-6850-6296

Professors
Fueno, Takayuki (Emeritus)
Hirata, Yushi (Emeritus)
Imanaka, Toshinobu (Emeritus)
Inoue, Yoshiro (Chair) 81-6-6850-6276
　　　inoue@cheng.es.osaka-u.ac.jp
　　　TP
Ito, Ryuzo (Emeritus)
Kaneda, Kiyotomi 81-6-6850-6260
　　　kaneda@cheng.es.osaka-u.ac.jp
Katayama, Takashi (Emeritus)
Komasawa, Isao (Emeritus)
Kuboi, Ryoichi 81-6-6850-6285
　　　kuboi@cheng.es.osaka-u.ac.jp
Kunugita, Eiichi (Emeritus)
Nakano, Masayoshi 81-6-6850-6268
　　　mnaka@cheng.es.osaka-u.ac.jp
Nitta, Tomoshige (Emeritus) 81-6-6850-6265
　　　nitta@cheng.es.osaka-u.ac.jp
Ohgaki, Kazunari 81-6-6850-6290
　　　ohgaki@cheng.es.osaka-u.ac.jp
Otake, Tsutao (Emeritus)
Taya, Masahito 81-6-6850-6251
　　　taya@cheng.es.osaka-u.ac.jp
Tone, Setsuji (Emeritus)
Ueyama, Korekazu 81-6-6850-6255
　　　ueyama@cheng.es.osaka-u.ac.jp

Associate Professors
Ebitani, Kohki 81-6-6850-6262
　　　ebitani@cheng.es.osaka-u.ac.jp
Egashira, Yasuyuki 81-6-6850-6259
　　　egashira@cheng.es.osaka-u.ac.jp
Kino-oka, Masahiro 81-6-6850-6253
　　　kino-oka@cheng.es.osaka-u.ac.jp
Nishiyama, Norikazu 81-6-6850-6256
　　　nisiyama@cheng.es.osaka-u.ac.jp
Sato, Hiroshi 81-6-6850-6291
　　　hsato@cheng.es.osaka-u.ac.jp
Shiraishi, Yasuhiro 81-6-6850-6271
　　　shiraish@cheng.es.osaka-u.ac.jp
Takahashi, Hideaki 81-6-6850-6267
　　　takahasi@cheng.es.osaka-u.ac.jp
Umakoshi, Hiroshi 81-6-6850-6287
　　　umakoshi@cheng.es.osaka-u.ac.jp

Accredited by:

Graduate advisor: Inoue, Yoshiro

Undergraduate advisor: Inoue, Yoshiro

Placement service: Chairman, Division of Chemical Engineering

Seikei University

Department of Applied Chemistry
Seikei University
Musashino, Tokyo 180-8633
Deliveries: Prof. T. Kojima

University phone...................... 81-422-37-3705
Department phone 81-422-37-3750
Department FAX 81-422-37-3871

Professors
Higuchi, Akon 0422-37-3748
　　　higuchi@ch.seikei.ac.jp

Katoh, Akira 0422-37-3747
    katoh@ch.seikei.ac.jp
Kojima, Toshinori 0422-37-3750
    kojima@ch.seikei.ac.jp
Kurita, Keisuke 0422-37-3745
    kurita@ch.seikei.ac.jp
Morita, Makoto 0422-37-3749
    morita@ch.seikei.ac.jp
Ozaki, Yoshiharu 0422-37-3751
    ozaki@ch.seikei.ac.jp
Totani, Yoichiro 0422-37-3744
    totani@ch.seikei.ac.jp
Tsubomura, Taro (Head) 0422-37-3752
    tsubo@ch.seikei.ac.jp

Associate Professors

Hara, Setsuko 0422-37-3753
    shara@ch.seikei.ac.jp

Accredited by: Ministry of Education Science and Culture

Degrees granted 2003-2004:
    B.S.: 92   M.S.: 21   Ph.D.: 1

Graduate advisor: Prof. A. Higuchi

Undergraduate advisor: Prof. A. Katoh

University of Tokyo

Department of Chemical System Engineering
Hongo, Bunkyo-ku, Tokyo 113-8656
University phone 03-3812-2111
Department FAX 03-5841-7321

Professors

Hirano, Toshisuke (Emeritus)
Iizuka, Yoshinori 5841-7298
Inoue, Hakuai (Emeritus)
Kawazoe, Kunitaro (Emeritus)
Kimura, Shoji (Emeritus)
Koda, Seiichiro (Emeritus)
Komiyama, Hiroshi 5841-7323
Koshi, Mitsuo 5841-7295
Kume, Hitoshi (Emeritus)
Kunii, Daizo (Emeritus)
Nakao, Shin-ichi (Chair) 5841-7326
Nishimura, Hajime (Emeritus)
Oshima, Yoshito 5841-3027
Sadakata, Masayoshi 5841-7344
Sakoda, Akiyoshi 5452-6350
Sekizawa, Ai 5841-7287
Takahashi, Hiroshi 5841-7751
Tamura, Masamitsu 5841-7291
Yamaguchi, Yukio 5841-7303
Yamamoto, Kenji 5841-7303
Yamashita, Koichi 5841-7228
Yanagisawa, Yukio 5841-7324
Yoshida, Kunio (Emeritus)

Associate Professors

Akutsu, Yoshiaki 5841-7384
Arai, Mitsuru 5841-7292
Asai, Yoshihiro 029-861-5373
Dobashi, Ritsu 5841-7304
Hirao, Masahiko 5841-7343
Hori, Keiichi 0427-59-8283
Miyoshi, Akira 5841-7296
Okada, Fumio 5841-7352
Okubo, Tatsuya 5841-7348
Sakai, Yasuyuki 5452-6352
Tsutsumi, Atsushi 5841-7336
Yamaguchi, Takeo 5841-7345
Yoshinaga, Jun 5841-8859

Assistant Professors

Kraines, Steven 5841-8837

Tokyo Institute of Technology

Department of Chemical Engineering
2-12-1, O-okayama, Meguro-ku,
Tokyo 152-8552
University phone 81-3-3726-1111
Department phone 81-3-5734-2475
Department FAX 81-3-5734-2475

Professors

Kawasaki, Junjiro .... 5734-3285 [†]
    jkawasak@chemeng.titech.ac.jp
Kuroda, Chiaki 5734-2115
    ckuroda@chemeng.titech.ac.jp
Masuko, Masabumi 5734-3036
    mmasuko@chemeng.titech.ac.jp
Ogawa, Kohei 5734-2117
    kogawa@chemeng.titech.ac.jp
Ohtaguchi, Kazuhisa 5734-2113
    kotaguch@chemeng.titech.ac.jp
Suzuki, Masaaki (Head) 5734-2112
    masaaki@chemeng.titech.ac.jp
Tsuda, Ken 5734-2116
    ktsuda@chemeng.titech.ac.jp

Associate Professors

Aida, Takashi 5734-2883
    taida@chemeng.titech.ac.jp
Fuchino, Tetsuo 5734-2474
    fuchino@chemeng.titech.ac.jp
Kosuge, Hitoshi 5734-2151
    hkosuge@chemeng.titech.ac.jp
Kubouchi, Masatoshi 5734-2119
    mkubouch@chemeng.titech.ac.jp
Sekiguchi, Hidetoshi 5734-2110
    hsekiguc@chemeng.titech.ac.jp
Taniguchi, Izumi 5734-2155
    itaniguc@chemeng.titech.ac.jp
Yoshikawa, Shiro 5734-3278
    syoshika@chemeng.titech.ac.jp

Assistant Professors

Asami, Kazuhiro 5734-3034
    kasami@chemeng.titech.ac.jp
Iwakabe, Koichi 5734-3287
    kiwakabe@chemeng.titech.ac.jp
Matsumoto, Hideyuki 5734-2115
    hmatsumo@ chemeng.titech.ac.jp
Mori, Shinsuke 5734-2629
    smori@chemeng.titech.ac.jp
Ookawara, Shinichi 5734-3035
    sokawara@chemeng.titech.ac.jp
Sakai, Tetsuya 5734-2124
    tsakai@chemeng.titech.ac.jp
Suzuki, Akihito 5734-2628
    asuzuki@chemeng.titech.ac.jp

[†] Faculty numbers have prefix 81-3.

Accredited by:

Degrees granted 2004-2005:
B.S.: 32   M.S.: 33   Ph.D.: 3

Graduate advisor: Suzuki, Masaaki

Undergraduate advisor: Fuchino, Tetsuo

Placement service: Kawasaki, Junjiro

Toyama University

Chemical Process Engineering Department
3190, Gofuku
Toyama, 930-8555

Department FAX .... (81) (76) 445-6697

Professors

Kumazawa, Hidehiro 445-6859
    kumazawa@eng.toyama-u.ac.jp
Miyabe, Kanji 445-6835
    miyabe@eng.toyama-u.ac.jp
Morohashi, Sho-ichi 445-6856
    morohasi@eng.toyama-u.ac.jp
Yamamoto, Ken-ichi 445-6831
    kyamamot@eng.toyama-u.ac.jp
Yamazaki, Ryohei (Chair) 445-6884
    ryamaza@eng.toyama-u.ac.jp

Associate Professors

Hoshino, Kazuhiro 445-6857
    khoshino@eng.toyama-u.ac.jp
Kawasaki, Hiroyuki 445-6860
    hkawasak@eng.toyama-u.ac.jp
Kurooka, Taketoshi 445-6829
    kurooka@eng.toyama-u.ac.jp
Takase, Hitoshi 445-6830
    htakase@eng.toyama-u.ac.jp
Yoshida, Masamichi 445-6836
    yoshida@parrot.eng.toyama-u.ac.jp

Accredited by:

Degrees granted 2004-2005:
B.S.: 40   M.S.: 8

Yamagata University

School of Chemical Engineering
4-3-16 Jonan, Yonezawa
Yamagata, Japan 992-8510

Department phone .... (81) 238-26-3160
Department FAX .... (81) 238-26-3414

Professors

Hasegawa, Masahiro (Chair) .... (81) 238-26-3162
    mhase@yz.yamagata-u.ac.jp
Kanda, Yoshiteru .... (81) 238-26-3161
    kanda@yz.yamagata-u.ac.jp
Kuriyama, Masafumi .... (81) 238-26-3150
    kuri@yz.yamagata-u.ac.jp
Sato, Shimio .... (81) 238-26-3145
    shimio@yz.yamagata-u.ac.jp
Takahashi, Koji .... (81) 238-26-3156
    koji@yz.yamagata-u.ac.jp
Toda, Masayuki .... (81) 238-26-3165
    toda@yz.yamagata-u.ac.jp
Yokota, Toshiyuki .... (81) 238-26-3155
    yokota@yz.yamagata-u.ac.jp

Associate Professors

Aita, Tadahiro .... (81) 238-26-3146
    aita@yz.yamagata-u.ac.jp
Katsuyama, Tetsuo .... (81) 238-26-3131
    katsuma@yz.yamagata-u.ac.jp
Shioi, Akihisa .... (81) 238-26-3164
    shioi@yz.yamagata-u.ac.jp
Shishido, Masahiro .... (81) 238-26-3166
    sisido@yz.yamagata-u.ac.jp

Department reports to: Prof. M. Haswgawa, Chair of Chemical Engineering

Placement service: Ms. H. Ando

Yokohama National University

Department of Chemical Engineering
79-5 Tokiwadai, Hodogaya-ku
Yokohama, 240-8501

Department phone 45-339-3991
Department FAX 45-339-4012

Professors

Habuka, Hitoshi .... 339-3998 †
    AM,RE
Hara, Takao 339-4189
Iida, Yoshihiro (Chair)
Kaminoyama, Meguru 339-3999
    kaminoyama@chemeng.bsk.ynu.ac.jp
    MO,PR
Kamiwano, Mitsuo (Emeritus) 339-3988
Koizumi, Jun-ichi 339-4266
    jikoizmi@ynu.ac.jp
    BT
Matsumoto, Kanji 339-4008
    SE
Ohya, Haruhiko (Emeritus) 339-3989
Okuyama, Kunito 339-4009
    EN
Tanisho, Shigeharu (Joint) 339-3996
    EN,EV
Watanabe, Masatoshi 339-3997
    BM

Associate Professors

Aihara, Masahiko 339-4229
    EN
Aramaki, Kenji 339-4190
    SU
Nishi, Kazuhiko 339-3988
    MO,NT,PO
Takeda, Minoru 339-4566
    BT
Tsunakawa, Hiroshi 339-4006
    NT

Assistant Professors

Misumi, Ryuta 339-3995
    MO,SE
Nakamura, Kazuho 339-3980
    BM,BT

Nittami, Tadashi 339-4025
  BT,SE
Suzuki, Ichiro 339-4266
  BT
Takeuchi, Takashi 339-3975
  MO

† Phone numbers have prefix 81-45.

Accredited by: JABEE

Graduate advisor: Minoru Takeda

Undergraduate advisor: Kazuhiko Nishi

Republic of Korea

Chonnam National University

Chemical Engineering Department
Kwang-ju, 500-757

University phone .... (062) 530-0114
Department phone .... (062) 530-1850
Department FAX .... (062) 530-1869

Professors

Cho, Chong-Hyun (Emeritus) .... 530-1850 †
Chough, Sung-Hyo 530-1825
Han, Neung-Won (Emeritus) 530-1850
Kang, Choon-Hyoung (Chair) 530-1818
Kang, Sung J. 530-1817
Kim, Jae-Hyung (Emeritus) 530-1850
Kim, Jae-Seung (Emeritus) 530-1850
Kim, Jin-Hwan 530-1815
Lee, Ki-Young 530-1843
Lee, Woo-Tai 530-1813
Park, Don-Hee 530-1841
Park, Nam-Cook 530-1826
Ryu, Hwa-Won 530-1842
Ryu, Min-Su 530-1816
Shin, Jae-Soon 530-1827
Sunwoo, Chang-Shin 530-1814
Yang, Jae-Ho 530-1812
Yu, Eui-Yeon 530-1811

Associate Professors

Kim, Do-Man 530-1844
Kim, Young-Chul 530-1828
Rhee, Jong-Il 530-1847

Assistant Professors

Han, Eun-Mi 530-1829

† Direct dial with prefix 062.

Accredited by: Ministry of Education Korea

Keimyung University

Chemical Engineering Department
1000 Sindangdong Dalseo-gu Taegu
704-701

University phone 81-53-580-5114
Department phone 81-53-580-5042
Department FAX 81-53-581-7686

Professors

Lee, Chun-Soo 81-53-580-5235
Kim, Jong-Sik 81-53-580-5236
Synn, Dong-Su 81-53-580-5238

Associate Professors

Choe, Seok-Bum 81-53-580-5239
Ha, KiRyong (Chief) 81-53-580-5263

Assistant Professors

Rho, Seung-Baik 81-53-580-5237
  rhosb@kmucc.keimyung.ac.kr
Suh, Soong-Hyuck 81-53-580-5545
Byun, Hong-Sik 81-53-580-5569

Korea Advanced Institute of Science and Technology

Department of Chemical and Biomolecular Engineering
373-1 Guseong-dong, Yuseong-gu
Daejeon 305-701, Republic of Korea

University phone 82-42-869-2114
University FAX 82-42-869-
Department phone 82-42-869-3902 4
Department FAX 82-42-869-3910

Professors

Chang, Ho N. 82-42-869-3912
  hnchang@kaist.ac.kr
Chang, Yong K. 82-42-869-3927
  ychang@kaist.ac.kr
Chung, In J. 82-42-869-3916
  chung@kaist.ac.kr
Hong, Won-Hi 82-42-869-3919
  whhong@kaist.ac.kr
Ihm, Son-Ki 82-42-869-3915
  skihm@kaist.ac.kr
Kim, Do H. 82-42-869-3929
  DoHyun.Kim@kaist.ac.kr
Kim, Jong-Duk (Director) 82-42-869-3921
  jdkim@kaist.ac.kr
Kim, Sang D. (Director) 82-42-869-3913
  kimsd@kaist.ac.kr
Kim, Sung C. (Director) 82-42-869-3914
  kimsc@kaist.ac.kr
Lee, Huen 82-42-869-3917
  h_lee@kaist.ac.kr
Lee, Sang Y. (Director) 82-42-869-3930
  leesy@kaist.ac.kr
Lee, Tai-yong 82-42-869-3926
  tylee@kaist.ac.kr
Park, Jung-Ki 82-42-869-3925
  jungpark@kaist.ac.kr
Park, O O. 82-42-869-3923
  ookpark@kaist.ac.kr
Park, Seung B. (Chair) 82-42-869-3928
  SeungBinPark@kaist.ac.kr
Park, Sunwon 82-42-869-3920
  sunwon@kaist.ac.kr
Woo, Seong I. (Director) 82-42-869-3918
  siwoo@kaist.ac.kr
Yang, Ji-Won 82-42-869-3924
  jwyang@kaist.ac.kr
Yang, Seung-Man 82-42-869-3922
  smyang@kaist.ac.kr

Associate Professors

Jung, Hee-Tae 82-42-869-3931
heetae@kaist.ac.kr

Assistant Professors

Park, Hyun G. 82-42-869-3932
hgpark@kaist.ac.kr

Accredited by: Ministry of Education

Degrees granted 2004-2005:
B.S.: 13   M.S.: 34   Ph.D.: 33

Graduate advisor: Seung Bin Park

Undergraduate advisor: Seung Bin Park

Student organization: Seung Bin Park
Advisor: Seung Bin Park

Department reports to: Sang Soo Kim, Dean, College of Engineering

Placement service: Seung Bin Park

Korea University

Department of Chemical and Biological Engineering
1, 5-Ka, Anam-Dong, Sungbuk-Ku
Seoul 136-701

University phone 82-2-3290-1114
Department phone 82-2-3290-3290
Department FAX 82-2- 926-6102

Professors

Chun, Hai S. (Emeritus) 82-2-3290-3400
Doh, Dong S. (Emeritus)
Hong, Suk I. 82-2-3290-3294
sihong@korea.ac.kr
Hyun, Jae C. 82-2-3290-3295
jchyun@grtrkr.korea.ac.kr
Kim, Chongyoup 82-2-3290-3302
cykim@grtrkr.korea.ac.kr
Kim, Seung W. 82-2-3290-3300
kimsw@korea.ac.kr
Kim, Sung H. 82-2-3290-3297
kimsh@korea.ac.kr
Kim, Woo N. 82-2-3290-3296
kimwn@korea.ac.kr
Kim, Yong J. (Emeritus)
Lee, Chul S. 82-2-3290-3293
cslee@korea.ac.kr
Lee, Kwan Y. (Chair) 82-2-3290-3299
kylee@korea.ac.kr
Yang, Dae R. 82-2-3290-3298
dryang@korea.ac.kr
Yoo, Young H. (Emeritus)

Associate Professors

Ahn, Dong J. 82-2-3290-3301
ahn@korea.ac.kr
Ha, Jeong S. 82-2-3290-3303
jeongsha@korea.ac.kr
Lee, Jee W. 82-2-3290-3304
leejw@korea.ac.kr

Assistant Professors

Jung, Hyun W. 82-2-3290-3306
hwjung@grtrkr.korea.ac.kr
Kang, Jung W. 82-2-3290-3305
jwkang@korea.ac.kr

Accredited by: Ministry of Education & Human Resources Development

Graduate advisor: Kwan Young Lee

Undergraduate advisor: Kwan Young Lee

Student organization: Chongyoup Kim
Advisor: Chongyoup Kim

Department reports to: Dean of Engineering

Pohang University of Science and Technology

San 31, Hyoja-dong, Nam-gu
Pohang, Korea 790-784
Deliveries: Prof. Kyung Hee Lee

University phone . . . . (82) 54-275-0900
Department phone . . . . (82) 54-279-2720
Department FAX . . . . (82) 54-279-2699

Professors

Chang, Kun S. (Emeritus) . . . . 279-2262 [†]
kschang@postech.ac.kr
Cho, Kil W. 279-2270
kwcho@postech.ac.kr
Chung, Jong S. 279-2267
jsc@postech.ac.kr
Kang, In S. 279-2273
iskang@postech.ac.kr
Kim, Jin K. 279-2276
jkkim@postech.ac.kr
Kim, Young G. (Emeritus) 279-2261
ygkim@postech.ac.kr
Lee, In B. 279-2274
iblee@postech.ac.kr
Lee, Jae S. 279-2266
jlee@postech.ac.kr
Lee, Kun H. 279-2271
ce20047@postech.ac.kr
Lee, Kyung H. (Head) 279-2263
kyunglee@postech.ac.kr
Lee, Sun B. 279-2268
sblee@postech.ac.kr
Nam, In-Sik 279-2264
isnam@postech.ac.kr
Park, Chan E. 279-2269
cep@postech.ac.kr
Park, Jong M. 279-2275
jmpark@postech.ac.kr
Rhee, Shi W. 279-2265
srhee@postech.ac.kr

Associate Professors

Cha, Hyung J. 279-2280
hjcha@postech.ac.kr
Yoon, Byung J. 279-2277
bjyoon@postech.ac.kr

Assistant Professors

Yong, Kijung 279-2278
kyong@postech.ac.kr

[†] Faculty numbers have prefix 82-54.

Accredited by: Ministry of Education
Degrees granted 2003-2004:
    B.S.: 18   M.S.: 34   Ph.D.: 14
Graduate advisor: Prof. Kilwon Cho
Undergraduate advisor: Prof. Jong Shik Chung

Seoul National University

School of Chemical Engineering
Seoul 151-744, Korea
University phone 82-2-880-5114
Department phone 82-2-880-7068
Department FAX 82-2-888-1604

Professors

Ahn, Tae-Oan (Emeritus)
Char, Kookheon 82-2-880-7431
    khchar@snu.ac.kr
Choi, Cha Y. 82-2-880-7071
    choicy@snu.ac.kr
Choi, Chang K. 82-2-880-7407
    ckchoi@snu.ac.kr
Choi, Ung (Emeritus)
Jang, Jyongsik 82-2-880-7069
    jsjang@plaza.snu.ac.kr
Kim, Byung-Gee 82-2-880-6774
    byungkim@snu.ac.kr
Kim, Hwayong 82-2-880-7406
    hwayongk@snu.ac.kr
Kim, Joon Y. (Emeritus)
Lee, Chai-sung (Emeritus)
Lee, Chung-Hak 82-2-880-7075
    leech@snu.ac.kr
Lee, Ho-In (Head) 82-2-880-7072
    hilee@snu.ac.kr
Lee, Hong H. (Director) 82-2-880-7403
    honghlee@plaza.snu.ac.kr
Lee, Jihwa 82-2-880-7076
    jiwhalee@plaza.snu.ac.kr
Lee, Ki-Jun 82-2-880-7402
    kijunlee@plaza.snu.ac.kr
Lee, Moon-Deuk (Emeritus)
Lee, Seung J. 82-2-880-7410
    sjlee@plaza.snu.ac.kr
Lee, Wha Y. 82-2-880-7404
    wyl@snu.ac.kr
Lee, Yoon S. 82-2-880-7073
    yslee@snu.ac.kr
Moon, Sang H. 82-2-880-7409
    shmoon@surf.snu.ac.kr
Oh, Seung M. 82-2-880-7074
    seungoh@plaza.snu.ac.kr
Rhee, Hyun-Ku 82-2-880-7405
    hkrhee@snu.ac.kr
Shim, Jyong S. (Emeritus)
Shin, Yoon-Kyoung (Emeritus)
Yoo, Young J. (Associate Dean) 82-2-880-7411
    yjyoo@plaza.snu.ac.kr
Yoon, En S. 82-2-880-7408
    esyoon@pslab.snu.ac.kr

Associate Professors

Hyeon, Taeghwan 82-2-880-7150
    thyeon@plaza.snu.ac.kr
Jho, Jae Y. 82-2-880-8346
    jyjho@snu.ac.kr
Kim, Young G. 82-2-880-8347
    ygkim@plaza.snu.ac.kr
Park, Tai H. 82-2-880-8020
    thpark@plaza.snu.ac.kr
Yi, Jongheop 82-2-880-7438
    jyiecerl@snu.ac.kr

Assistant Professors

Ahn, Kyung H. 82-2-880-8322
    ahnnet@snu.ac.kr
Kim, Jae J. 82-2-880-8863
    jjkimm@snu.ac.kr
Lee, Jong-Chan 82-2-880-7070
    jongchan@snu.ac.kr
Yoon, Jeyong 82-2-880-8927
    jeyong@snu.ac.kr

Accredited by: Ministry of Education Korea
Degrees granted 2003-2004:
    B.S.: 109   M.S.: 80   Ph.D.: 32
Department reports to: Jang-Moo Lee, Dean of Engineering

Kuwait

Kuwait University

Chemical Engineering Department
College of Engineering and Petroleum
P.O. Box 5969
13060, Safat
University phone 965 4811188
Department phone 965 4817662
Department FAX 965 4839498

Professors

AL-Haddad, Amir A. .... x5221 [†]
AL-Sahhaf, Taher x5673
AL-Saif, Walid (Visiting)
Alatiqi, Imad M. x5776
Bouhmra, Walid S. x5613
Ettouney, Hisham x5619
Fahim, M.A x5127
Riazi, Mohammed R. x5772
Shabhan, Habib I. (Head) x5109

Associate Professors

AL-Ameeri, Rashid S. x5148
AL-Enezy, Ghazi 965-4810272
Baker, Christofer G.J x5121
Bishara, Ahmed E. x5617
Lababidi, Haitham M.S. x5778
Mahdi, Khaled A. x5618
    k_mahdi@kuc01.kuniv.edu.kw

Assistant Professors

AL-Adwani, Hamad A.H. x7460
Al-Bahri, Tareq A. x7459
Al-Bofersn, Osama A. x5856
Al-Mulla, Adam x5775
Al-Musallam, Abdulwahhab x5612
Al-Ramadan, Hamida A. x5694
AL-Roomi, Yousef M. x7016

Al-Sahali, Mohammad B. x5783
AL-Shayji, Khawla x5108
Ali, Sammi H. G. x5702

† Faculty numbers are extensions of University phone.

Accredited by: ABET

Placement service: Eng. Mubarak Ghanim

Malaysia

University of Malaya

Chemical Engineering Department
Faculty of Engineering
50603 Kuala Lumpur

University phone 60 3 7967-7022
Department phone 60 3 7967-5206
Department FAX 60 3 7967-5319

Professors

Hasan, Masitah (Chair) .... 7967 5295 †
    masitahhasan@um.edu.my
Hashim, Mohd A. 7967 4522
    alihashim@um.edu.my
Sulaiman, Nik M. 7967 5299
    meriam@um.edu.my

Associate Professors

Abdullah, Ezzat C. 7967 5301
    ezzat@um.edu.my
Aroua, Mohamed K. 7967 5313
    mk_aroua@um.edu.my
Hussain, Mohamed A. 7967 5214
    mohd_azlan@um.edu.my
Leong, Yub C. 7967 5294
    ycleong@um.edu.my
Wan Daud, Wan M. A. 7967 5297
    ashri@fk.um.edu.my

Assistant Professors

Chua, Adeline S. M.
Ghasem, Nayef M. 7967 5360
    nayef@um.edu.my
Hashim, Nur A. 7967 6892
    awanis@um.edu.my
Hassan, Che R. 7967 5314
    rosmani@um.edu.my
Hisham, Badrul 7967 6869
    badrules@um.edu.my
Mohamed Nor, Mohamed I. 79675347
    misk@um.edu.my
Raman, Abdul A. 7967 5300
    azizraman@um.edu.my
Tahir, Faridah 7967 5219
    faridah2002@um.edu.my
Yeoh, Hak K.
Yusoff, Rozita 7967 6891
    ryusoff@um.edu.my

† Direct dial with prefix 03.

Accredited by: IChemE (UK)

Degrees granted 2004-2005:
    B.S.: 45   M.S.: 5   Ph.D.: 2

Graduate advisor: Ezzat Chan Abdullah

Department reports to: Prof. Dr. Masitah Hasan, Head of Department

Universiti Sains Malaysia

School of Chemical Engineering
Engineering Campus
Seri Ampangan
14300 Nibong Tebal
Seberang Perai Selatan
Pulau Pinang

University phone .... (604) 5937788
Department phone .... (604) 5937788x6400
Department FAX .... (604) 5941013

Professors

Bhatia, Subhash .... x6409 †
    chbhatia@eng.usm.my

Associate Professors

Abdullah Aziz, Jalal x6411
    chjalal@eng.usm.my
Ahmad, Abdul L. x6418
    chlatif@eng.usm.my
Fernando, W.J.N x6428
    chnoel@eng.usm.my
Kamaruddin, Azlina H. x6417
    chazlina@kimia.eng.usm.my
Mohamed, Abdul R. (Dean) x6400
    chrahman@eng.usm.my
Najafpour, Ghasem x6412
    chghasem@eng.usm.my

Assistant Professors

Bakar, Mohd Z. A. x6402
    chmohdz@eng.usm.my
bin Aziz, Norashid x6457
    chnaziz@eng.usm.my
bin Othman, Mohd R. x6426
    chroslee@eng.usm.my
Choudhury, J.P. x6424
    chjyoti@eng.usm.my
Daud, Mohd Z. M. x6414
    chzulk@eng.usm.my
Hameed, Bassim H. x6422
    chbassim@eng.usm.my
Hussein, Ye L. M. x6460
    chhussin@eng.usm.my
Qassim, Mohd. Q. H. x6423
    chqassim@eng.usm.my
Srinivasakannan, C. x6425
    chkannan@eng.usm.my
Zakaria, Ridzuan x6403
    chduan@eng.usm.my

† Dial (604) 5937788 and ask for extension.

Accredited by: Board of Engineering

Department reports to: Abdul Rahman Mohamed, Dean of the School

Mexico

Instituto Tecnologico y de Estudios Superiores de Monterrey

Chemical Engineering Department
Sucursal de Correos J
Monterrey, N.L. 64849
Deliveries: Dr. Joaquin Acevedo

University phone .... (8) 358-2000
Department phone .... (52) 8158-2034
Department FAX .... (52) 8328-4250

Professors
Garcia, Jorge
    jorge.garcia@itesm.mx
Mejia, Gerardo
    gmejia@itesm.mx
Rito, Marco
    mrito@itesm.mx
Romero, Miguel A. (Associate Dean)
    mromero@itesm.mx
    EN,MO,RE
Trevino, Belzahet
    btrevino@itesm.mx
Yague, Omar
    oyague@itesm.mx

Associate Professors
Acevedo, Joaquin (Head)
    jacevedo@itesm.mx
    MO,PR
Lozano, Francisco J.
    fjlozano@itesm.mx
Martinez, Jeronimo
    jemartin@itesm.mx
Ortiz, Enrique
    eortiz@itesm.mx

Assistant Professors
Garza, Vicente J.
    vicente.garza@itesm.mx
Lapizco, Blanca
    blapizco@itesm.mx
    AM,BT
Mendoza, Alberto
    mendoza.alberto@itesm.mx
Patino, Veronica
    vpatino@itesm.mx
Ramirez, Darinka
    darinka@itesm.mx
Soriano, Ricardo
    rsoriano@itesm.mx

Accredited by: SACS, CACEI, ABET

Degrees granted 2005-2006:
    B.S.: 100   M.S.: 10

Graduate advisor: Dr. Jorge García

Undergraduate advisor: Veronica Patiño

Student organization: IMIQ
Advisor: Veronica Patiño

Department reports to: Arturo Molina, Dean, Div. of Engineering

Placement service: Arturo Molina, Dean, Div. of Engineering

Universidad Iberoamericana

Coordinacion de Ingenieria Quimica
Prolongacion Paseo de la Reforma 880,
Col. Lomas de Santa Fe, Mexico D. F. 01210
Deliveries: Quim. Juan Abud Saint Martin

University phone .... (52) 55 5950 4000
Department phone .... (52) 55 5950 4033
Department FAX .... (52) 55 5950 4279

Professors
Abud Saint Martín, Juan (Associate Head) x4033
    juan.abud@uia.mx
Hernández Esparza, Margarita .... x 4629
    margarita.hernandez@uia.mx

Associate Professors
Chaparro Mercado, Ma. C. x4098
    carmen.chaparro@uia.mx
Flores Tlacuahuac, Antonio x7367
    antonio.flores@uia.mx
Huerta Cevallos, Rene x7088
    rene.huerta@uia.mx
López Rubio, Javier F. x7091
    javier.lopez@uia.mx
Ortiz Estrada, Ciro x4028
    ciro.ortiz@uia.mx
Pedraza Segura, Lorena .... x 4624
    lorena.pedraza@uia.mx
Rivera Toledo, Martin x4078
    martin.river@uia.mx
Silva Beard, Andrea (Head) x4534
    andrea.silva@uia.mx
Vasquez Medrano, Ruben x7321
    ruben.vasquez@uia.mx

Accredited by: CACEI

Degrees granted 2004-2005:
    B.S.: 52   M.S.: 7

Undergraduate advisor: Quim. Juan Abud

Student organization: SAICQ
Advisor: Eduardo Alvarez F.

Universidad de las Americas, Puebla

Chemical and Food Engineering Department
Ex-hacienda Santa Catarina Martir
San Andres Cholula, Puebla 72820

University phone 52 (222) 229-2000
University FAX 52 (222) 229-2009
Department phone 52 (222) 229-2126
Department FAX 52 (222) 229-2727

Professors
Argaiz-Jamet, Alvaro 52 (222) 229-2000x4092
    aargaiz@mail.udlap.mx
Fonseca-Sandoval, Raul G. 52 (222) 229-2026
    fonseca@mail.udlap.mx
Lopez-Malo, Aurelio 52 (222) 229-2409
    amalo@mail.udlap.mx
Palou-Garcia, Enrique 52 (222) 229-2658
    epalou@mail.udlap.mx

Reyes-Mazzoco, Rene 52 (222) 229-2660
rreyes@mail.udlap.mx
Velez-Ruiz, Jorge F. 52 (222) 229-2648
jfvelezr@mail.udlap.mx
Vergara-Balderas, Fidel 52 (222) 229-2146
fvergara@mail.udlap.mx
Welti-Chanes, Jorge S. 52 (222) 229-2728
jwelti@mail.udlap.mx

Associate Professors
Barcenas-Pozos, M. E. 52 (222) 229-2126
mbar@mail.udlap.mx
Lara-Diaz, Rene A. 52 (222) 229-2000x4353
rlara@mail.udlap.mx
Maza-Aranda, J. G. 52 (222) 229-2659
gmaza@mail.udlap.mx
Rios-Casas, Luis G. (Head) 52 (222) 229-2661
lrios@mail.udlap.mx
San Martin-Gonzalez, Fernanda 52 (222) 229-2927
fernanda@mail.udlap.mx

Accredited by: SACS

Degrees granted 2003-2004:
    B.S.: 35   M.S.: 19

Graduate advisor: Dr. Aurelio Lopez-Malo (Food Sc.) and Dr. Rene Reyes (Chem. Eng.)

Undergraduate advisor: Ma. Elena Sosa (Food Sc.), René Lara (Chem. Eng.)

Student organization: CEUDLA
Advisor: M. Sc. Luis Rios

Department reports to: Dr. Juan Manuel Ramirez C., Dean of Engineering

Placement service: Silvia Carreño Castillo

Universidad Autonoma Metropolitana-Iztapalapa

Department of Chemical Engineering
San Rafael Atlixco No. 186
Col. Vicentina, Del. Iztapalapa
09340 Mexico, D. F. MEXICO
University phone .... (525) 804-4600
Department phone .... (525) 804-4642/43
Department FAX .... (525) 804-4900

Professors
Alvarez, Jesús ....(525) 804-46-48
jac@xanum.uam.mx
Alvarez-Ramirez, José ....(525) 804-46-48
jjar@xanum.uam.mx
de los Reyes, José A. ....(525) 804-46-48
jarh@xanum.uam.mx
Fuentes, Gustavo A. ....(525) 804-46-48
gfuentes@xanum.uam.mx
Lapidus, Gretchen ....(525) 804-46-48
gtll@xanum.uam.mx
Lobo, Ricardo ....(525) 804-46-48
lobo@xanum.uam.mx
López, Felipe ....(525) 804-46-48
felipe@xanum.uam.mx
Ochoa, J. A. ....(525) 804-46-48
jaot@xanum.uam.mx
Revah, Sergio ....(525) 804-46-48
srevah@xanum.uam.mx
Ruiz, Richard S. (Associate Head) ....(525) 804-46-48
rmr@xanum.uam.mx
Soria, Alberto .... (525) 804-4648
asor@xanum.uam.mx
Vernon, Jaime ....(525) 804-46-48
jvc@xanum.uam.mx
Viveros, Tomás (Dean) ....(525) 804-46-48
tvig@xanum.uam.mx
Vizcarra, Mario ....(525) 804-46-48
mgvm@xanum.uam.mx

Associate Professors
Aréchiga, Uriel ....(525) 804-46-48
uriel@xanum.uam.mx
Escobar, Angel ....(525) 804-46-48
ange@xanum.uam.mx
Gómez, Sergio .... (525) 804-4648
sgomez@xanum.uam.mx
Jarquín, Hugo ....(525) 804-46-48
hja@xanum.uam.mx
Martínez, Carlos ....(525) 804-46-48
cmv@xanum.uam.mx

Accredited by: CACEI Undergraduate Program, CONACyT Graduate Program

Degrees granted 2004-2005:
    B.S.: 24   M.S.: 10   Ph.D.: 3

Graduate advisor: Dr. Mario Vizcarra

Undergraduate advisor: Dr. Gretchen Lapidus

Student organization: Marco Antonio Loza
Advisor: Dr. Gretchen Lapidus

Department reports to: Dr. Eduardo S. Pérez Cisneros, Head of the Process Engineering and Hydraulics Department

Placement service: gtll@xanum.uam.mx

Netherlands

University of Amsterdam

Department of Chemical Engineering
Nieuwe Achtergracht 166
1018 WV Amsterdam
University phone 31 20 525 9111
Department phone 31 20 525 5265
Department FAX 31 20 525 5604

Professors
Bliek, Alfred (Chair) .... 525 6479 [†]
bliek@science.uva.nl
Das, Hendrik A. (Emeritus)
das@ecn.nl
Fasolino, Annalisa 525 6448
fasolino@science.uva.nl
Fortuin, Jan M. H. (Emeritus) 525 6478
Frenkel, Daan 525 5265
frenkel@amocf3.amolf.nl
Iedema, Piet D. 525 6484
piet@science.uva.nl
Krishna, Rajamani 525 7007
krishna@science.uva.nl

Poppe, Hans (Emeritus) 525 5265
Schoenmakers, Peter J. 525 6642
                                pjschoen@science.uva.nl
Smilde, Age K. 525 5062
                                  asmilde@science.uva.nl
Smit, Berend 525 5067
                                  B.Smit@science.uva.nl
Tijssen, Robert 525 6550
                                  rtijssen@science.uva.nl

Associate Professors
Dimian, Alexandre C. 525 6034
                                  alexd@science.uva.nl
Hoefsloot, Huub C. J. 525 5867
                                  huubh@science.uva.nl
Jérôme, Blandine 525 6992
                                  jerome@science.uva.nl
Kok, Wim T. 525 6539
                                    wkok@science.uva.nl
van den Heuvel, Johannes C. 525 5078
                                  heuvel@science.uva.nl

Assistant Professors
Bolhuis, Peter G. 525 6447
                                  bolhuis@science.uva.nl
Eiser, Erika 525 6916
                                   eiser@science.uva.nl
Lowe, Christopher P. 525 6485
                                    lowe@science.uva.nl
Meijer, Evert J. 525 6448
                                  ejmeijer@science.uva.nl
Rothenberg, Gadi 525 6963
                                    gadi@science.uva.nl
Westerhuis, Johan A. 525 6546
                                  westerhu@science.uva.nl

† Direct dial with prefix 31 20.

Accredited by: IChemE (UK)

Delft University of Technology

Faculty of Applied Sciences
Department of Chemical Engineering & Bioproces Technology
Julianalaan 136
2628 BL-Delft

University phone 31 (15) 278-9111
Department phone 31 (15) 278-4841
Department FAX 31 (15) 278-5868

Professors
Akker van den, H.E.A.vanden A. (Director)
Coppens, M.O.Coppens
                            M.O.Coppens@tnw.tudelft.nl
Dedem van, G.W.K. van D.
                            G.W.K.vanDedem@tnw.tudelft.nl
Frens, G.Frens (Emeritus)
Grievink, J. .... 278-4351 †
Hagen, W.R 278-5051
                            W.R.Hagen@tnw.tudelft.nl
Harmsen, G. J. 278-5006
Heijnen, J. J. 278-2341
Jansens, P. P. J. 278-6678
Kapteijn, F 278-4384
Katgerman, L. 278-2249
Kuenen, J. G. 278-5308
Loosdrecht, M.C.M.vanLoosdrecht
                          M.C.M.vanLoosdrecht@tnw.tudelft.nl
Luyben, K. C. A. M. (Dean) 278-2353
Moulijn, J.A. 278-5008
Odijk, Th. 278-5590
Pasman, J 278-5006
Picken, S 278-6946
Pronk, J 278-3214
Schmidt-Ott, A.Schmidt-Ott
                          A.Schmidt-Ott@tnw.tudelft.nl
Schoonman, J 278-2647
Sheldon, R. A. 278-2675
Swaan Arons, J.deSwaanArons (Emeritus)
                          J.deSwaanArons@tnw.tudelft.nl
van den Bleek, C. M. 278-4390
van Loosdrecht, M. 278-1618
van Turnhout, J. 278-2623
Wielen v.d., L.A.M. 278-2361

Associate Professors
de Loos, Th. W. 278-2619
de Loos-Vollebregt, Mrs. M. T. C. 278-6379
Goossens, A 278-4919
Jongejan, J 278-2371
Koper, G.J.M. 2788218
                          G.J.M.koper@tnw.tudelft.nl
Lameris, G. H. 278-2648
Lemkowitz, S. M. 278-4394
Makkee, M.Makkee
                          M.Makkee@tnw.tudelft.nl
Nieswaag, H. 278-2234
Nijenhuis, K. te 278-2630
Peters, C. J. 278-2660
Peters, J. A. 278-5892
Schapink, F. W. 278-2272
Steensma, H. Y. 278-2412
van den Ham, A. G. J. 278-4374
van Dijken, J. P. 278-2412
Verheijen, P 278-4326

Assistant Professors
Arends, I.W.C.E.Arends
                          I.W.C.E.Arends@tnw.tudelft.nl
Gotsis, A.D.Gotsis
                          A.D.Gotsis@tnw.tudelft.nl
Hanefeld, U.Hanefeld
Kleeff van, B.H.A. van K.
                          B.H.A.vanKleeff@tnw.tudelft.nl
Ommen van, J.R.vanOmmen
                          J.R.vanOmmen@tnw.tudelft.nl

† Direct dial with prefix 31 15.

Accredited by: IChE, ABET, VSNU

Undergraduate advisor: Mrs ir A.Schaap

Student organization: Technologisch Gezelschap

Eindhoven University of Technology

Department of Chemical Engineering
Den Dolech 2
5600 MB Eindhoven

University phone 31 (40) 247911
Department phone 31 (40) 2473000
Department FAX 31 (40) 2444321

Professors

Benthem, R. van
Broer, D.J.
Bruin, S.
Cramers, C.A. (Emeritus)
Dautzenberg, F.M. (Emeritus)
de Jeu, W.H.
de With, G.
Drinkenburg, A.A.H. (Emeritus)
German, A. L. (Emeritus)
Haan, A.B. de
Herk, A. M. van
Hulshof, L.A.
Janssen, R. A. J.
Kerkhof, P. J. A. M.
Keurentjes, J.T.F.
Koning, C.E.
Kramer, G.J.
Leegwater, H.
Leeuwen, P.W.N.M. van
Lemstra, P. J.
Linde, R. van der (Emeritus)
Loo, F.J.J. van (Emeritus)
Meijer, E. M.
Meijer, E. W.
Metselaar, R. (Emeritus)
Niemantsverdriet, J.W. (Dean)
Nieuwenhuys, B.
Nolte, R.J.M.
Notten, P.
Santen, R. van
Schouten, J.C. (Associate Dean)
Schubert, U.
Sijbesma, R.
Veen, J. A. R. van
Vogt, D.

Associate Professors

Abbenhuis, E.
Bastiaansen, C.W.M.
Coumans, W. J.
Hintzen, H. T.
Janssen, A.P.J.
Janssen, L. J. J. (Emeritus)
Klumperman, L.
Kuster, B. F. M.
Meuldijk, J.
Nies, E.
Ptasinski, K.
Rastogi, S.
Reijenga, J. C.
Vekemans, J.A.J.M.
Wijers, J. G.

Accredited by: ABET

Department reports to: Executive Board of University

New Zealand

University of Auckland

Department of Chemical and Materials Engineering
Private Bag 92019
Auckland
Deliveries: Gate 3, Grafton Road, Auckland City
University phone . . . . (64) (9) 373 7599
Department phone . . . . (64) (9) 373 7599x88135
Department FAX . . . . (64) (9) 373 7463

Professors

Chen, John J. J. . . . . 88137 †
        j.chen@auckland.ac.nz
        HT,MO,PR
Chen, X. D. x87004
        d.chen@auckland.ac.nz
        BT,HT,MO
Duffy, Geoffrey G. (Associate Dean) x87805
        gg.duffy@auckland.ac.nz
        TP
Farid, Mohamed M x84807
        m.farid@auckland.ac.nz
        BT,EN,HT,MO
Ferguson, W. G. (Head) x88133
        wg.ferguson@auckland.ac.nz
        AM,SU
Gao, Wei (Associate Dean) x88175
        w.gao@auckland.ac.nz
        AM,ME,MO,NT,SU

Associate Professors

Broom, Neil D. x88974
        nd.broom@auckland.ac.nz
        BM
Hossain, Monwar x83962
        m.hossain@auckland.ac.nz
        BT,EV,SE
Hyland, Margaret M. x87865
        m.hyland@auckland.ac.nz
        AM,PR,SU
Kirkpatrick, Rob 85128
        r.kirkpatrick@auckland.ac.nz
        PR
Young, Brent 85606
        b.young@auckland.ac.nz
        MO,PR,SE

Assistant Professors

Chiu, Yu L. 86924
        yl.chiu@auckland.ac.nz
        AM,NT
Hodgson, Michael A. x88218
        ma.hodgson@auckland.ac.nz
        AM,ME,NT,SU
James, Bryony J. x85813
        b.james@auckland.ac.nz
        AM,NT,SU
Jones, Mark 84548
        mark.jones@auckland.ac.nz
        AM,BM,NT,SU
Patterson, Darrell 85027
        Darrell.Patterson@auckland.ac.nz
        BT,EV,RE,SE
Zhang, Lu
        BT,MO,RE

† Dial (64) (9) 373 7599 and ask for extension.

Accredited by: IChemE (UK), IPENZ

Degrees granted 2005-2006:
    B.S.: 43   M.S.: 12   Ph.D.: 2

Undergraduate advisor: Dr M. A. Hodgson

Department reports to: Dean of Engineering

Placement service: Jim Bohm (7488), Careers Advisory Service

University of Canterbury

Chemical and Process Engineering
Private Bag 4800
Christchurch 8001
Deliveries: Corner Creyke & Forestry Rd, Ilam
University phone 64 3 366-7001
Department phone 64 3 364-2543
Department FAX 64 3 364-2063

Professors

Abrahamson, John .... 364-2318 †
    john.abrahamson@canterbury.ac.nz
    NT
Keey, Roger B. (Emeritus) 364-2543
    r.keey@cape.canterbury.ac.nz
Marsh, Kenneth N. 364-2140
    ken.marsh@canterbury.ac.nz
Pang, Shusheng 364-2538
    shusheng.pang@canterbury.ac.nz
Williamson, Arthur G. (Emeritus) 364-2543
    a.williamson@cape.canterbury.ac.nz

Associate Professors

Gilmour, Ian A. 364-2137
    ian.gilmour@canterbury.ac.nz
Gostomski, Peter A. 364-2141
    peter.gostomski@canterbury.ac.nz
Jordan, Patrick J. 364-2864
    pat.jordan@canterbury.ac.nz
Morison, Ken R. 364-2578
    ken.morison@canterbury.ac.nz
Williamson, Chris J. 364-2865
    chris.williamson@canterbury.ac.nz

Assistant Professors

Chu, Khim H 364-2217
    khim.chu@canterbury.ac.nz

† Direct dial with prefix 64 3.

Accredited by: IChemE (UK), Institution of Professional Engineers

Graduate advisor: C J Fee

Undergraduate advisor: K Morison

Department reports to: P Jackson, Pro Vice Chancellor (Engineering)

Norway

Norwegian University of Science and Technology

Department of Chemical Engineering
Institutt for kjemisk prosessteknologi
NO-7491 Trondheim
Norway
Deliveries: Sem Saelands vei 4
University phone 47 73 5950000
Department phone 47 73 594030
Department FAX 47 73 594080

Professors

Berge, Arvid (Emeritus) 94138
    arvid.berge@nt.ntnu.no
Blekkan, Edd A. 94157
    edd.anders.blekkan@nt.ntnu.no
Chen, De 93149
    de.chen@nt.ntnu.no
Christensen, Per K. (Emeritus) 94029
    pkochchr@nt.ntnu.no
Erga, Olav (Emeritus) 94120
    olav.erga@nt.ntnu.no
Friedemann, John D. (Adjunct) 97022
    johndani@nt.ntnu.no
Genzer, Jan (Adjunct) 97022
    jan.genzer@nt.ntnu.no
Gregersen, Øyvind 94029
    gregersen@nt.ntnu.no
Helle, Torbjorn (Emeritus) 94031
    torbjoern.helle@nt.ntnu.no
Hertzberg, Terje 94113
    terje.hertzberg@nt.ntnu.no
Holmen, Anders 94151
    anders.holmen@nt.ntnu.no
Hägg, May-Britt 94033
    may-britt.hagg@nt.ntnu.no
Jakobsen, Hugo A. 94132
    hugo.jakobsen@nt.ntnu.no
Lovland, Jorgen (Emeritus) 94124
    jolovland@hotmail.com
Malthe-Sørenssen, Didrik (Adjunct) 94119
    dms@nt.ntnu.no
Moljord, Kjell (Adjunct) 94147
    kmol@statoil.com
Mørk, Preben 94148
    preben.cato.moerk@nt.ntnu.no
Nesse, Norvald 94133
    norvald.nesse@nt.ntnu.no
Preisig, Heinz 92807
    preisig@nt.ntnu.no
Rytter, Erling (Adjunct) 94147
    err@statoil.no
Sjoblom, Johan 95505
    johsj@nt.ntnu.no
Skogestad, Sigurd (Head) 94154
    sigurd.skogestad@nt.ntnu.no
Stenius, Per (Adjunct) 94030
    per.stenius@nt.ntnu.no
Stöcker, Michael (Adjunct) 97022
    michael.stocker@nt.ntnu.no
Svendsen, Hallvard 94100
    hallvard.svendsen@nt.ntnu.no
Thorsen, Gunnar (Emeritus) 94115
    gunnar.thorsen@nt.ntnu.no

Associate Professors

Andreassen, Jens-Petter	94209
	jens-petter.andreassen@nt.ntnu.no
Haanæs, Egil (Emeritus)	94137
	egil.haanas@nt.ntnu.no
Haug-Warberg, Tore	94108
	tore.haug-warberg@nt.ntnu.no
Moe, Stoerker T.	94032
	stoerker.moe@nt.ntnu.no
Rønning, Magnus	94121
	magnus.ronning@nt.ntnu.no
Venvik, Hilde	92831
	hilde.venvik@nt.ntnu.no
Øye, Gisle	94135
	gisle.oyent.ntnu.no

Accredited by:

Degrees granted 2004-2005:
M.S.: 25   Ph.D.: 14

Department reports to: David Nicholson (47 73596204)

Philippines

Bicol University

College of Engineering
Chemical Engineering Department
Campus 1
EM's Bo.
Legazpi City 4500 Philippines

Professors
Padre, Natividad (Chair)

Associate Professors
Macam, Anelia
annmacam@yahoo.com

Undergraduate advisor: Natividad M.Padre,Ph.D.

De La Salle University

Chemical Engineering Department
2nd Floor, Velasco Building
De La Salle University
2401 Taft Avenue, Manila

University phone	(632) 524-4611
University FAX	(632)526-1403
Department phone	(632) 536-0257
Department FAX	(632) 524-0563

Professors
Gallardo, Susan M.
gallardos@dlsu.edu.ph
Gaspillo, Pag-Asa D.
gaspillop@dlsu.edu.ph
Olano, Servillano, Jr. S.B.
olanos@dlsu.edu.ph
Roces, Susan A.
rocess@dlsu.edu.ph
Salazar, Carlito M.
salazarc@dlsu.edu.ph

Associate Professors
Abella, Leonila C. (Chair)
abella@dlsu.edu.ph
Auresenia, Joseph
aureseniaj@dlsu.edu.ph
Bacani, Florinda T.
bacanif@dlsu.edu.ph
Brondial, Yolanda P. (Associate Chair)
brondialy@dlsu.edu.ph
Cabigon, Noel P.
cabigonn@dlsu.edu.ph
Maridable, Julius B. (Dean)
maridablej@dlsu.edu.ph
Razon, Luis F.
razonl@dlsu.edu.ph
Tan, Raymond R.
tanr_a@dlsu.edu.ph

Assistant Professors
Baraoidan, Wilheliza A.
baraoidanw@dlsu.edu.ph
Uy, Marylou M.
uym@dlsu.edu.ph

Accredited by: PAASCU, CHED

Degrees granted 2003-2004:
B.S.: 29   M.S.: 6   Ph.D.: 1

Graduate advisor: Dr. Pag-asa Gaspillo

Undergraduate advisor: Dr. Leonila Abella

Student organization: CHEN Society
Advisor: Mr. Ted Monroy

Department reports to: Dean of Engineering

University of Santo Tomas

Department of Chemical Engineering
Roque Ruaqo Buliding
Espaqa, Manila 1008

University phone	00632 7313101
University FAX	00632 406 1611
Department phone	00632 7313101 loc. 8275
Department FAX	00632 7314041

Professors
Laurito, Evelyn (Head) .... 00632 7313101 loc. 8275
odgs@ust.edu.ph
Mabini, Marilyn (Dean) .... 00632 7313101 loc. 8275
odgs@ust.edu.ph

Accredited by:

Graduate advisor: Christina Castro Cabral

Undergraduate advisor: Dr. Marilyn C. Mabina

Poland

Technical University of Lodz

Faculty of Process and Environmental Engineering
ul. Wolczanska 213/215
93-005 Lodz

University phone	48 42 6312005
University FAX	48 42 6365014
Department phone	48 42 6313700
Department FAX	48 42 6365663

Professors

Gorak, Andrzej (Visiting)	
Heim, Andrzej (Chair)	48 42 631 37 30
Kaminski, Wladyslaw (Chair)	48 42 631 37 08
Ledakowicz, Stanislaw (Dean)	48 42 6313715
Mucha, Maria	48 42 631 37 85
Piddubniak, Oleksa	
Strumillo, Czeslaw (Emeritus)	48 42 631 37 35
Swiatkowski, Witold (Emeritus)	
Tyczkowski, Jacek	48 42 631 37 23
Wodzinski, Piotr	48 42 631 37 40
Zarzycki, Roman (Chair)	48 42 631 37 42
Zbicinski, Ireneusz (Chair)	48 42 631 37 73

Associate Professors

Chacuk, Andrzej	48 42 631 37 46
Dziubinski, Marek (Chair)	48 42 631 37 34
Kazimierski, Piotr (Adjunct)	48 42 631 36 94
Krzystek, Liliana (Adjunct)	48 42 631 37 38
Kuncewicz, Czeslaw (Associate Dean)	48 42 631 37 27
Nowicki, Lech (Associate Dean)	48 42 631 37 81
Pakowski, Zdzislaw (Associate Dean)	48 42 631 37 31
Petera, Jerzy	42 48 631 37 07
Skrzypski, Jerzy	48 42 631 37 09

Accredited by: PAKA Polish Accreditation Committee, UKA University Accreditation Committee

Degrees granted 2004-2005:

B.S.: 5   M.S.: 46   Ph.D.: 14

Department reports to: Dean of the Faculty of Process and Environmental Engineering

Technical University of Szczecin

Department of Chemical Engineering
and Environmental Protection Processes
Al. Piastow 42
71-065 Szczecin

University phone	91 4346751
Department phone	91 4494472
Department FAX	91 4494642

Professors

Jaworski, Zdzislaw	4494020 [†]
	jaworski@carbon.tuniv.szczecin.pl
Karcz, Joanna (Head)	4494335
	joanka@carbon.tuniv.szczecin.pl
Kosmider, Joanna	4494519
	jakos@carbon.tuniv.szczecin.pl
Masiuk, Stanislaw	4494847
	smasiuk@carbon.tuniv.szczecin.pl
Paderewski, Mscislaw (Emeritus)	4494426
	mpader@carbon.tuniv.szczecin.pl
Strek, Fryderyk (Emeritus)	4494560
	stef@carbon.tuniv.szczecin.pl

Associate Professors

Dudczak, Jan	4494922
	dudus@carbon.tuniv.szczecin.pl
Haba, Alfred (Emeritus)	4494528
	ahaba@carbon.tuniv.szczecin.pl
Nastaj, Jozef	4494084
	jonas@carbon.tuniv.szczecin.pl
Szaniawska, Daniela	4494688
	dszan@carbon.tuniv.szczecin.pl

Assistant Professors

Ambrozek, Bogdan	4494622
	ambog@carbon.tuniv.szczecin.pl
Derecki, Wladyslaw	4494023
	wder@carbon.tuniv.szczecin.pl
Gabrus, Elzbieta	4494925
	elga@carbon.tuniv.szczecin.pl
Kawecka-Typek, Julita	4494183
	Julita.Kawecka-Typek@ps.pl
Kuzniewska-Lach, Irena	4494528
	drikula@carbon.tuniv.szczecin.pl
Lach, Krzysztof	4494436
	klach@carbon.tuniv.szczecin.pl
Lacki, Henryk	4494155
	hlacki@carbon.tuniv.szczecin.pl
Majkut, Aleksander	4494437
	almat@carbon.tuniv.szczecin.pl
Major, Marta	4494243
	mmajor@carbon.tuniv.szczecin.pl
Michalska, Marzena	4494243
	mamich@carbon.tuniv.szczecin.pl
Paterkowski, Wojciech	4494399
	wpater@carbon.tuniv.szczecin.pl
Suszek, Ewa (Associate Dean)	4494731
	evelinas@carbon.tuniv.szczecin.pl
Zakrzewska, Barbara	4494159
	zakrzewska@carbon.tuniv.szczecin.pl

[†] Faculty numbers have prefix 91.

Department reports to: Prof. Eugeniusz Milchert, Dean of College

Warsaw University of Technology

Faculty of Chemical and Process Engineering
ul. Warynskiego 1
00-645 Warszawa

University phone	(4822) 660 7220
Department phone	(4822) 660 6369
Department FAX	(4822) 825 1440

Professors

Baldyga, Jerzy (Associate Dean)	660 6376 [†]
	baldyga@ichip.pw.edu.pl
Bin, Andrzej	660 6373
	bina@ichip.pw.edu.pl
Chmielewski, Andrzej	660 6249
	chmielewski@ichip.pw.edu.pl
Gawronski, Roman	660 6276
	gawronski@ichip.pw.edu.pl

Gradon, Leon (Dean) 660 6279
gardon@ichip.pw.edu.pl
Marcinkowski, Ryszard (Emeritus) 660 6419
marcinkowski@ichip.pw.edu.pl
Pohorecki, Ryszard 660 6314
pohorecki@ichip.pw.edu.pl
Sieniutycz, Stanislaw 825 6340
sieniutycz@ichip.pw.edu.pl
Szewczyk, Krzysztof 660 6413
szewczyk@ichip.pw.edu.pl
Szwast, Zbigniew (Associate Dean) 825 6340
szwast@ichip.pw.edu.pl
Urbanek, Andrzej (Emeritus) 660 6352
urbanek@ichip.pw.edu.pl
Warych, Jerzy 660 6372
warych@ichip.pw.edu.pl
Wolny, Andrzej 660 6350
wolny@ichip.pw.edu.pl
Wronski, Stanislaw (Emeritus) 660 6295
wronski@ichip.pw.edu.pl

Associate Professors

Molga, Eugeniusz 660 6293
molga@ichip.pw.edu.pl
Piatkiewicz, Wojciech 660 5352
piatkiew@ichip.pw.edu.pl
Podgorski, Albert 660 6351
podgorsa@ichip.pw.edu.pl

† Faculty numbers have prefix 4822.

Accredited by: Conference Rectors Academic Schools

Degrees granted 2004-2005:

     M.S.: 62   Ph.D.: 10

Graduate advisor: Dr. Machniewski Piotr

Undergraduate advisor: Dr. Moniuk Wladyslaw

Department reports to: Kurnik Wlodzimierz, Rector of Warsaw University of Technology

Portugal

Universidade do Porto

Departamento de Engenharia Quimica
Faculdade de Engenharia
Rua Dr Roberto Frias s/n
4200-465 Porto
University phone .... (351) 22 508 1400
Department phone .... (351) 22 508 1884
Department FAX .... (351) 22 508 1449

Professors

Azevedo, Sebastiao J.C.F. (Chair) ...... (351) 22 508 1647
sfeyo@fe.up.pt
MO,PR
Costa, Carlos .... (351) 22 508 1670
ccosta@fe.up.pt
EN,EV,MO
de Carvalho, João R. G. .... (351) 22 508 1640
jrguedes@fe.up.pt
HT,TP
Figueiredo, José L. C. .... (351) 22 508 1663
jlfig@fe.up.pt
AM,RE,SU

Medina, A. G. .... (351) 22 508 1656
amedina@fe.up.pt
SE
Melo, Luís .... (351) 22 508 1588
lmelo@fe.up.pt
BT,HT,TP
Rodrigues, Alírio E. .... (351) 22 508 1671
arodrig@fe.up.pt
MO,PR,RE,SE
Salcedo, Romualdo L. R. (Associate Chair) (351) 22 508 1644
rsalcedo@fe.up.pt
EV,MO,PR

Associate Professors

Alves, Arminda .... (351) 22 508 1883
aalves@fe.up.pt
BT
Campos, João M. .... (351) 22 508 1692
jmc@fe.up.pt
HT,TP
Costa, Mário R. .... (351) 22 508 1666
mrcosta@fe.up.pt
PO,RE
Dias, Madalena .... (351) 22 508 1661
dias@fe.up.pt
HT,MO,TP
Faria, Joaquim L. .... (351) 22 508 1645
jlfaria@fe.up.pt
AM,SU
Gonçalves, Maria P. .... (351) 22 508 1684
pilarg@fe.up.pt
SU
Lopes, Jose C. .... (351) 22 508 1667
lopes@fe.up.pt
HT,TP
Loureiro, José M. .... (351) 22 508 1672
loureiro@fe.up.pt
MO,RE,SE
Macedo, Eugénia .... (351) 22 508
eamacedo@fe.up.pt
TH
Martins, J. I. .... (351) 22 508 1643
jipm@fe.up.pt
AM,SU
Mendes, Adélio .... (351) 22 508 1695
mendes@fe.up.pt
SE
Pinto, Alexandra .... (351) 22 508 1675
apinto@fe.up.pt
EN,TP
Sereno, Alberto C. .... (351) 22 508 1655
sereno@fe.up.pt
BT,MO
Órfão, José .... (351) 22 508 1665
jjmo@fe.up.pt
RE,SU

Assistant Professors

Alves, Manuel .... (351) 22 508 1680
mmalves@fe.up.py
TP
Barbosa, Domingos .... (351) 22 508 1660
dbarbosa@fe.up.pt
SE,TH
Bastos, João .... (351) 22 508 1658
jbastos@fe.up.pt
MO,SE

Botelho, Cidália ....(351) 22 508 1885
    cbotelho@fe.up.pt
    BT,EV
Coelho, Manuel ....(351) 22 508 1679
    mcoelho@fe.up.pt
    MO,NT
Ferraz, Maria C. A. ....(351) 22 508 1688
    aferraz@fe.up.pt
    EV
Ferreira, Palmira O. ....(351) 22 508 1668
    pof@fe.up.pt
    SE
Madeira, L. M. ....(351) 22 508 1519
    mmadeira@fe.up.pt
    RE
Magalhães, Fernão D. ....(351) 22 508 1601
    fdmagalh@fe.up.pt
    PO,SE
Martins, Fernando G. ....(351) 22 508 1974
    fgm@fe.up.pt
    MO
Mendonça, João ....(351) 22 508 1654
    mendonca@fe.up.pt
    SE
Morgado, José
    jmorgado@citeve.pt
Nunes, Olga P. ....(351) 22 508 1917
    opnunes@fe.up.pt
    BM,BT
Pereira, M. do C. ....(351) 22 508 1590
    mccsp@fe.up.pt
    EV,NT
Pereira, M. F. ....(351) 22 508 1468
    fpereira@fe.up.pt
    EV,RE
Rocha, Fernando A. N. ....(351) 22 508 1678
    frocha@fe.up.pt
    SE,TP
Santos, Lúcia ....(351) 22 508 1682
    lsantos@fe.up.pt
    EV
Soares, Helena ....(351) 22 508 1650
    hsoares@fe.up.pt
    EV
Tavares, Manuel
    manuel.leao@petrogal.pt

Accredited by: Ordem dos Engenheiros

Degrees granted 2005-2006:
    B.S.: 6   M.S.: 12   Ph.D.: 14

Undergraduate advisor: Sebastião Feyo de Azevedo

Department reports to: Carlos A. V. da Costa, Dean, Fac. of Engineering

Qatar

University of Qatar

College of Engineering
Department of Chemical Engineering
P. O. Box 2713
Doha, Qatar
Deliveries: Head of Department
University phone 974 485-2466
University FAX 974 485-2491
Department phone 974-485-2986
Department FAX 974-485-2101

Professors
Wilson, Andrew J. .... 485-1119 [†]
    awilson@qu.edu.qa

Associate Professors
Al-Asheh, Sameer 485-1748
    alasheh@qu.edu.qa
Haghtalab, Ali 485-2122
    ahaghtalab@qu.edu.qa

Assistant Professors
Al-Sayegh, Abdulreda 485-2123
    ralsaygh@qu.edu.qa
Alfadalah, Hassan (Acting Head) 485-2100
    alfadala@qu.edu.qa
Almohanadi, Nasser 485-2100
    nalmohan@qu.edu.qa

[†] Faculty numbers have prefix 974.

Accredited by: ABET

Degrees granted 2004-2005:
    B.S.: 9

Student organization: ChE Majlis
Advisor: Head of Department

Saudi Arabia

King Fahd University of Petroleum & Minerals

Chemical Engineering Department
PO Box 5050
Dhahran 31261
University phone 966-3-860-0000
Department phone 966-3-860-2205
Department FAX 966-3-860-4234

Professors
Abul-Hamayel, Mohammad A. .... 860-3903 [†]
    hamayel@kfupm.edu.sa
    EN
Al-Amer, Adnan M. (Chair) 860-2353
    alamer@kfupm.edu.sa
    RE
Al-Saleh, Muhammad A. 860-2029
    masaleh@kfupm.edu.sa
    EN,RE

208 Saudi Arabia

Al-Shalabi, Mazen A. 860-2198
mshalabi@kfupm.edu.sa
RE
Amin, Mohamed B. 860-2205
mbamin@kfupm.edu.sa
EV,MO,PO
Kahraman, Ramazan 860-4987
kahraman@kfupm.edu.sa
AM,EN,NT
Maadhah, Ali G. 860-3882
amaadhah@kfupm.edu.sa
EN,SE
Shaikh, Abdullah A. 860-2257
aashaikh@kfupm.edu.sa
MO,RE

Associate Professors

Abbas, Nureddin M. 860-3667
abbasnm@kfupm.edu.sa
SU
Abu-Sharkh, Basel 860-2744
sharkh@kfupm.edu.sa
NT,PO
Al-Ali, Habib H. 860-2599
hhali@kfupm.edu.sa
EN
Al-Harbi, Dulaihan K. 860-2375
dharbi@kfupm.edu.sa
MO
Al-Khattaf, Sulaiman S. 860-1429
skhattaf@kfupm.edu.sa
RE
Al-Naafa, Mohammad A. 860-3688
manaafa@kfupm.edu.sa
SE
Hussein, Ibnelwaleed A. 860-2235
ihussein@kfupm.edu.sa
NT,PO
Rahman, S.U. 860-2219
srahman@kfupm.edu.sa
EV,NT,RE

Assistant Professors

Al-Arfaj, Muhammad A. 860-1694
maarfaj@kfupm.edu.sa
MO,PR
Al-Baghli, Nadhir A. 860-1476
nalbaghli@kfupm.edu.sa
EN,MO
Al-Mubaiyedh, Usamah A. 860-1528
usamah@kfupm.edu.sa
MO
Zaidi, S.M. J. 860-1242
zaidismj@kfupm.edu.sa
AM,NT,RE

[†] Direct dial with prefix 03.

Accredited by: ABET

Degrees granted 2005-2006:
    B.S.: 51   M.S.: 5

Graduate advisor: Dr. Basel F. Abu-Sharkh

Undergraduate advisor: Dr. Ramazan Kahraman

Student organization: ChE Club
Advisor: Dr. Nadhir A. Al-Baghli

Department reports to: Dr. Samir A. Al-Baiyat, Dean, College of Engineering

King Saud University

Chemical Engineering Department
P.O. Box 800
Riyadh 11421
Kingdom of Saudi Arabia

University phone 9661-467-4000
University FAX 9661-467-7580
Department phone 966-1-4676851
Department FAX 966-1-4678770

Professors

Abasaeed, E. A. .... 76856 [†]
abasaeed@ksu.edu.sa
BT,MO,PO,RE
Abashar, E. E. M. 75843
mabashar@ksu.edu.sa
MO,RE
Ajbar, M. A. 76843
aajbar@ksu.ed.sa
MO,PR
Al-Fariss, F. T. 76875
fariss@ksu.edu.sa
PO,SU
Al-Humaizi, l. K. (Chair) 76813
humaizi@ksu.edu.sa
MO,PR
Al-Mutaz, S. I. 76870
almutaz@ksu.edu.sa
EN,EV,NT
Al-Zahrani, M. S. 76873
szahrani@ksu.edu.sa
PO,RE
Ali, M. E. 76871
amkamal@ksu.edu.sa
MO,PR
Asif, Mohamed 76849
masif@ksu.edu.sa
SE,SU
Boumaaza, M. M. 79151
mouradb@ksu.edu.sa
EN,EV,SU
Fakeeha, H. A. 76847
anishf@ksu.edu.sa
EN,RE,SE
Mustafa, M. H. 76854
hmohm@ksu.edu.sa
SE
Wagialla, M. K. 76846
wagialla@ksu.edu.sa
MO

Associate Professors

Al-Ahmed, I. M. 76874
malahmad@ksu.edu.sa
EN,EV
Al-Habdan, M. F. 76842
habdan@ksu.edu.sa
EN,MO,SE
Al-Hazzaa, I. M. 76845
masai@ksu.edu.sa
AM,SU

Al-Odan, A. M. 76869
    alodan@ksu.edu.sa
    AM,EV,PO,SU
Almasry, A. W. 76853
    walmasry@ksu.edu.sa
    BT,RE

Assistant Professors

Al-Hussaini, A. A. 76738
    amalik@ksu.edu.sa
    EN,SE
Al-Mubaddel, S. F. 76848
    mubaddel@ksu.edu.sa
    AM,EN,EV,PO
Al-Mutlaq, M. A. A. 76738
    mutlaq@ksu.edu.sa
    EN,MO
Al-Nashef, Inas 76865
    alnashef@ksu.edu.sa
    EN,EV
Al-Rabiah, A. A. 76844
    arabiah@ksu.edu.sa
    EN,MO,SE
Al-Sugair, A. K. 76857
    alsugair@ksu.edu.sa
Al-Zaghyer, S. Y. 76855
    yszs@ksu.edu.sa
    RE

† Direct dial with prefix 966-1-46.

Accredited by: ABET

Degrees granted 2005-2006:
    B.S.: 40   M.S.: 9   Ph.D.: 1
Graduate advisor: Prof. Anis Fakeeha

Undergraduate advisor: Prof. Emad Ali

Student organization: Saudi Chemical Society
Advisor: Abdul-Razzaq Al-Shammary

Placement service: Dr. Khalid Al-Humaizi

Singapore

National University of Singapore

Department of Chemical and Biomolecular Engineering
Blk E5 #02-09, 4 Engineering Drive 4
Singapore 117576

University phone .... (65) 6516 6666
Department phone .... (65) 6516 2186
Department FAX .... (65) 6779 1936

Professors

Chung, Tai-Shung N. .... (65) 6516-6645
    chencts@nus.edu.sg
    AM,EN,EV,MO,NT,PO,SE,SU
Farooq, Shamsuzzaman .... (65) 6516-6545
    chesf@nus.edu.sg
    EN,MO,SE
Kang, En-Tang .... (65) 6516-2189
    cheket@nus.edu.sg
    AM,BM,ME,NT,PO,SU
Karimi, Iftekhar A. .... (65) 6516-6359
    cheiak@nus.edu.sg
    PR
Krantz, William B. (Visiting) .... (65) 6516-1067
    chekwb@nus.edu.sg
Lee, Jim Y. .... (65) 6516-2899
    cheleejy@nus.edu.sg
    AM,EN,NT,PO,RE,SU
Neoh, Koon G. .... (65) 6516-2176
    chenkg@nus.edu.sg
    AM,BM,NT,PO,SU
Rajagopalan, Raj (Head) .... (65) 6516-2186
    chehead@nus.edu.sg
Sandler, Stanley I. (Visiting)
    chessi@nus.edu.sg
    SE,TH
Tan, Thiam C. .... (65) 6516-2192
    chetantc@nus.edu.sg
    TH
Yap, Gek S. M. .... (65) 6516-6369
    cheyapm@nus.edu.sg

Associate Professors

Chen, Shing B. .... (65) 6516-5237
    checsb@nus.edu.sg
    MO,NT,PO,SE,TP
Chiu, Min-Sen .... (65) 6516-2223
    checms@nus.edu.sg
    PR
Feng, Si-Shen .... (65) 6516-3835
    chefss@nus.edu.sg
    AM,BM,BT,MO,NT,PO,SU
Foo, Swee C. .... (65) 6516-8721
    chefoosc@nus.edu.sg
Hidajat, Kus .... (65) 6516-2191
    chehidak@nus.edu.sg
Hong, Liang .... (65) 6516-5029
    chehongl@nus.edu.sg
    AM,EN,PO,RE,SE,SU
Kawi, Sibudjing .... (65) 6516-6312
    chekawis@nus.edu.sg
    AM,EN,EV,NT,RE,SE,SU
Krishnaswamy, Peruvemba R. .... (65) 6516-2177
    chekrish@nus.edu.sg
    PR
Li, Zhi .... (65) 6516-5083
    chelz@nus.edu.sg
    BM,BT,EV,PO,PR,RE
Loh, Kai C. .... (65) 6516-2174
    chelohkc@nus.edu.sg
    BT,EV,MO,SE
Rangaiah, Gade P. .... (65) 6516-2187
    chegpr@nus.edu.sg
    MO,PR,RE
Srinivasan, Madapusi P. .... (65) 6516-2171
    chesmp@nus.edu.sg
    AM,EV,ME,NT,PO,SU
Srinivasan, Rajagopalan .... (65) 6516-8041
    chergs@nus.edu.sg
    BT,EN,EV,MO,PR
Tan, Reginald B. H. .... (65) 6516-6360
    chetanbh@nus.edu.sg
    BM,EV,NT,PR,SE,TH,TP
Ti, Hwei C. .... (65) 6516-2188
    chetihc@nus.edu.sg
    MO
Ting, Yen P. .... (65) 6516-2190
    chetyp@nus.edu.sg
    BT,EV,SU

Uddin, Mohammad S. .... (65) 6516-2886
  cheshahb@nus.edu.sg
  SE
Wang, Chi-Hwa .... (65) 6516-5079
  chewch@nus.edu.sg
  BM,MO,NT,TP
Zeng, Hua C. .... (65) 6516-2896
  chezhc@nus.edu.sg
  NT
Zhao, Xiu S. G. .... (65) 6516-4727
  chezxs@nus.edu.sg
  AM,EN,EV,NT,RE,SU

Assistant Professors

Birgersson, Karl E. .... (65) 6516-7132
  chebke@nus.edu.sg
  TP
Gunawan, Rudiyanto .... (65) 6516-6617
  chegr@nus.edu.sg
  MO,PR
Jiang, Jianwen .... (65) 6516-5083
  chejj@nus.edu.sg
  EV,MO,PO,SE,TH
Khan, Saif A. .... (65) 6516-5133
  chesakk@nus.edu.sg
  AM,MO,NT,RE
Lee, Dong-Yup .... (65) 6516-6907
  cheld@nus.edu.sg
  BT,MO,PR
Liu, Bin .... (65) 6516-8049
  cheliub@nus.edu.sg
Saeys, Mark .... (65) 6874-5826
  chesm@nus.edu.sg
Samavedham, Lakshminarayanan .... (65) 6516-8484
  chels@nus.edu.sg
  BM,EV,PR
Tong, Yen W. .... (65) 6516-8467
  chetyw@nus.edu.sg
  AM,BM,BT,EV,NT,PO,SU
Trau, Dieter W. .... (65) 6516-8052
  chetrau@nus.edu.sg
Yang, Kun-Lin .... (65) 6516-6614
  cheyk@nus.edu.sg
Yung, Lin-Yue L. .... (65) 6516-1699
  cheyly@nus.edu.sg
  BM,BT,EV,NT,PO,SU

Accredited by: IChemE(UK), IES (Singapore)

Degrees granted 2005-2006:
  B.S.: 252   M.S.: 59   Ph.D.: 30

Undergraduate advisor: G. P. Rangaiah

Student organization: Chemical and Biomolecular Engineering Students Society

Department reports to: Seeram Ramakrishna, Dean of Engineering

Slovenia

University of Ljubljana

Faculty of Chem. and Chem. Technology
Askerceva 5
1000 Ljubljana
Deliveries: P.O.B. 537, SLO-1001 Ljubljana

University phone 386 1 2419 100
University FAX 386 1 2419 220
Department phone 386 1 2419 350
Department FAX 386 1 2419 144

Professors

Golob, Janvit .... x522 [†]
  janvit.golob@fkkt.uni-lj.si
Koloini, Tine x500
  tine.koloini@fkkt.uni-lj.si
Levec, Janez x502
  janez.levec@fkkt.uni-lj.si
Modic, Roman (Emeritus) x520
Pejovnik, Stanislav (Dean) x202
  stane.pejovnik@fkkt.uni-lj.si
Zumer, Miha x504
  miha.zumer@fkkt.uni-lj.si

Associate Professors

Berovic, Marin x510
  marin.berovic@fkkt.uni-lj.si
Krajnc, Matjaz (Associate Dean) x524
  matjaz.krajnc@fkkt.uni-lj.si
Macek, Jadran x200
  jadran.macek@fkkt.uni-lj.si
Pavko, Aleksander x506
  saso.pavko@fkkt.uni-lj.si
Plazl, Igor x512
  igor.plazl@fkkt.uni-lj.si
Zagorc-Koncan, Jana x537
  jana.zagorc@fkkt.uni-lj.si

Assistant Professors

Lakota-Druzina, Ana x514
  ana.lakota-druzina@fkkt.uni-lj.si
Marinsek, Marjan x204
  marjan.marinsek@fkkt.uni-lj.si
Poljansek, Ida x544
  ida.poljansek@fkkt.uni-lj.si
Zgajnar-Gotvajn, Andreja x518
  andreja.zgajnar@fkkt.uni-lj.si
Zupancic-Valant, Andreja x529
  andreja.valant@fkkt.uni-lj.si

[†] Dial 386 1 2419 and extension.

Accredited by:

Degrees granted 2004-2005:
  B.S.: 41   M.S.: 8   Ph.D.: 5

Department reports to: Prof. Dr. Matjaz Krajnc

University of Maribor

Fakulteta za kemijo in kemijsko tehnologijo
Faculty of Chemistry and Chemical Engineering
Smetanova 17
SI, 2000 Maribor
Deliveries: Prof.Dr. Peter Glavic
University phone .... (386 2) 23 55 800
Department phone .... (386 2) 22 94 400
Department FAX .... (386 2) 25 27 774

Professors

Dolecek, Valter (Dean) .... (386 2) 22 94 441
    valter.dolecek@uni-mb.si
Drofenik, Mihael .... (386 2) 22 94 416
    miha.drofenik@uni-mb.si
Glavic, Peter .... (386 2) 22 94 451
    glavic@ijs.si
Knez, Zeljko .... (386 2) 22 94 461
    zeljko.knez@uni-mb.si
Kravanja, Zdravko .... (386 2) 22 94 481
    kravanja@uni-mb.si
Krope, Jurij .... (386 2) 22 94 475
    jurij.krope@uni-mb.si

Associate Professors

Butinar, Branko .... (386 2) 22 94 489
    branko.butinar@uni-mb.si
Stropnik, Crtomir .... (386 2) 22 94 421
    crtomir.stropnik@uni-mb.si

Assistant Professors

Ban, Irena .... (386 2) 22 94 417
    irena.ban@uni-mb.si
Brodnjak Voncina, Darinka .... (386 2) 22 94 432
    darinka.brodnjak@uni-mb.si
Dobcnik, Danilo (Emeritus) .... (386 2) 22 94 431
    danilo.dobcnik@uni-mb.si
Goricanec, Darko .... (386 2) 22 94 475
    darko.goricanec@uni-mb.si
Gorek, Andreja .... (386 2) 22 94 453
    andreja.gorsek@uni-mb.si
Habulin, Maja .... (386 2) 22 94 462
    maja.habulin@uni-mb.si
Korpar, Samo .... (386 2) 22 94 490
    samo.korpar@ijs.si
Krajnc, Majda .... (386 2) 22 94 452
    majda.krajnc@uni-mb.si
Krajnc, Peter .... (386 2) 22 94 422
    peter.krajnc@uni-mb.si
Novak Pintaric, Zorka .... (386 2) 22 94 482
    zorka.novak@uni-mb.si
Oreki, Severina .... (386 2) 22 94 454
    severina.oreski@uni-mb.si
Ozim, Vojko (Emeritus) .... (386 2) 22 94 471
    vojko.ozim@uni-mb.si
Petek, Aljana .... (386 2) 22 94 443
    aljana.petek@uni-mb.si
Simonic, Marjana .... (386 2) 22 94 472
    marjana.simonic@uni-mb.si
Volavek, Bogdan (Emeritus) .... (386 2) 22 94 415
    bogdan.volavsek@uni-mb.si
kerget, Mojca .... (386 2) 22 94 463
    mojca.skerget@uni-mb.si

Graduate advisor: Zdravko Kravanja, prof. dr.

Undergraduate advisor: Peter Glavic, prof. dr.

Student organization: Vanja Bogadi

Department reports to: Valter Dolecek, Dean

Placement service: Franc Purkeljc, M.Sc.

Republic of South Africa

University of Cape Town

University of Cape Town
Department of Chemical Engineering
Private Bag X3
Rondebosch 7701
University phone 27 21 650 9111
Department phone 27 21 650 2518
Department FAX 27 21 650 5501

Professors

Dry, Mark
    med@chemeng.uct.ac.za
Hansford, Geoff S.
    gsh@chemeng.uct.ac.za
Harrison, Sue T. L.
    stlh@chemeng.uct.ac.za
van Steen, Eric (Head) 27 21-6502509
    evs@chemeng.uct.ac.za

Associate Professors

Burton, Stephanie
    sburton@chemeng.uct.ac.za
Fletcher, Jack C.Q.
    jcqf@chemeng.uct.ac.za
Fraser, Duncan M.
    dmf@chemeng.uct.ac.za
Gaylard, Peter
    pgaylard@chemeng.uct.ac.za
Harris, Peter
    pharris@chemeng.uct.ac.za
Lewis, Alison E.
    alison@chemeng.uct.ac.za
Moller, Klaus
    km@chemeng.uct.ac.za
Petrie, Jim G.
    petrie@chemeng.uct.ac.za

Assistant Professors

Bradshaw, Dee J.
    db@chemeng.uct.ac.za
Case, Jenni M
    jcase@chemeng.uct.ac.za
Chakraborty, Aninda
    aninda@chemeng.uct.ac.za
Deglon, Dave A.
    dad@chemeng.uct.ac.za
Harris, Martin C
    mch@chemeng.uct.ac.za
Musonge, Paul
    pmusonge@chemeng.uct.ac.za
von Blottnitz, Harro
    hvb@chemeng.uct.ac.za

Accredited by: ECSA

Graduate advisor: S Burton

Undergraduate advisor: K Möller

Department reports to: C. T. O'Connor, Dean

University of KwaZulu-Natal

School of Chemical Engineering
University of KwaZulu-Natal
Howard College Campus
Durban, 4041
Deliveries: University of Kwazulu-Natal, Howard College Campus, King George V Avenue, DURBAN, 4001

University phone 27-31-2609111
Department phone 27-31-2603115
Department FAX 27-31-2601118

Professors

Buckley, Christopher A. 27-31-2603131
    Buckley@ukzn.ac.za
    BT,EV
Carsky, Milan 27-31-2603544
    carskym@ukzn.ac.za
    EN,RE
Loveday, Brian K. (Emeritus) 27-31-2603121
    Loveday@ukzn.ac.za
Mulholland, Mike 27-31-2603123
    Mulholland@ukzn.ac.za
    PR
Raal, J. D. (Emeritus) 27-31-2603124
    Raal@ukzn.ac.za
    SE
Ramjugernath, Deresh (Head) 27-31-2603128
    Ramjuger@ukzn.ac.za
    SE

Associate Professors

Arnold, David R. 27-31-2601228
    Arnoldd@ukzn.ac.za
    RE,SE
Starzak, Maciej J. 27-31-2603117
    Starzak@ukzn.ac.za
    MO,RE

Assistant Professors

Pocock, Jon 27-31-2603377
    Pocockj@ukzn.ac.za

Accredited by: IChemE (UK), ECSA

Degrees granted 2005-2006:
    B.S.: 66   M.S.: 14

Graduate advisor: Deresh Ramjugernath

Department reports to: Prof. N. Ijumba, Dean

Placement service: Deresh Ramjugernath, Head of School

North-West University, Potchefstroom Campus

School of Chemical and Minerals Engineering
Private Bag X6001
Potchefstroom, 2520

University phone 27 (18) 299-1111
Department phone 27 (18) 299-1656
Department FAX 27 (18) 299-1535

Professors

Bruinsma, O.S.L. D. 27 18 299-1669
    chioslb@puk.ac.za
Marx, Sanette 27 18 299-1995
    chism@puk.ac.za
Neomagus, HWJP H. 27 18 299 1991
    chihwn@puk.ac.za
Waanders, Frans B. (Director) 27 18 299-1994
    chifbw@puk.ac.za

Assistant Professors

Campbell, Quentin P. 27 18 299-1993
    chiqpc@puk.ac.za
le Roux, Marco 27 18 299-1990
    chimlr@puk.ac.za
van der Gryp, Percy 27 18 299-1953
    chipvdg@puk.ac.za

Accredited by: IChemE(SA), ECSA, IChemE(UK)

Degrees granted 2004-2005:
    B.S.: 15   Ph.D.: 2

Graduate advisor: Frans Waanders

Undergraduate advisor: Retha Potgieter

Student organization: Engineering Student Society
Advisor: Retha Potgieter

Department reports to: Dean of Faculty of Engineering

University of Pretoria

Department of Chemical Engineering
Pretoria 0001

University phone 27 (12) 420-4111
Department phone 27 (12) 420-2475
Department FAX 27 (12) 362-5173

Professors

Chirwa, Evans .... 27 (12) 420-5894 [†]
    evans.chirwa@up.ac.za
    BT,EV
deVaal, Philip (Head) 27 (12) 420-2197
    philip.devaal@up.ac.za
    MO,PR
Focke, Walter 27 (12) 420-2588
    xyris@mweb.co.za
    AM,PO
Grimsehl, Uys (Emeritus) 27 (12) 420-2475
    uys.grimsehl@up.ac.za
    EN
Heydenrych, Michael 27 (12) 420-2199
    mike.heydenrych@up.ac.za
    MO,PR
Majozi, Thokozani 27 (12) 420-4130
    thoko.majozi@up.ac.za
    MO,PR
Mandersloot, Wim (Honorary) 27 (12) 803-6863
    wmanders@eng.up.ac.za
    RE
Morgan, Dave (Honorary) 27 (12) 420-2856
    dave.morgan@up.ac.za
    AM
Pretorius, W. A. (Emeritus) 27 (12) 804-3070
    BT,EV

Schoeman, Japie 27 (12) 420-3569
japie.schoeman@up.ac.za
BT,EV
Schutte, Frik 27 (12) 420-3571
cschutte@eng.up.ac.za
BT,EV
Skinner, W (Emeritus) 27 (12) 665-0200
AM

Associate Professors

Christopher, Lew (Honorary) 27 (12) 420-2593
lew.christopher@up.ac.za
BT
duPlessis, Barend 27 (12) 420-3740
barend.duplessis@up.ac.za
SE
Friend, Francois 27 (12) 420-3741
francois.friend@up.ac.za
EV
Nicol, Willie 27 (12) 420-3796
willie.nicol@up.ac.za
RE
Tolmay, Andries 27 (12) 420-3020
dries.tolmay@up.ac.za
EN,MO

Assistant Professors

du Toit, Elizbe L. 27 (12) 420-3641
elizbe.dutoit@up.ac.za
RE
Rolfes, Heidi 27 (12) 420 2588
heidi.rolfes@up.ac.za
AM,BM,PO
Sandrock, Carl
MO,PR

† The prefix for all numbers is 27 (12).

Accredited by: IChemE (SA), IChemE (UK), ECSA

Degrees granted 2005-2006:
    B.S.: 34   M.S.: 9

Department reports to: R.F. Sandenbergh (Dean)

University of the Witwatersrand, Johannesburg

School of Chemical and Metallurgical Engineering
Private Bag 3
WITS
2050
Deliveries: Richard Ward Building, Room 321

University phone .... (11) 717-7510
University FAX .... (11) 403 1471
Department phone .... (11) 717-7510
Department FAX .... (11) 403 1471

Professors

Bryson, A.W. (Emeritus) .... (11) 717 7528 †
awb@prme.wits.ac.za
EV,RE,SE
Cornish, Lesley (Visiting) .... (11) 709 4474
lesleyc@mintek.co.za
AM,MO
Eric, R. H. .... (11) 717 7537
rhe@prme.wits.ac.za
AM,HT,MO,RE
Falcon, Rosemary (Visiting) .... (11) 717 7535
falcons@icon.co.za
EN
Glasser, David (Emeritus) .... (11) 717 7512
david.glasser@comps.wits.ac.za
EV,HT,MO,PR,TH
Hildebrandt, Diane .... (11) 717 7527
diane.hildebrandt@comps.wits.ac.za
BT,EV,MO,PR,TH
Koursaris, Andreas .... (11) 717 7530
koursaris@prme.wits.ac.za
AM,SU
Luyckx, Silvana (Visiting) .... (11) 717 7524
silvana.luyckx@prme.wits.ac.za
AM,SU
Moys, Michael .... (11) 717 7518
mhm@comps.wits.ac.za
EN,MO,PR
Potgieter, Herman (Head) .... (11) 717 7546
hermanp@prme.wits.ac.za
AM,EN,EV
Sigalas, Jack .... (11) 717 7502
isigalas@prme.wits.ac.za
AM,NT,SU
Te Riele, Wolter (Emeritus) .... (11) 717 7538
wolter.teriele@prme.wits.ac.za

Associate Professors

Iyuke, Sunny .... (11) 717 7594
siyuke@prme.wits.ac.za
BT,EV,NT,TP
Kucukkaragoz, Serdar (Visiting) .... (11) 717 7514
cskucuk@prme.wits.ac.za
HT,MO,RE
Williams, Donald .... (11) 717 7531
dfw@prme.wits.ac.za

Assistant Professors

Aoyi, Ochieng .... (11) 717 7552
EN,EV
Jewell, Linda .... (11) 717 7507
ljewell@prme.wits.ac.za
BT,EN,EV,RE
Kauchali, Shehzaad .... (11) 717 7533
skauchali@prme.wits.ac.za
EN,EV,MO,TH
Ndlovu, Sehliselo .... (11) 717 7516
sendlovu@prme.wits.ac.za
BT,RE,SE
Sacks, Natasha .... (11) 717 7523
nsacks@prme.wits.ac.za
AM,SU
Wagner, Nikki 011 717 7540
nicola.wagner@wits.ac.za
EN,EV
Woollacott, Laurie .... (11) 717 7560
lwool@prme.wits.ac.za

† The country code for South Africa is 0027

Accredited by: ECSA [Engineering Council of South Africa]IChemE (UK)

Degrees granted 2005-2006:
    B.S.: 55   M.S.: 5   Ph.D.: 2

Graduate advisor: M. Raghununan

Undergraduate advisor: D. Williams

Department reports to: Prof. R. Nkado, Dean

Placement service: Prof. J.H. Potgieter

Spain

University of Alcala de Henares
Ingenieria Quimica

Chemical Engineering
Faculty of Sciences
28871 Alcala de Henares
Madrid
Deliveries: Crta. Barcelona, km 33,600
University phone 34 91 885 49 41
Department phone 34 91 885 49 41
Department FAX 34 91 885 50 88

Professors
García Calvo, Eloy 918 854939
eloy.garcia@uah.es

Associate Professors
Guardiola, Jesus 918 854976
jesus.guardiola@uah.es
Leton, Pedro 918 854974
pedro.leton@uah.es
Ramos, Guadalupe 918 855099
guadalupe.ramos@uah.es
Rodriguez, Antonio 918 854974
antonio.rodriguez@uah.es
Rosal, Roberto
roberto.rosal@uah.es

Assistant Professors
Arranz, Miguel A. 918 854950
miguelan.arranz@uah.es
Boltes, Karina 918 854974
karina.boltes@uah.es
Cortes, Candido 918 855099
candido.cortes@uah.es
Elvira, Rosario 918 854950
rosario.elvira@uah.es

Department reports to: Eloy Garcia Calvo

University of Alicante

Ingenieria Quimica
Facultad de Ciencias
Apartado 99
03080 Alicante
University phone 34-96-5903400
Department phone 34-96-5903867
Department FAX 34-96-5903826

Professors
Fernandez-Sempere, Julio
julio.fernandez@ua.es
Font-Montesinos, Rafael
Rafael.Font@ua.es
Gomis-Yagües, Vicente (Director)
vgomis@ua.es
Marcilla-Gomis, Antonio
antonio.marcilla@ua.es
Prats-Rico, Daniel
prats@ua.es
Ruiz-Bevia, Francisco
ruiz.bevia@ua.es

Associate Professors
Asensi-Steegmann, J. C.
JC.Asensi@ua.es
Beltran-Rico, Maribel
maribel.beltran@ua.es
Boluda-Botella, Nuria
Nuria.Boluda@ua.es
Caballero-Suárez, José A.
caballer@ua.es
Conesa-Ferrer, Juan A.
JA.Conesa@ua.es
Fernández-Torres, María J.
fernandez@ua.es
Garcia-Cortes, Angela N.
angela.garcia@ua.es
Gomez-Siurana, Amparo
amparo.gomez@ua.es
Martin-Gullon, Ignacio
gullon@ua.es
Olaya-Lopez, M. M.
maria.olaya@ua.es
Reyes-Labarta, Juan A.
ja.reyes@ua.es
Varó-Galvañ, Pedro J.
Pedro.Varo@ua.es

Accredited by:

Student organization: Aiqua Asociacion de Ingenieros Quimicos de Alicante

Autonoma University of Barcelona

Enginyeria Quimica
Edifici C
08193 Bellaterra
Barcelona
University phone 34-93-5811000
Department phone 34-93-5811018
Department FAX 34-93-5812013

Professors
Gòdia, Francesc . . . . 5812692 [†]
francesc.godia@uab.es
Lafuente, Javier 5812143
javier.lafuente@uab.es
Lopez, Josep (Director) 5811806
josep.lopez@uab.es
Solà, Carles 5812002
carles.sola@uab.es

Associate Professors
Benaiges, Dolors 5812144
mariadolors.benaiges@uab.es
Cairó, Jordi 5812694
jordijoan.cairo@uab.es
Casas, Carles 5811809
carles.casas@uab.es
de Mas, Carles 5811019
carles.demas@uab.es
Gabarrell, Xavier 5812789
xavier.gabarrell@uab.es

González, Glòria 5812791
 gloria.gonzalez@uab.es
Montesinos, José L. 5812142
 joseluis.montesinos@uab.es
Sarra, Montserrat 5812789
 montserrat.sarra@uab.es
Valero, Francisco 5811809
 francisco.valero@uab.es
Vicent, M. T. 5812142
 teresa.vicent@uab.es

Assistant Professors

Baeza, Juan 5812695
 juanantonio.baeza@uab.es
Blánquez, Francisca 5812141
 paqui.blanquez@uab.es
Cabello, Fernando 5811879
 fernando.cabello@uab.es
Carrera, Julian 5811808
 julian.carrera@uab.es
Cos, Oriol 5812695
 oriol.cos@uab.es
Gabriel, David 5812141
 david.gabriel@uab.es
Huix, Josep 5811018
Lecina, Martí 5811808
 marti.lecina@uab.es
Mach, Josep M. 5811018
Marín, Antoni 5811018
 antonio.marin@uab.es
Montràs, Anna 5811808
 anna.montras@uab.es
Pérez, Julio 5812141
 julio.perez@uab.es
Rieradevall, Joan 5811018
Sendra, Cristina 5811808
 cristina.sendra@uab.es
Suau, Trinitat 5811879
 trinitat.suau@uab.es

† Faculty numbers have prefix 34 93.

Accredited by:

Degrees granted 2003-2004:

 B.S.: 35 M.S.: 6 Ph.D.: 3

Department reports to: Joan Sorribes, Director, Escola Tècnica Superior d'Enginyeria

University of Cantabria

Dpto. Ingeniería Química y Quimica Inorganica
Universidad de Cantabria ETSIIT
Avenida de Los Castros, s/n.
39005 Santander Spain

University phone 34 942 20 15 00
University FAX 34 942 20 11 03
Department phone 34 942 20 15 90
Department FAX 34 942 20 15 91

Professors

Blanco Delgado, Carmen 34 942 20 14 71
 blancoc@unican.es
Ortiz Uribe, Inmaculada (Head) 34 942 20 15 85
 ortizi@unican.es
Viguri Fuente, Javier R. 34 942 20 15 89
 vigurij@unican.es

Associate Professors

Fernández Ferreras, Josefa 34 942 20 20 26
 fernandj@unican.es
Fernández Olmo, Ignacio 34 942 20 15 86
 fernandi@unican.es
Galán Corta, Berta 34 942 20 15 98
 galanb@unican.es
Garea Vázquez, Aurora 34 942 20 15 88
 gareaa@unican.es
González Martínez, Fernando 34 942 20 14 73
 gonzalfe@unican.es
Herrero Romero, Juana 34 942 20 15 80
 herreroj@unican.es
Ibáñez Mendizábal, Raquel 34 942 20 14 70
 ibanezr@unican.es
Oliván Martínez, Olga 34 942 20 15 96
 olivano@unican.es
Otero Hermida, Jose A. 34 942 20 15 95
 oteroj@unican.es
Pesquera González, Carmen 34 942 20 17 72
 pesquerc@unican.es
Renedo Omaechevarría, Josefina 34 942 20 15 80
 renedomj@unican.es
Rico Gutierrez, Jose L. 34 942 20 15 99
 ricoj@unican.es
San Roman San Emeterio, M F. 34 942 20 15 83
 sanromm@unican.es
Urtiaga Mendía, Ana M. (Associate Dean) 34 942 20 15 87
 urtiaga@unican.es

Assistant Professors

Aldaco, Rubén 34 942 201588
 aldacor@unican.es
Arce Recio, Roberto 34 942 20 15 93
 arcer@unican.es
Ayerbe de Aragón Aguilera, Asuncion 34 942 84 65 45
 ayerbea@unican.es
Colina Pérez, Jose M. 34 942 20 15 94
 colinajm@unican.es
Coz Fernández, Alberto 34 942 20 13 59
 coza@unican.es
García Posadas, Hipolito 34 942 20 14 70
 garciah@unican.es
Lasa Díaz, Cristina 34 942 20 15 88
 lasac@unican.es
Lena López, Gumersindo 34 942 20 15 92
 lenag@unican.es
Poza Fernández, Angel 34 942 20 15 92
 pozaa@unican.es
Rivero Martínez, María J. 34 942 20 15 82
 riveromj@unican.es
Ruiz Gutierrez, Gema 34 942 20 15 83
 gema.ruiz@unican.es

Accredited by: Spanish Education and Science Ministry

Degrees granted 2003-2004:

 B.S.: 30 Ph.D.: 3

Graduate advisor: Gorri Cirella, Eugenio Daniel

Undergraduate advisor: Garea Vázquez, Aurora

Placement service: COIE

216 Spain

Universidad de Castilla-La Mancha

Dpto. de Ingenieria Quimica
Av. Camilo Jose Cela s/n
13004 Ciudad Real

University phone	(34) 902.204.100
Department phone	(34) 902/204.100 x 3416
Department FAX	(34) 926-295318

Professors

Canizares, Pablo (Head)	x3412 [†]
	Pablo.Canizares@uclm.es
de Lucas, Antonio (Associate Chair)	x3426
	alucas@inqu-cr.uclm.es
Valverde, Jose L. (Director)	x3415
	JoseLuis.Valverde@uclm.es

Associate Professors

Alonso, Miguel A.	(34) 926-710577
	MiguelAngel.Alonso@uclm.es
Carnicer, Angel	(34) 926-710577
	Angel.Carnicer@uclm.es
Dorado, Fernando	x3411
	Fernando.Dorado
Duran, Antonio (Associate Dean)	x3814
	Antonio.Duran@uclm.es
Frades, Jesus M.	(34) 926-710577
	Jesús.Frades@uclm.es
Fuertes, Juan	x5414
	Juan.Fuertes@uclm.es
Gomez, Rocio (Associate Dean)	x3768
	Rocio.Gomez@uclm.es
Gracia, Ignacio	x3418
	Ignacio.Gracia@uclm.es
Lobato, Justo	x3418
	Justo.Lobato@uclm.es
Monteagudo, Jose M.	(34) 926-710577
	JoseMaria.Monteagudo@uclm.es
Perez, Angel	x3416
	Angel.Perez@uclm.es
Ricon, Jesusa	x5412
	Jesusa.Rincon@uclm.es
Rodrigo, Manuel A.	x3419
	Manuel.Rodrigo@uclm.es
Rodriguez, Joaquin	x5414
	Joaquin.Rodriguez@uclm.es
Rodriguez, Juan F.	x3416
	juanfran@inqu-cr.uclm.es
Rodriguez, Lourdes (Director)	x3413
	Lourdes.Rodriguez@uclm.es
Rodriguez, Luis	x 6021
	Luis.RRomero@uclm.es
Sanchez, Paula	x3418
	Paula.Sanchez@uclm.es
Villasenor, Jose	x3419
	Jose.Villasenor

Assistant Professors

Asencio, Isaac	x3417
	Isaac.Asencio@uclm.es
Cabra, Luis	x3411
	lcabrad@repsol-ypf.es
Carmona, Manuel S.	x3814
	carmonaf@inqu-cr.uclm.es
Fernández, Fco. J.	x3417
	FcoJesus.FMorales@uclm.es
Garcia, Jesus	x3506
	Jesus.GGomez@uclm.es
Ramos, Manuel	x3416
	manuel.ramos.go@teleline.es
Ramos, Maria J.	x3417
	mjramos@inqu-cr.uclm.es

[†] Faculty numbers are extensions of (34) 26-295300.

Graduate advisor: Juan F. Rodríguez

Undergraduate advisor: A. de Lucas

Department reports to: Pablo Cañizares

Complutense University of Madrid

Departamento de Ingeniería Química
Facultad de Ciencias Químicas
Universidad Complutense
E-28040 Madrid

Department phone	(91) 394 4115
Department FAX	(91) 394 4114

Professors

Aracil, José	(91) 394-4175
	jam1@quim.ucm.es
Corella, José	(91) 394 4164
	narvaez@quim.ucm.es
García-Ochoa, Félix	(91) 394 4176
	fgochoa@quim.ucm.es
López, Federico	(91) 394 4249
Ovejero, Gabriel	(91) 394 4111
	govejero@quim.ucm.es
Rodríguez, Francisco (Associate Dean)	(91) 394 4246
	frsomo@quim.ucm.es
Romero, Arturo	(91) 394 4170
	arturo@emducms1.sis.ucm.es
Sotelo, José L. (Head)	(91) 394 4117
	jose.sotelo@quim.ucm.es
Tijero, Julio	(91) 394 4250
	jtijero@quim.ucm.es
Uguina, M. Ángeles	(91) 394 4113
	uguinama@quim.ucm.es

Associate Professors

Aragón, José M.	(91) 394-4173
	jomar@quim.ucm.es
Blanco, Ángeles	(91) 394 4247
	ablanco@quim.ucm.es
Delgado, José A.	(91) 394 4119
	jadeldob@quim.ucm.es
García, Julián	(91) 394 5119
	jgarcia@quim.ucm.es
Guardiola, Elita	(91) 394 4248
	guforedo@quim.ucm.es
Martínez, Mercedes	(91) 394 4167
	mmr1@quim.ucm.es
Mirada, Fernando	(91) 394 4251
	fmirada@quim.ucm.es
Negro, Carlos	(91) 394 4242
	cnegro@quim.ucm.es
Oliet, Mercedes	(91) 394 4241
	moliet@quim.ucm.es
Palancar, M. C.	(91) 394 4169
	mcpalanc@quim.ucm.es
Rodríguez, Araceli	(91) 394 4182
	arodri@quim.ucm.es
Romero, M. D.	(91) 394 4118
	mdolores@quim.ucm.es

Santos, Aurora .... (91) 394 4171
aursan@quim.ucm.es

Assistant Professors
Alba, Cristina
  crisalba@quim.ucm.es
Alonso, Virginia .... (91)394 42 47
  valonso@quim.ucm.es
Bautista, Fernando .... (91) 394 4167
  lfbausan@quim.ucm.es
Calvo, Lourdes .... (91) 394 4185
  lcalvo@quim.ucm.es
Espínola, Ascensión .... (91)394 4112
Fuente, Elena 91 394 4245
  helenafg@quim.ucm.es
Garcia, Juan .... (91)394 41 12
  juangcia@quim.ucm.es
Guijarro, Isabel
  migg@quim.ucm.es
Gómez, Emilio
  emgomez@quim.ucm.es
Gómez, José M. .... (91) 394 4185
  segojmgm@quim.ucm.es
Ladero, Miguel A.
  mladero@quim.ucm.es
Lodares, Carmen .... (91)394 4240
  clodares@quim.ucm.es
Monte, Concepción .... (91) 394 4245
  cmonte@quim.ucm.es
Perez, Alejandro 91 394 4115
  alperez@quim.ucm.es
Pérez, Ponciano .... (91) 394 4112
Santos, Victoria .... (91) 394 4179
  vesantos@quim.ucm.es
Tijero, Antonio
  atijero@quim.ucm.es
Torrecilla, Santiago .... (91) 394 4240
  jstorre@quim.ucm.es
Águeda, Ismael .... (91) 394 4115
  viam@quim.ucm.es

Accredited by:

Degrees granted 2003-2004:
  B.S.: 83   M.S.: 25   Ph.D.: 13

University of Granada

Department of Chemical Engineering
Faculty of Sciences
c/ Fuentenueva s/n
18071, Granada
University phone 34 958 243025
Department phone 34 958 243308
Department FAX 34 958 248992

Professors
Bravo, V. .... 3310 [†]
  vbravo@ugr.es
  EV,SU
Camacho, F. 3309
  fcamacho@ugr.es
  BT,SE
Gonzalez, P. 3310
  pgtello@ugr.es
  BT,SE
Jurado, E. (Chair) 3307
  ejurado@ugr.es
  EV,SU

Associate Professors
Blazquez, G. 0770
  gblazque@ugr.es
  EV,SU
Calero, M. 3311
  mcaleroh@ugr.es
  EV,SU
Fernandez, M. 3311
  mferse@ugr.es
  EV,SU
Galvez, A. 0770
  agalvez@ugr.es
  EV,SU
Garcia, A. I. 9018
  anaigl@ugr.es
  EV,SU
Gomez, M. 3313
  mggarzon@ugr.es
  BT,SE
Guadix, E. M. 2925
  eguadix@ugr.es
  BT,SE
Hernainz, F. 3315
  hernainz@ugr.es
  EV,SU
Luzon, G. (Associate Dean) 2925
  german@ugr.es
  EV,SU
Martinez, L. 8991
  lmartin@ugr.es
  EV,PR
Martinez, Ma. E. (Associate Chair) 3315
  meugenia@ugr.es
  BT,PR
Paez, Ma. P. 3307
  mppaez@ugr.es
  BT,SE
Reyes, A. 9018
  areyesr@ugr.es
  EV,SU
Rodriguez, S. 3313
  srodrig@ugr.es
  EV,PR

Assistant Professors
Bailon, R. 4075
  bailonm@ugr.es
  EV,SU
Jimenez, J. M. 8991
  mjimenez@ugr.es
  BT,PR
Luque, A. 4075
  antluque@ugr.es
  EV
Martínez, A 3314
  amferez@ugr.es
  BT,SE
Nunez, J. 4075
  jnolea@ugr.es
  EV,SU
Vicaria, J.M. 0445
  vicaria@ugr.es
  EV,SU

[†] Faculty numbers have extension 34 958-24.

218 Spain

Accredited by: Spanish Ministry of Education

Degrees granted 2005-2006:
B.S.: 60   M.S.: 4   Ph.D.: 3

Department reports to: E. Jurado, Departamento de Ingenieria Quimica

University of Illes Balears

Quimica Tecnica
College of Sciences
Palma de Mallorca

Department phone .... (971) 20-7111x283

Professors
Bergueiro, Jose R. .... (971) 17-3240
Rossello, Carmen .... (971) 17-3239

Associate Professors
Canellas, J. .... (971) 17-3241

Department reports to: Felix Grases, Dean of Sciences College

Universidad Politecnica de Madrid-ETSI Industriales

Departamento de Ingenieria Quimica Industrial y Medio Ambiente
c/Jose Gutierrez Abascal, 2
28006 Madrid

University phone 34 913366035
Department phone 34 913363026
Department FAX 34 913363009

Professors
Larena, A. 34 913363181
    alarena@iqi.etsii.upm.es
Laso, M. 34 913363015
    laso@diquima.upm.es
Losada, J. (Head) 34 913363184
    jlosada@iqi.etsii.upm.es
Ramos, M. A. 34 913365344
    mramos@diquima.upm.es

Associate Professors
De Maria, M R. 34 913363189
    rdemaria@diquima.upm.es
Fernandez, A. 34 913363183
    afernandez@iqi.etsii.upm.es
Fuente, M. M. de la 34 913363185
    mmfuente@iqi.etsii.upm.es
Galan, S. 34 913363185
    sgalan@diquima.upm.es
Garcia, P. 34 913363185
    pigarcia@iqi.etsii.upm.es
Gonzalez, I. 34 913365341
    aaee.ig@gised.com
Lumbreras, Julio 34913363189
    jlumbreras@etsii.upm.es
Martinez, J. 34 913363183
    jmartinez@iqi.etsii.upm.es
Matias, M. C. 34 913363182
    mcmatias@iqi.etsii.upm.es
Molina, M. J. 34 913363182
    mjmolina@iqi.etsii.upm.es
Narros, A. 34 913363186
    anarros@iqi.etsii.upm.es
Paz, I. 34 913365343
    i.paz@iqi.etsii.upm.es
Peso, M. I. del 34 913363185
    ipeso@iqi.etsii.upm.es
Pinto, G. 34 913363183
    gpinto@iqi.etsii.upm.es
Quintana, F. J. 34 913363188
    fquintanam@clh.es
Quintanilla, J.E. 34 913363185
    je.quintanilla@iqfr.csic.es
Rodríguez, M. 34 913363189
    mrod@diquima.upm.es
Santos, A. 34 913365341
    asantos@diquima.upm.es
Zubizarreta, J. I. 34 913363182
    jizubi@iqi.etsii.upm.es

Assistant Professors
Díez de Garay, E. 34 913365341
    ediez@diquima.upm.es
Gayoso, J. A. 34 913365341
    jagayoso@tecnicasreunidas.es
Jimeno, N. 34 913363015
    njimeno@diquima.upm.es
Llorente, V. 34 915451192
    vllorente@cyii.es
Rojas, R. 34 913365343
    rrojas@iqi.etsii.upm.es
Soto, A. 34 913363189
    asoto@uee.es

Accredited by:

University of Murcia

Departamento de Ingenieria Quimica
Facultad de Quimica
Campus de Espinardo
Aptdo. Correos. 4.021
30071 Murcia

University phone .... (968) 363600
Department phone .... (968) 367359
Department FAX .... (968) 364148

Professors
Bodalo, Antonio 968 367354
    abodalo@um.es
Minana, Agustin 968 367353
    minana@um.es
Rubio, Manuel (Director) 968 367357
    mrubio@um.es
Saez, Jose 968 367358
    saezmer@um.es
Soler, Antonio 968 367355
    ansoler@um.es

Associate Professors
Bastida, Josefa 968 367361
    jbastida@um.es
Cabanes, A. L. 968 367362
    cabanes@um.es
Gomez, Demetrio 968 367356
    demetrio@um.es

Gomez, Elisa 968 367352
    egomez@um.es
Gomez, Jose L. 968 367351
    carrasco@um.es
Gonzalez, Enrique 968 367364
    ferradas@um.es
Llorens, Mercedes 968 367349
    llorens@um.es
Meseguer, Victor 968 367333
    vzapata@um.es
Máximo, Fuensanta 968 367367
    fmaximo@um.es
Ortuno, Juan F. 968 367360
    jfortuno@um.es
Tomas, Francisca 968 367331
    ptomas@um.es
Villora, Gloria (Dean) 968 367363
    gvillora@um.es

Assistant Professors

Aguilar, M. I. 968 367091
    maguilar@um.es
Hidalgo, Asuncion M. 968 367219
    iqasocia@um.es
Montiel, Claudia 968 367219
    cmontiel@um.es
Quesada, Joaquín 968 367228
    quesamed@um.es

University of Pais Vasco

Chemical Engineering
Faculty of Chemistry
P. O. Box 1072
20080, San Sebastian

University phone 34-943-018000
Department phone 34-943-015328
Department FAX 34-943-212236

Professors

Asua, Jose M. .... 018181 [†]
    jmasua@sq.ehu.es
Montes, Mario 018183
    qppmoram@sq.ehu.es

Associate Professors

Barandiaran, Maria J. 015330
    qppbasam@sq.ehu.es
Canton, Lourdes 018169
    qppcaorl@sq.ehu.es
de la Cal, José C. 015331
    qppdedej@sq.ehu.es
Forcada, Jacqueline 018182
    qppfogaj@sq.ehu.es
Legorburu, Iñigo (Associate Dean) 018215
    qpplefai@sq.ehu.es
Leiza, Jose R. 015329
    qpplerej@sq.ehu.es

Assistant Professors

Antxustegi, Mirari 015325
    qppanbem@sq.ehu.es

[†] Faculty numbers have prefix 34 943.

Department reports to: J.M. Asua

Rovira i Virgili University

Departament d'Enginyeria Quimica
Escola Tècnica Superior d'Enginyeria Química
Avinguda dels Països Catalans, 26
Campus Sescelades
43007 Tarragona

Department phone 34 977 559603
Department FAX 34 977 559621

Professors

Castells, F. 34 977 559644
    fcastell@etseq.urv.es
Fabregat, A. 34 977 559643
    afabrega@etseq.urv.es
Farriol, X. 34 977 559642
    xfarriol@etseq.urv.es
Giralt, F. 34 977 559638
    fgiralt@etseq.urv.es
Giralt, J. 34 977 559650
    jgiralt@etseq.urv.es
Sueiras, J. 34 977 558701
    jsueiras@etseq.urv.es

Associate Professors

Alabart, J. R. 34 977 559658
    jalabart@etseq.urv.es
Bañares-Alcantara, R 34 977 559673
    rbanares@etseq.urv.es
Bonet, J. 34 977 559645
    jbonet@etseq.urv.es
Ferrando, M. 34 977 558505
    mferrand@etseq.urv.es
Font, J. 34 977 559646
    jfont@etseq.urv.es
Garcia, R. 34 977 559611
    rgarcia@etseq.urv.es
Gavaldà, J. 34 977 559647
    jgavalda@etseq.urv.es
Grifoll, J. 34 977 559639
    jgrifoll@etseq.urv.es
Güell, C. 34 977 558504
    cguell@etseq.urv.es
Herrero, J. 34 977 559649
    jherrero@etseq.urv.es
Katakis, I. 34 977 559655
    ikatakis@etseq.urv.es
Lopez, F. 34 977 558503
    flopez@etseq.urv.es
Mackie, A. 34 977 559674
    amackie@etseq.urv.es
Mateo, J. M. 34 977 559676
    jmateo@etseq.urv.es
Medina, F. 34 977 559787
    fmedina@etseq.urv.es
Medir, M. 34 977 558177
    mmedir@etseq.urv.es
Montané, D. 34 977 559652
    dmontane@etseq.urv.es
Renau, J. 34 977 558550
    jrenau@etseq.urv.es
Rolán, A. 34 977 559651
    arolan@etseq.urv.es
Salvadó, J. 34 977 559641
    jsalvado@etseq.urv.es
Schuhmacher, M. 34 977 559653
    mschuh@etseq.urv.es

Stüber, F. 34 977 559671
fstuber@etseq.urv.es

Assistant Professors

Bes, E. 34 977 559604
Blanco, M. 34 977 558549
mblanco@etseq.urv.es
Blasco, A. 34 977 559604
Boix Rita, B. 34 977 559661
Boix Sabater, R. 34 977 559661
rboix@etseq.urv.es
Borrajo, M.A. 34 977 559661
Cabré, J. 34 977 559661
Chillida Rabada, J. 34 977 559661
Claramonte, V. 34 977 559661
vclaram@etseq.urv.es
Cuesta, J. 34 977 558549
Guardiola, S. 34 977 250000
sguardio@etseq.urv.es
Guitérrez, A. 34 977 559661
aguti@etseq.urv.es
Lloret, R. 34 977 559661
rlloret@etseq.urv.es
Mañé, E. 34 977 559661
emanye@etseq.urv.es
Merino, J. 34 977 558549
jmerino@etseq.urv.es
Moreno Cabello, M. 34 977 558549
mmoreno@etseq.urv.es
Moreno Garcia, M.E. 34 977 250000
emoreno@etseq.urv.es
Pemán, J.M. 34 977 559604
jpeman@etseq.urv.es
Rabadà, J. 34 977 250000
Ramírez, B. 34 977 250000
bramirez@etseq.urv.es
Robles, M. 34 977 559661
mrobles@etseq.urv.es

Department reports to: A. Fabregat

University of Salamanca

Departamento de Ingenieria Quimica y Textil
Facultad de Ciencias Quimicas
Plaza de los Caidos, 1-5
37008 Salamanca

University phone 34-923294400
Department phone 34-923294479
Department FAX 34-923294574

Professors

Galan, Miguel A. 34-923294479
magalan@usal.es

Associate Professors

Catalan, Jacinto 34-923294479
jcatalan@usal.es
Costa, Carlos 34-923294479
ccosta@usal.es
Cuellar, Jorge (Chair) 34-923294479
cuellar@usal.es
Estevez, Angel M. 34-923294479
estevez@usal.es
Fernandez, Angel 34-923294479
aftena@usal.es
Marquez, M. C. 34-923294479
mcm@usal.es
Martin, Jose L. 34-923294479
jolmasa@usal.es
Montes, Francisco J. 34-923294479
javimon@usal.es
Ramos, Pedro 34-923294479
pramos@usal.es
Rodriguez, Jesus M. 34-923294479
jesusr@usal.es

Assistant Professors

Alvaro, Audelino 34-923294479
audea@usal.es
Arranz, Juan L. 34-923294479
juluar@usal.es
Domingo, Antonio M. 34-923294479
tonidm@usal.es
Ruiz, Cesar A. 34-923294479
car@usal.es
Sánchez, José M. 34-923294479
chemasal@usal.es
Torrente, Carmen 34-923294479
carmina@usal.es
Tovar, Tomas R. 34-923294479
manana@usal.es

Graduate advisor: Francisco J. Montes

Department reports to: Faculty of Chemical Sciences

Universidad de Santiago de Compostela

Departamento de Ingenieria Quimica
Lope Gómez de Marzoa, s/n
15782 Santiago de Compostela

University phone 34 981 563100
University FAX 34 981 528050
Department phone 34 981 563100
Department FAX 34 981 528050

Professors

Antorrena, Gervasio .... 16787 [†]
eqgantor@usc.es
Arce, Alberto 16790
eqaaarce@usc.es
Bao, Manuel 16789
eqbao@usc.es
Casares, Juan J. (Chair) 16794
eqjjcl@usc.es
Lema, Juan M. (Dean) 16793
eqjmlema@usc.es
Méndez, Ramón 16791
eqrmndzp@usc.es
Nunez, M. J. 16792
eqmajose@usc.es
Vazquez, Gonzalo 16788
eqgvazq@usc.es

Associate Professors

Blanco, Antonio 16796
eqanblas@usc.es
Chenlo, Francisco 16797
eqchenlo@usc.es
Feijoo, Gumersindo 24073
eqfeijoo@lugo.usc.es

Fernandez, Eugenio 24075
    usciqefc@cesga.es
Gonzalez, Julia 24102
    eqjulia@lugo.usc.es
Navaza, Jose M. 16795
    eqnavaza@usc.es
Omil, Francisco 24074
    eqomil@usc.es
Roca, Enrique 24074
    eqeroca@lugo.usc.es
Soto, Ana M. 24134
    eqequilf@usc.es
Veiga, Jose L. 24136
    jlveiga@lugo.usc.es
Vidal, Isabel 16798
    eqivteef@lugo.usc.es

Assistant Professors

Campos, Jose L. 16777
    eqluis@usc.es
Freire, Sonia 16758
    sfreire@usc.es
Garrido, Juan 16778
    equenlla@usc.es
Martinez, Jose M. 24125
    eqageito@lugo.usc.es
Moreira, Ramon 16759
    eqmoncho@usc.es
Moreira, Teresa 16767
    tmoreira@usc.es
Méndez, M. R. 24077
    usciqmmg@cesga.es
Pereira, Gerardo 24140
    gepegon@usc.es
Rodil, Eva 16760
    erodil@usc.es
Sineiro, Jorge 24133
    eqxurxo@lugo.usc.es
Vázquez, M. J. 24077
    eqmjvv@usc.es

† Dial 34 981 563100 and ask for extension.

Accredited by:

Escuela Superior de Ingenieros Industriales de Sevilla

Department of Chemical and Environmental Engineering
Camino de los Descubrimientos s/n
41092 Sevilla
Deliveries: José F. Vale Parapar

University phone 34 95 4486100
Department phone 34 95 4487260
Department FAX 34 95 4461775

Professors

Alcalde Moreno, Manuel 34 95 4487270
    alcalde@esi.us.es
Cañadas Serrano, Luis 34 95 4487267
    canadas@esi.us.es
Cortes Galeano, Vicente J. 34 95 4487260
    vjcortes@esi.us.es
Fernandez Pereira, Constantino 34 95 4487271
    pereira@esi.us.es
Garcia Lopez, Angel 34 95 4487271
    angelgl@esi.us.es
Gracia Manarillo, Ignacio 34 95 4487273
    gracia@esi.us.es
Olivares del Valle, Joaquin 34 95 4487275
    jov@esi.us.es
Ollero de Castro, Pedro (Associate Dean) ... 34 95 4487260
    ollero@esi.us.es
Plumed Rubio, Antonio 34 95 4487262
    plumed@esi.us.es
Salvador Martinez, Luis 34 95 4487260
    salvador@esi.us.es
Sanchez Pena, Eduardo J. 34 95 4487266
    eduardoj@esi.us.es
Usero Garcia, Jose 34 95 4487274
    usero@esi.us.es
Vale Parapar, José F. (Head) 3494 4487269
    vale@esi.us.es

Associate Professors

Arjona Antolin, Ricardo 34 95 4487223
    arjona@esi.us.es
Gutierrez Ortiz, Fco. J. 34 95 4487268
    fjgo@esi.us.es
Morillo Aguado, José 34 95 4487276
    jmorillo@esi.us.es
Navarrete Rubia, Benito 34 95 4487280
    navarre@esi.us.es
Rodriguez Piñero, Miguel A. 34 95 4487270
    marp@esi.us.es
Vidal Barrero, Fernando 34 95 4487222
    vidal@esi.us.es
Vilches Arenas, Luis 34 95 4487282
    vilches@esi.us.es
Villegas Sanchez, Rosario 34 95 4487282
    villegas@esi.us.es

Accredited by:

Degrees granted 2003-2004:
    B.S.: 40   Ph.D.: 2

Department reports to: José F. Vale Parapar

Universidad de Sevilla

Departamento de Ingenieria Quimica
Facultad de Quimica
41071 Sevilla

University phone 34 95 4551000
Department phone 34 95 4557180
Department FAX 34 95 4556447

Professors

Alfaro Rodríguez, M. C. 34 95 4557179
    alfaro@us.es
Alvarez Mateos, Paloma 34 95 4552846
    palvarez@us.es
Berjano Núñez, Manuel 34 95 4557179
    mberjano@us.es
Carranza Mora, Francisco 34 95 4557182
    fcarranza@us.es
Carrillo de la Fuente, Francisco (Adjunct) ... 34 95 4557178
    cfuente@us.es
Cordobés Carmona, Felipe 34 95 4557179
    fcordobe@us.es
Cota Galán, Juan 34 95 4557178/3877
    cota@us.es

De la Fuente, Julia 34 95 4557183/2845
jfferia@us.es
Flores Luque, Vicente (Head) 34 95 4557179
vfloresluque@us.es
Guerrero Conejo, Antonio 34 95 4557179
aguerrero@us.es
Iglesias González, Nieves 34 95 4557182
mnieves@us.es
Martín Aguilar, Ana 34 95 4557178
amartina@us.es
Mazuelos Rojas, Alfonso 34 95 4557182
mazuelos@us.es
Moraga Borrell, Jose M. 34 95 4557178
moraga@us.es
Muñoz García, José 34 95 4557179
jmunoz@us.es
Palencia Pérez, Inmaculada 34 95 4557182
palencia@us.es
Pereda Marín, Juan 34 95 4557183
jpmarin@us.es
Rodriguez Patino, Juan M. 34 95 4556446
jmrodri@us.es
Rodríguez Niño, Rosario 34 95 4556446
mdrrodri@us.es
Romero Guzmán, Fernando 34 95 4557183
fromero@us.es
Roselló Segado, Antonio 34 95 4557178
rosello@us.es
Ruiz Dominguez, Manuela 34 95 4557183
manuela@us.es

Associate Professors

Carrera Sánchez, Ceciclio 34 95 4556446/7147
cecilio@us.es
Durán Barrantes, Montaña 34 95 4557183
mmduran@us.es

Accredited by:

Degrees granted 2004-2005:

    B.S.: 42   M.S.: 6   Ph.D.: 3

Department reports to: dingquimica@us.es

Placement service: dingquimica@us.es

Universitat de Valencia

Departamento de Ingenieria Quimica
Escola Tècnica Superior d'Enginyeria
46100 Burjassot (Valencia)

University phone...................... 34 96 38 64100
Department phone 34 96 35 44325
Department FAX 34 9635 44898

Professors

Aucejo, Antonio 34 9635 44318
Antonio.Aucejo@uv.es
PR,SE
Martinez-Andreu, Antoni (Chair) 34 9635 44319
Antoni.Martinez@uv.es
SE
Monton, Juan B. 34 9635 44317
Juan.B.Monton@uv.es
SE

Associate Professors

Berna, Angel 34 9635 44316
Angel.Berna@uv.es
BT
Burguet, M. C. 34 9635 44318
Cruz.Burguet@uv.es
SE
de la Torre, Javier 34 9635 44326
Javier.Torre@uv.es
PR,SE
Dejoz, Ana M. 34 9635 43436
Ana.M.Dejoz@uv.es
RE
Gabaldon, M. C. 34 9635 43331
Carmen.Gabaldon@uv.es
EV
Gonzalez-Alfaro, M. V. 34 9635 43436
Vicenta.Gonzalez@uv.es
RE,SE
Llopis, Francisco (Associate Chair) 34 9635 43130
Francisco.Llopis@uv.es
RE
Loras, Sonia 34 9635 43169
Sonia.Loras@uv.es
SE
Martinez-Soria, Vicente 34 9635 43169
Vicente.Mtnez-soria@uv.es
EV
Marzal, Paula (Associate Dean) 34 9635 43331
Paula.Marzal@uv.es
EV
Miguel, Pablo J. 34 9635 43130
Pablo.J.Miguel@uv.es
RE,SE
Munoz, Rosa 34 9635 44319
Rosa.Munoz@uv.es
SE
Orchilles, A. V. 34 9635 43170
Vicent.Orchilles@uv.es
RE,SE
Pena, M. P. 34 9635 44316
Maria.P.Pena@uv.es
Penarrocha, Josep M. 34 9635 43169
Josep.Penarrocha@uv.es
EV
Sanchotello, Margarita 34 9635 44317
Margarita.Sanchotello@uv.es
SE
Seco, Aurora 34 9635 44326
Aurora.Seco@uv.es
EV
Vazquez, M. I. 34 9635 43170
Isabel.Vazquez@uv.es
RE,SE
Vercher, Ernesto 34 9635 43131
Ernesto.Vercher@uv.es
SE

Assistant Professors

Bouzas, Alberto 34 9635 43434
Alberto.Bouzas@uv.es
EV
Chafer, Amparo 34 9635 43434
Amparo.Chafer@uv.es
SE
Ribes, Josep 34 9635 43169
Josep.Ribes@uv.es
EV

Accredited by: Council of Universities of Spain

Degrees granted 2005-2006:
B.S.: 48   Ph.D.: 3
Department reports to: P. Marzal, Vice-Dean, ETSE

University of Valladolid

Chemical Engineering and Environmental Technology Department
Faculty of Sciences
C/ Prado de la Magdalena s/n
47011 Valladolid
University phone .... (34) 983 423250
University FAX .... (34)983423234
Department phone .... (34) 983 423166
Department FAX .... (34) 983 423013

Professors

Carton, Angel .... (34) 983 423167
 carton@iq.uva.es
Cocero, Maria J. .... (34) 983 423174
 mjcocero@iq.uva.es
Fdz-Polanco, Fernando .... (34) 983 423172
 ffp@iq.uva.es
Gonzalez, Gerardo (Dean) .... (34) 983 423170
 gerardo@iq.uva.es
Mato, Fidel (Emeritus) .... (34) 983 423175
 fmato@iq.uva.es

Associate Professors

Alonso, Gloria E. .... (34) 983 423166
 ealonso@iq.uva.es
Antolin, Gregorio .... (34) 983 423362
 greant@eis.uva.es
Bolado, Silvia .... (34) 983 423958
 silvia@iq.uva.es
Espinel, Manuela .... (34) 983 423508
 espinel@sid.eup.uva.es
Fdz-Polanco, Maria .... (34) 983 424506
 maria@iq.uva.es
Garcia, Maite .... (34) 983 423237
 maite@iq.uva.es
Garcia, Pedro (Head) .... (34) 983 423171
 pedro@iq.uva.es
Irusta, Ruben .... (34) 983 423693
 rubiru@eis.uva.es
Martinez, Bernardo .... (34) 983 423506
 bern@iq.uva.es
Mato, Fidel A. .... (34) 983 423169
 pepa@iq.uva.es
Mato, Rafael B. .... (34) 983 423177
 rbmato@iq.uva.es
Pena, Maria M. .... (34) 983 423176
 pena@iq.uva.es
Sobron, Francisco .... (34) 983 423168
 sobron@iq.uva.es
Torio, Isabel .... (34) 983 423507
 misato@sid.eup.uva.es
Uruena, Miguel A. .... (34) 983 423506
 uru@iq.uva.es
Villaverde, Santiago .... (34) 983 423656
 svillave@iq.uva.es

Assistant Professors

García Serna, Juan .... (34) 983 423166
 jgserna@iq.uva.es

Lucas, Susana .... (34) 983 423166
 susana@iq.uva.es

Accredited by: IChemE

Degrees granted 2004-2005:
M.S.: 56   Ph.D.: 3
Department reports to: Jesus M. Sanz Serna, Rector of the University

University of Zaragoza

Department of Chemical and Environmental Engineering
Pedro Cerbuna, 12
E50009, Zaragoza
University phone .... (34) 976 761000
Department phone .... (34) 976 761154
Department FAX .... (34) 976 762142

Professors

Bilbao, Rafael (Head) .... (34) 976 761150
 rbilbao@unizar.es
Menendez, Miguel .... (34) 976 761152
 qtmiguel@unizar.es
Ovelleiro, J. L. .... (34) 976 761156
 oveleiro@unizar.es
Santamaria, Jesus M. (Dean) .... (34) 976 761153
 iqcatal@unizar.es

Associate Professors

Alzueta, M. U. .... (34) 976 761876
 uxue@unizar.es
Arauzo, Jesus .... (34) 976 761878
 qtarauzo@unizar.es
Aznar, M. P. .... (34) 976 762391
 paznar@unizar.es
Calvo, Manuel .... (34) 976 762191
 calvom@unizar.es
Ceamanos, Jesus .... (34) 976 762160
 ceamanos@unizar.es
Coronas, Joaquin .... (34) 976 762471
 coronas@unizar.es
Frances, Eva .... (34) 976 762550
 efrances@unizar.es
Garcia-Bacaicoa, Pedro .... (34) 976 761880
 bacaicoa@unizar.es
Garcia-Nieto, Lucia .... (34) 976 762194
 luciag@unizar.es
Herguido, Javier .... (34) 976 762393
 jhergui@unizar.es
Iranzo, Carlos .... (34) 976 762190
 ciranzo@unizar.es
Lazaro, Luisa .... (34) 976 761877
 llazaro@unizar.es
Mastral, Jose F. .... (34) 976 761876
 pepe@unizar.es
Matute, Rosa P. .... (34) 976 762550
 rmatute@unizar.es
Millera, Angela .... (34) 976 761875
 amillera@unizar.es
Monzon, Antonio .... (34) 976 761157
 amonzon@unizar.es
Murillo, M. B. .... (34) 976 761880
 murillo@unizar.es
Ormad, M. P. .... (34) 976 761877
 mpormad@unizar.es

Oro, M. A. .... (34) 976 761151
maoro@unizar.es
Pena, J. A. (Associate Dean) .... (34) 976 762390
jap@unizar.es
Perez, J. I. .... (34) 976 762194
jiperezs@unizar.es
Revuelta, Gillermo .... (34) 976 762190
grevuelt@unizar.es
Royo, Carlos J. .... (34) 974 761334
cjroyo@unizar.es
Salvador, M. L. .... (34) 976 761584
mlsalva@unizar.es
Sanchez, Emilio (Associate Dean) .... (34) 974 761335
blas@unizar.es
Sarasa, Judith .... (34) 976 762392
jsarasa@unizar.es
Tellez, Carlos .... (34) 976 762471
ctellez@unizar.es
Villacampa, J. I. .... (34) 974 761336
villacam@unizar.es

Assistant Professors

Callejas, Alicia .... (34) 976 733977
acalleja@carbon.icb.csic.es
Gea, Gloria .... (34) 976 763308
gloria.gea@unizar.es
Mallada, Reyes .... (34) 974 761461
rmallada@unizar.es
Oliva, Miriam .... (34) 976 761339
miroliva@unizar.es
Pina, M. P. .... (34) 976 761155
mapina@unizar.es
Romeo, Eva .... (34) 974 761461
evaromeo@unizar.es
Romero, Enrique .... (34) 976 762392
eromero@unizar.es
Ruiz, Joaquin .... (34) 976 761339
jruizp@unizar.es

Accredited by: Ministry of Education

Student organization: AIQA

Department reports to: Prof. M. Menéndez

Placement service: Mrs. P. Munoz

Sweden

University of Lund

Chemical Engineering
Lund Institute of Technology
P.O. Box 124
S-221 00 Lund

University phone...................... (46) 46 2220000
Department phone .... (46) 46 2228285
Department FAX .... (46) 46 2224526

Professors

Andersson, Arne .... (46) 46 2228280
arne.andersson@chemeng.lth.se
Axelsson, Anders .... (46) 46 2228282
Anders.Axelsson@chemeng.lth.se
Jonsson, Ann S. .... (46) 46 2228291
Ann-Sofi.Jonsson@chemeng.lth.se
Karlsson, Hans .... (46) 46 2228244
hans.karlsson@chemeng.lth.se
La Cour Jansen, Jes .... (46) 46 2228999
Jes.la_Cour_Jansen@vateknik.lth.se
Lidén, Gunnar (Associate Chair) .... (46) 46 2220826
gunnar.liden@chemeng.lth.se
Odenbrand, Ingemar .... (46) 46 2228284
Ingemar.Odenbrand@chemeng.lth.se
Stenstrom, Stig .... (46) 46 2228292
Stig.Stenstrom@chemeng.lth.se
Sverdrup, Harald .... (46) 46 2228274
Harald.Sverdrup@chemeng.lth.se
Warfvinge, Per .... (46) 46 2223626
per.warfvinge@chemeng.lth.se
Zacchi, Guido (Chair) .... (46) 46 2228297
Guido.Zacchi@chemeng.lth.se

Associate Professors

Dolby, Ingemar (Emeritus) .... (46) 46 2228290
Ingemar.Dolby@chemeng.lth.se

Assistant Professors

Alveteg, Mattias .... (46) 46 2223627
Mattias.Alveteg@chemeng.lth.se
Jonsson, Karin .... (46) 46 2228607
Karin.Jonsson@vateknik.lth.se
Nilsson, Bernt .... (46) 46 2228299
Bernt.Nilsson@chemeng.lth.se

Accredited by:

Graduate advisor: Gunnar Lidén

Royal Institute of Technology

Department of Chemical Engineering and Technology
Teknikringen 28
S-100 44
Stockholm

University phone...................... 46 8 790 6000
Department phone 46 8 790 8229
Department FAX 46 8 105 228

Professors

Björnbom, Pehr (Chair) 46 8 790 8255
pehr@ket.kth.se
Järås, Sven (Chair) 46 8 790 8917
svenj@ket.kth.se
Lindbergh, Göran (Chair) 46 8 790 8143
goeran.lindbergh@ket.kth.se
Neretnieks, Ivars (Head) 46 8 790 8229
niquel@ket.kth.se
Rasmuson, Åke (Chair) 46 8 790 8227
rasmuson@ket.kth.se
Wennersten, Ronald (Chair) 46 8 790 6347
rw@ket.kth.se
Westermark, Mats 46 8 790 6220
mw@ket.kth.se
Yan, Jinyue 46 8 790 6528
yanjy@ket.kth.se

Associate Professors

Alvfors, Per (Chair) 46 8 790 6526
alvfors@ket.kth.se
Berendson, Jaak 46 8 790 7028
jaak@admin.kth.se
Björnbom, Emilia 46 8 790 8256
emilia@ket.kth.se

Boutonnet, Magali 46 8 790 8245
    magali@ket.kth.se
Liu, Jinsong 46 8 790 6346
    liuv@ket.kth.se
Martinez, Joaquin 46 8 790 6570
    jmc@ket.kth.se
Moreno, Luis 46 8 790 6412
    lm@ket.kth.se
Pettersson, Lars 46 8 790 8259
    larsp@ket.kth.se
Sjöström, Krister 46 8 790 8248
    krister@ket.kth.se
Sylwan, Christopher 46 8 790 8258
    sylwan@ket.kth.se
Uusi-Penttilä, Marketta 46 8 790 8228
    marketta@ket.kth.se

Accredited by: The Swedish Government

Degrees granted 2003-2004:
    B.S.: 19   M.S.: 42   Ph.D.: 13

Graduate advisor: Prof Göran Lindbergh

Undergraduate advisor: Ass Prof Per Alvfors

Department reports to: Prof Anders Flodström, President of the Royal Institute of Technology (KTH)

Placement service: Ms Ulla Schött

Switzerland

Swiss Federal Institute of Technology

ETH Hoenggerberg HCI
Institute for Chemical and Bioengineering ICB
CH-8093 Zurich

University phone 41 44 632 1111
Department phone 41 44 632 3048
Department FAX 41 44 632 1509

Professors

Baiker, Alfons 41 44 632 31 53
    alfons.baiker@chem.ethz.ch
    AM,EN,MO,NT,PR,RE,SU
Fussenegger, Martin (Head) 41 44 633 34 48
    martin.fussenegger@chem.ethz.ch
    BM,BT
Hungerbuehler, Konrad 41 44 632 60 98
    konrad.hungerbuehler@chem.ethz.ch
    EV
Morbidelli, Massimo 41 44 632 30 34
    massimo.morbidelli@chem.ethz.ch
    AM,PO,PR,RE,SE,TH,TP
Prins, Roel (Emeritus) 41 44 632 54 90
    roel.prins@chem.ethz.ch
Wokaun, Alexander 41 44 632 71 46
    alexander.wokaun@chem.ethz.ch
    EN,EV

Assistant Professors

Stark, Wendelin J. 41 44 632 09 80
    wendelin.stark@chem.ethz.ch
    AM,BM,NT
van Bokhoven, Jeroen 41 44 632 55 42
    jeroen.vanbokhoven@chem.ethz.ch
    AM,MO,PR,RE

Accredited by:

Degrees granted 2005-2006:
    M.S.: 9   Ph.D.: 35

Graduate advisor: Dr. Otmar Dossenbach

Department reports to: Department of Chemistry and Applied Biosciences D-CHAB

Swiss Federal Institute of Technology, Ecole Polytechnique Federale de Lausanne

Section of Chemistry and Chemical Engineering
Faculty of Sciences
CH, 1015 Lausanne

University phone .... (4121) 693 1111
Department phone .... (4121) 693 9850
Department FAX .... (4121) 693 9855

Professors

Bodenhausen, Geoffrey .... 41 21 693 9431 [†]
    geoffrey.bodenhausen@epfl.ch
Girault, Hubert (Head) 41 21 693 3151
    hubert.girault@epfl.ch
Graetzel, Michael 41 21 693 3112
    michael.graetzel@epfl.ch
Hubbell, Jeffrey 41 21 693 96 81
    jeffrey.hubbell@epfl.ch
Merbach, André E. 41 21 693 9871
    andre.merbach@epfl.ch
Mutter, Manfred 41 21 693 94 11
    manfred.mutter@epfl.ch
Renken, Albert 41 21 693 3181
    albert.renken@epfl.ch
Rizzo, Thomas (Dean) 41 21 693 3073
    thomas.rizzo@epfl.ch
Roulet, Raymod (Emeritus) 41 21 693 9861
    raymond.roulet@epfl.ch
Schlosser, Manfred (Emeritus) ... 41 21 693 9351
    manfred.schlosser@epfl.ch
Stockar, Urs von 41 21 693 3191
    urs.vonstockar@epfl.ch
Vogel, Horst 41 21 693 3155
    horst.vogel@epfl.ch
Vogel, Pierre 41 21 693 03 1
    pierre.vogel@epfl.ch

Associate Professors

Comninellis, Christos 41 21 693 3674
    christos.comninellis@epfl.ch
Friedli, Claude 41 21 693 3121
    claude.friedli@epfl.ch
Johnsson, Kai 41 21 693 9356
    kai.johnsson@epfl.ch
Moser, Jacques-E. 41 21 693 3628
    je.moser@epfl.ch
Röthlisberger, Ursula 41 21 693 0321
    ursula.roethlisberger@epfl.ch
Stoessel, Francis 41 21 693 3671
    francis.stoessel@epfl.ch

Assistant Professors

Dyson, Paul 41 21 693 9854
    paul.dyson@epfl.ch

Krossing, Ingo 41 21 693 93 15
    ingo.krossing@epfl.ch
Pitsch, Stefan 41 21 693 9380
    stefan.pitsch@epfl.ch
Pohnert, Georg 41 21 693 93 01
    georg.pohnert@epfl.ch
Severin, Kay 41 21 693 9302
    kay.severin@epfl.ch

[†] Direct dial with prefix 4121.

Accredited by:

Degrees granted 2004-2005:
    M.S.: 30   Ph.D.: 30

Graduate advisor: Prof. H. Girault

Undergraduate advisor: Prof. H. Girault

Student organization: AGEPOLY

Department reports to: Marendaz Jean-Luc ¡jean-luc.marendaz@epfl.ch¿

Taiwan

Chang Gung University

Chemical and Materials Engineering Department
259, Wen-Hwa 1st Road
Kwei-Shan, Tao-Yuan
333

University phone 886-3-2118800
Department phone .... 886-3-2118800 ext 5286
Department FAX 886-3-2118668

Professors

Chen, Jyh-Ping .... x5298 [†]
    jpchen@mail.cgu.edu.tw
Shiau, Lie-Ding (Chair) x5291
    shiau@mail.cgu.edu.tw
Yang, Jen-Ming x5290
    jmyang@mail.cgu.edu.tw

Associate Professors

Chiu, Fang-Chyou x5297
    maxson@mail.cgu.edu.tw
Hsu, Ruey-Chi x5288
    rcHsu@mail.cgu.edu.tw
Hua, Mu-Yi x5289
    huamy@mail.cgu.edu.tw
Lin, Andrew x5756
    Andrew@mail.cgu.edu.tw
Lu, Hsin-Chun x5292
    hsinchun@mail.cgu.edu.tw
Lu, Tsan-Sheng x5675
    tsheng@mail.cgu.edu.tw
Lue, Shing-Jiang x5489
    jessie@mail.cgu.edu.tw
Wang, Gow-Bin x5757
    gbwang@mail.cgu.edu.tw
You, Jiann-Hwa x5287
    you@mail.cgu.edu.tw

Assistant Professors

Kuo, Hsiu-Po x5488
    hpkuo@mail.cgu.edu.tw
Lin, Chia-Chang x5760
    higee@mail.cgu.edu.tw
Liu, Yu-Kuo x5328
    ykliu@mail.cgu.edu.tw

[†] Faculty numbers are extensions of University phone.

Accredited by: Ministry of Education

Degrees granted 2004-2005:
    B.S.: 50   M.S.: 60

Department reports to: Lie-Ding Shiau, Chairman

Placement service: Lie-Ding, Shiau

Feng-Chia University

Chemical Engineering Department
100 Wenhwa Road
Taichung, Taiwan 40724

University phone 886-4-24517250
Department phone .... 886-4-24517250 ext. 3651
Department FAX 886-4-24510890

Professors

Chan, Chih-Chieh .... ext. 3688 [†]
    ccchan@fcu.edu.tw
Chang, Hsin-Fu .... ext. 3687
    hfchang@fcu.edu.tw
Chao, Yun-Peng (Chair) .... ext. 3677
    ypshao@fcu.edu.tw
Chen, Chyi-Tsong (Associate Dean) .... ext. 3691
    ctchen@fcu.edu.tw
Chu, Hou-Hsien .... ext. 3686
    hhchu@fcu.edu.tw
Hwang, Jen-Tai .... ext. 3651
    jthwang@npf.org.tw
Lin, Ping J. .... ext. 3682
    pjlin@fcu.edu.tw
Wu, Der-Chang .... ext. 3689
    dcwu@fcu.edu.tw
Wu, Shu-Yii .... ext. 3679
    sywu@fcu.edu.tw

Associate Professors

Sheu, Jiang S. .... ext. 3690
    chsu@fcu.edu.tw
Ting, Hsing-Yie .... ext. 3079
    hyting@fcu.edu.tw

Assistant Professors

Chang, AlexC.-C. .... ext. 3676
    acchang@fcu.edu.tw
Lin, Yuang-Sen .... ext. 3659
    yslin@fcu.edu.tw
Shih, C. J. .... ext. 3680
    chshih@fcu.edu.tw
Yuan, Wei-Li .... ext. 6892
    wyuan@fcu.edu.tw

[†] Faculty numbers are extensions of 886-4-24517250

Accredited by: Ministry of Education

Department reports to: Chao, Yun-Peng

Placement service: Yen-Hsu Kuan

National Central University

Department of Chemical and Materials Engineering
Chung-Li, Taoyuan
Taiwan 32054

University phone .... (03) 4227151
Department phone .... (03) 4227151x4200
Department FAX .... (03) 4252296

Professors

Chang, Feg-Wen x4202
Chen, Hui x4216
    huichen@sparc20.ncu.edu.tw
Chen, Teng-Ko x4215
    T3100064@ncu865.ncu.edu.tw
Chen, Wen-Yih (Chair) x4222
    T313165.ncu865.ncu.edu.tw
Chen, Yin-Zu x4219
    YnZuChem@cc.ncu.edu.tw
Chen, Yu-Wen x4203
Chiang, Shiaw-Tseh x4214
    STChiang@cc.ncu.edu.tw
Fey, Ting-Kuo x4206
Kao, Chengheng R. x4225
    crkao@twncu865.ncu.edu.tw
Kuo, Kung-Tu (Emeritus) x4205
    T312310@twncu865.ncu.edu.tw
Lee, Liang-Sun x4208
    T3100206@twncu865.ncu.edu.tw
Lin, Hsiao-Tsung x4207
Ruaan, Rouh-Chyu x4232
    ruaan@cc.ncu.edu.tw
Shyu, Shin-Shing x4204
Tsao, Heng-Kwong x4226
    HKTsao@colloid.Chem.ncu.edu.tw
Yang, Sze-Ming x4217
    Szeyang@tpts6.seed.net.tw

Associate Professors

Shu, Chin-Hang x4227
    chinshu@cc.ncu.edu.tw
Wang, Ten-Tsai x4211

Assistant Professors

Liu, Cheng-Yi x4228

Accredited by: Ministry of Education

Graduate advisor: Cheng-Tung Chou

Undergraduate advisor: Chengheng R. Kao

Department reports to: Dean, College of Engineering

Placement service: Student Bureau

National Cheng Kung University

Department of Chemical Engineering
1 University Road
Tainan, Taiwan 70148

University phone 886-6-2757575
Department phone 886-6-2757575x62600
Department FAX 886-6-2344496

Professors

Chang, Chien-Hsiang x62671
    changch@mail.ncku.edu.tw
Chang, Chuei-Tin x62663
    ctchang@mail.ncku.edu.tw
Chang, Jo-Shu x62651
    changjs@mail.ncku.edu.tw
Chen, Chin-Cheng x62655
    ccchen@mail.ncku.edu.tw
Chen, Chuh-Yuan x62643
    ccy7@mail.ncku.edu.tw
Chen, Dong-Hwang x62680
    chendh@mail.ncku.edu.tw
Chen, Huey-Ing x62667
    hueying@mail.ncku.edu.tw
Chen, Ling-Yuan (Adjunct)
Chen, Teh-Liang x62660
    t62660@mail.ncku.edu.tw
Chen, Yun x62657
    yunchen@mail.ncku.edu.tw
Chiang, Chien-Lih x62648
    clchiang@mail.ncku.edu.tw
Chou, Tse-Chuan x62639
    tcchou@mail.ncku.edu.tw
Chu, Tzong-Jeng (Adjunct) x62681x304
    z7108017@email.ncku.edu.tw
Chung, Shyan-Lung x62654
    slchung@mail.ncku.edu.tw
Hong, Chau-Nan x62662
    hong@mail.ncku.edu.tw
Huang, Ting-Chia (Adjunct) x62630
    tchuang@mail.ncku.edu.tw
Hwang, Shyh-Hong x62661
    shhwang@mail.ncku.edu.tw
Kao, Chen-Feng x62642
    cfkao@mail.ncku.edu.tw
Kuo, Jen-Feng (Adjunct) x62638
    jenfkuo@mail.ncku.edu.tw
Kuo, Ping-Lin x62658
    plkuo@mail.ncku.edu.tw
Lee, Yuh-Lang x62693
    yllee@mail.ncku.edu.tw
Lin, Jui-Che x62665
    jclin@mail.ncku.edu.tw
Liu, Chung-Chiun (Adjunct)
Liu, Jui-Hsiang (Chair) x62646
    jhliu@mail.ncku.edu.tw
Maa, Jer-Ru (Adjunct) x62632
    jerrumaa@mail.ncku.edu.tw
Syu, Mei-Jywan x62631
    z8108094@email.ncku.edu.tw
Teng, Hsisheng x62640
    hteng@mail.ncku.edu.tw
Tsay, Sun-Yuan (Adjunct) x62681x341
    sytsay@mail.ncku.edu.tw
Wang, Chi x62645
    chiwang@mail.ncku.edu.tw
Wang, Chun-Shan (Adjunct) x62649
    cswang@mail.ncku.edu.tw
Wen, Ten-Chin x62656
    tcwen@mail.ncku.edu.tw
Weng, Hung-Shan x62637
    z5408008@email.ncku.edu.tw
Woo, Eamor x62670
    emwoo@mail.ncku.edu.tw
Yang, Ming-Chang x62666
    mcyang@mail.ncku.edu.tw
Yang, Yu-Min x62633
    ymyang@mail.ncku.edu.tw

Associate Professors

Chen, Bing-Hung x62695
    bkchen@mail.ncku.edu.tw
Cheng, Chu-Yuan x62664
    cycheng@mail.ncku.edu.tw
Lee, Jeng-Cheng (Adjunct)
Ling, Han-Chern x62659
    hcling@mail.ncku.edu.tw
Wu, Jih-Jen x62694
    wujj@mail.ncku.edu.tw

Assistant Professors

Chao, Ying-Chen (Adjunct)
Hou, Sheng-Shu x62641
    sshou@mail.ncku.edu.tw
Huang, Yao-Hui x62636
    yhhuang@mail.ncku.edu.tw
Juang, Yi-Je
    yjjuang@mail.ncku.edu.tw
Wei, Hsien-Hung x62691
    hhwei@mail.ncku.edu.tw

Accredited by: Ministry of Education

Degrees granted 2004-2005:
    B.S.: 135   M.S.: 85   Ph.D.: 20

Department reports to: Jhing-Fa Wang, Dean of Engineering

National Chung Cheng University

Department of Chemical Engineering
San-Hsing, Min-Hsiung
Chaia-Yi, 621-02

University phone 886-5-2720411
Department phone 886-5-2720411x23400
Department FAX 886-5-2721206

Professors

Chang, Jen-Ray .... 886-5-2720411 ext 33455
    chmjrc@ccu.edu.tw
Chen, Chien-Chong .... 886-5-2720411 ext 33462
    chmccc@ccu.edu.tw
Hu, Chi-Chang .... 886-5-2720411 ext 33411
    chmhcc@ccu.edu.tw
Hua, Chi-Chung .... 886-5-2720411 ext 33412
    chmcch@ccu.edu.tw
Hwang, Chyi (Honorary) .... 886-5-2720411 ext 33456
    chmch@ccu.edu.tw
Lee, Wen-Chien .... 886-5-2720411 ext 33406
    chmwcl@ccu.edu.tw
Tsiang, Raymond C. (Chair) .... 886-5-2720411 ext 33454
    chmcct@ccu.edu.tw
Wang, Feng-Sheng .... 886-5-2720411 ext 33404
    chmfsw@ccu.edu.tw
Wang, Maw-Ling (Honorary) ..... 886-5-2720411 ext 33461
    chmmlw@ccu.edu.tw

Associate Professors

Kuo, Yung-Chih .... 886-5-2720411 ext 33459
    chmyck@ccu.edu.tw
Lin, Tsao-Jen .... 886-5-2720411 ext 33405
    chmtjl@ccu.edu.tw
Tsai, Jing-Cherng .... 886-5-2720411 ext 33460
    chmjct@ccu.edu.tw

Assistant Professors

Lee, Tai-Chou .... 886-5-2720411 ext 33409
    chmtcl@ccu.edu.tw
Huang, Kuang-Tse .... 886-5-2720411 ext 33407
    chmkth@ccu.edu.tw
Li, Yuan-Yao .... 886-5-2720411 ext 33403
    chmyyl@ccu.edu.tw

Accredited by:

Degrees granted 2004-2005:
    B.S.: 35   M.S.: 44   Ph.D.: 3

National Taiwan University of Science and Technology

Department of Chemical Engineering
43 Keelung Road, Section 4
Taipei, 106-07
Taiwan

University phone 886-2-2733-3141
Department phone 886-2-2737-6610
Department FAX 886-2-2737-6644

Professors

Chen, Hsiu-Mei 886-2-2737-6651
    hsiumei@ch.ntust.edu.tw
Chern, Chorng-Shyan 886-2-2737-6649
    chern@ch.ntust.edu.tw
Chern, Yaw-Terng 886-2-2737-6646
    cyt@ch.ntust.edu.tw
Chien, I-Lung 886-2-2737-6652
    Chien@ch.ntust.edu.tw
Chou, Yi-Shyong 886-2-2737-6622
    yschou@ch.ntust.edu.tw
Hong, Lu-Sheng 886-2-2737-6650
    hong@ch.ntust.edu.tw
Huang, Yan-Jyi 886-2-2737-6625
    yjhuang@ch.ntust.edu.tw
Hwang, Bing-Joe 886-2-2737-6624
    bjh@ch.ntust.edu.tw
Ju, Yi-Hsu 886-2-2737-6612
    ju@ch.ntust.edu.tw
Ku, Young 886-2-2737-6621
    Ku@ch.ntust.edu.tw
Lee, Cheng-Kang 886-2-2737-6629
    cklee@ch.ntust.edu.tw
Lee, Chiapyng 886-2-2737-6623
    cl@ch.ntust.edu.tw
Lee, Ming-Jer (Chair) 886-2-2737-6626
    mjl@ch.ntust.edu.tw
Liaw, Der-Jang 886-2-2737-6638
    liaw@ch.ntust.edu.tw
Lin, Chun-I 886-2-2737-6614
    cilin@ch.ntust.edu.tw
Lin, Ho-Mu 886-2-2737-6643
    hml@ch.ntust.edu.tw
Lin, Shi-Yow 886-2-2737-6648
    ling@ch.ntust.edu.tw
Liu, Chin-Hsin 886-2-2737-6647
    cjl@ch.ntust.edu.tw
Liu, Jhy-Chern 886-2-2737-6627
    liu@ch.ntust.edu.tw
Liu, Tuan-Chi 886-2-2737-6617
    tuan@ch.ntust.edu.tw
Shiau, Ching-Yeh 886-2-2737-6619
    cys@ch.ntust.edu.tw

Tsai, Dah-Shyang 886-2-2737-6618
tsai@ch.ntust.edu.tw

Associate Professors

Jiang, Jyh-Chiang 886-2-2737-6653
jcjiang@ch.ntust.edu.tw
Liaw, Ben-Ruey 886-2-2737-6628
ruey@ch.ntust.edu.tw
Tseng, Wen-Chi 886-2-2730-1078
tsengwc@ch.ntust.edu.tw

Assistant Professors

Ho, Ming-Hua
mhho@mail.ntust.edu.tw

Accredited by: Ministry of Education

Degrees granted 2004-2005:
 B.S.: 163　　M.S.: 74

Department reports to: Ming-Jer Lee

Placement service: Ming-Jer Lee

National Taiwan University

Department of Chemical Engineering
No.1, Sec. 4, Roosevelt Road, Taipei, Taiwan 106-17

University phone 886-2-23630231
Department phone 886-2-2363-5230
Department FAX 886-2-2362-3040

Professors

Chao, Yung-Cheng (Emeritus)
Chen, Cheng-Ching (Emeritus)
Chen, Cheng-Liang 886-2-2363-6194
ccl@ntu.edu.tw
Chen, Leo-Wang (Emeritus) 886-2-2363-9884
Chen, Li-Jen 886-2-2362-3296
ljchen@ntu.edu.tw
Chen, Wen-Chang 886-2-2362-8398
chenwc@ntu.edu.tw
Chen, Yan-Ping 886-2-2366-1661
ypchen@ntu.edu.tw
Chiu, Wen-Yen 886-2-2362-3259
ycchiu@ntu.edu.tw
Ho, Kuo-Chuan 886-2-2366-0739
kcho@ntu.edu.tw
Hsieh, Kuo-Huang 886-2-2362-7688
khhsieh@ntu.edu.tw
Hsu, Jyh-Ping 886-2-2363-7448
jphsu@ntu.edu.tw
Huang, Hsiao-Ping 886-2-2363-8999
husnaghpc@ntu.edu.tw
Huang, Shih-Yow (Emeritus) 886-2-2363-2542
hsyo@ntu.edu.tw
Keh, Huan-Jang (Dean) 886-2-2363-5462
huan@ntu.edu.tw
Lan, Chung-Wen 886-2-2363-3917
cwlan@ntu.edu.tw
Lee, Duu-Jong 886-2-2362-5632
djlee@ntu.edu.tw
Lee, Keh-Chyang 886-2-2362-2530
ericlee@ntu.edu.tw
Lee, Min-Dar (Emeritus)
Leu, Lii-Ping 886-2-2365-7200
lleulii@ntu.edu.tw
Liu, Hwai-Shen 886-2-2362-7499
hsliu@ntu.edu.tw
Lu, Chung-Hsin 886-2-2365-1428
chlu@ntu.edu.tw
Lu, Shaw-Mei (Emeritus)
Lu, Wei-Ming (Emeritus) 886-2-2362-2707
wmlu@ntu.edu.tw
Sheng, Yu-Jane 886-2-2366-0454
yjsheng@ntu.edu.tw
Shih, Shin-Min 886-2-2363-3974
smshih@ntu.edu.tw
Tai, Clifford Y. 886-2-2362-0832
cytai@ntu.edu.tw
Wan, Ben-Zu 886-2-2363-0737
benzuwan@ntu.edu.tw
Wang, Da-Ming 886-2-2366-0433
daming@ntu.edu.tw
Wu, Chi-Sheng 886-2-2363-1994
cswu@ntu.edu.tw
Wu, Nae-Lih (Chair) 886-2-2362-7158
nlw001@ntu.edu.tw
Yen, Shi-Chern 886-2-2363-0397
scyen@ntu.edu.tw
Yu, Cheng-Ching 886-2-3365-1759
ccyu@ntu.edu.tw

Associate Professors

Hsieh, Hsyue-Jen 886-2-2363-3097
hjhsieh@ntu.edu.tw
Tsai, Wei-Bor 886-2-2369-8627
weibortsai@ntu.edu.tw

Assistant Professors

Dai, Chi-An 886-2-2364-3378
polymer@ntu.edu.tw
Lin, Shiang-Tai 886-2-33663065
stlin@ntu.edu.tw
Wang, Sheng-Shih

Accredited by: Ministry of Education

Degrees granted 2004-2005:
 B.S.: 100　　M.S.: 65　　Ph.D.: 19

Graduate advisor: Prof. Leu, Lii-Ping; Prof. Hsieh, Hsyue-Jen

Department reports to: H. J. Keh, Dean, College of Engineering

Placement service: Staff: Tseng,Sui-hwa

National Tsing Hua University

Department of Chemical Engineering
Hsinchu, Taiwan 300

University phone 886-3-5715131
Department phone 886-3-5719036
Department FAX 886-3-5715408

Professors

Chang, Rong-Yu 886-3-5718344
rychang@moldex3d.com
Chen, Hsin-Lung 886-3-5721714
hslchen@mx.nthu.edu.tw
Chen, Show-An 886-3-5710733
sachen@che.nthu.edu.tw

Chen, Sinn-Wen (Chair) 886-3-5721734
    swchen@che.nthu.edu.tw
Chin, Wei-Kuo 886-3-5713721
    wkc@che.nthu.edu.tw
Chou, Kan-Sen 886-3-5713691
    kschou@che.nthu.edu.tw
Chu, I-Ming 886-3-5713704
    imchu@che.nthu.edu.tw
Hsiue, Ging-Ho 886-3-5719956
    ghhsiue@che.nthu.edu.tw
Huang, Ta-Jen 886-3-5716260
    tjhuang@che.nthu.edu.tw
Hwang, Shyh-Jye 886-3-5723221
    sjhuang@che.nthu.edu.tw
Jang, Shi-Shang 886-3-5713697
    ssjang@che.nthu.edu.tw
Ku, Chia-Soon (Visiting) 886-2-23126100
    kucs@cpc.com.tw
Lee, Yu-Der 886-3-5713204
    ydlee@che.nthu.edu.tw
Liu, Ta-Jo 886-3-5723380
    tjliu@che.nthu.edu.tw
Lu, Shih-Yuan 886-3-5714364
    sylu@mx.nthu.edu.tw
Ma, Chen-Chi 886-3-5713058
    ccma@che.nthu.edu.tw
Su, An-Chung 886-7-5254055
    acsu@mail.nsysu.edu.tw
Sung, Hsing-Wen 886-3-5742504
    hwsung@che.nthu.edu.tw
Tan, Chung-Sung 886-3-5721189
    cstan@che.nthu.edu.tw
Wan, Chi-Chao 886-3-5721664
    ccwan@mx.nthu.edu.tw
Wang, I-Kai 886-3-5713763
    ikwang@che.nthu.edu.tw
Wang, Yung-Yun 886-3-5713690
    yywang@mx.nthu.edu.tw
Wong, David S. 886-3-5721694
    dshwong@che.nthu.edu.tw
Wu, Wen-Teng (Visiting) 886-6-2757575ext62000
    wtwu@mail.ncku.edu.tw

Associate Professors

Ho, Rong-Ming 886-3-5738349
    rmho@mx.nthu.edu.tw
Hsu, John T. (Adjunct) 886-3-7246166ext35717
    tsuanhsu@nhri.org.tw
Hu, Yu-Chen 886-3-5718245
    ychu@mx.nthu.edu.tw
Mei, Marty T. (Visiting) 886-2-27134242ext517
    marty@harvester.com.tw

Assistant Professors

Tang, Shiue-Cheng 886-3-5715131ext3649
    sctang@che.nthu.edu.tw

Accredited by: Ministry of Education R.O.C.

Degrees granted 2004-2005:
    B.S.: 59   M.S.: 65   Ph.D.: 15

Graduate advisor: Sinn-wen Chen

Undergraduate advisor: Sinn-wen Chen

Student organization: Student Association of Chemical Engineering
Advisor: Sinn-wen Chen

Tamkang University

Department of Chemical and Materials Engineering
Tamsui, Taipei Hsien, Taiwan, 25137
    University phone 886-2-2621-5656
    Department phone 886-2-26215656x2614
    Department FAX 886-2-2620-9887

Professors

Chang, Yu-Chi .... x2722 †
    040791@mail.tku.edu.tw
Chen, Hsi-Jen x2721
    c1952814@ms12.hinet.net
Cheng, Liao-Ping x2725
    lpcheng@mail.tku.edu.tw
Cheng, Tung-Wen (Chair) x2725
    twcheng@mail.tku.edu.tw
Chi, Jung-Chang x2723
    jcchi@mail.tku.edu.tw
Han, Kwong-Wing x2601
    043467@mail.tku.edu.tw
Ho, Chii-Dong x2724
    cdho@mail.tku.edu.tw
Hwang, Kuo-Jen x2726
    kjhwang@mail.tku.edu.tw
Yeh, Ho-Ming x2601
    hmyeh@mail.tku.edu.tw

Associate Professors

Chang, Cheng-Liang x2723
    chlchang@mail.tku.edu.tw
Chang, Hsuan x2721
    nhchang@mail.tku.edu.tw
Chen, Ching-Chung x2722
    047927@mail.tku.edu.tw
Don, Trong-Ming x2670
    tmdon@mail.tku.edu.tw
Lin, Dar-Jong x2724
    djlin@mail.tku.edu.tw
Lin, Gwo-Geng x2726
    gglin@ms4.hinet.net
Yu, Hseng-Fu x2728
    hfyu@mail.tku.edu.tw

† Faculty numbers are extensions of 886-2-26215656.

Accredited by: Ministry of Education

Graduate advisor: Tung-Wen Cheng

Student organization: ChIChE
Advisor: Tung-Wen Cheng

Placement service: Ching-Sheng Shu, Public Relations, Section Secretariate

Tunghai University

Chemical Engineering Department
Tunghai University
P.O. Box 833.
Taichung, Taiwan R.O.C. 40704
    University phone 886-4-23590121
    Department phone 886-4-23590262-108
    Department FAX 886-4-23590009

Professors

Chang, You-Im 886-4-23590262
    yichang@thu.edu.tw
    EV,NT,SU
Chiao, Shu-Min 886-4-23590262
    smchiao@thu.edu.tw
    PO
Do, Jing-Shan 886-4-23590262
    jsdo@thu.edu.tw
    EV,RE
Huang, Chi-Tsung 886-4-23590262
    huangct@thu.edu.tw
    MO,PR,SE
Li, Kuo-Tseng (Head) 886-4-23590262
    ktli@thu.edu.tw
    EV,MO,RE,SU
Wang, Yeh 886-4-23590262
    yehwang@thu.edu.tw
    MO,PO
Yang, Fan-Chiang 886-4-23590262
    fcyang@thu.edu.tw
    BT,RE
Yang, I-Kuan 886-4-23590262
    ikyang@thu.edu.tw
    AM,PO

Associate Professors

Cheng, Shueh-Hen 886-4-23590262
    shcheng@thu.edu.tw
    MO,PR,SE
Gu, Yesong 886-4-23590262
    yegu@thu.edu.tw
    BM,BT,SE
Hsieh, Shu-Mu 886-4-23590262
    smhsieh@thu.edu.tw
    MO,PR,SE

Assistant Professors

Ho, Chih S. 886-04-23590262
    csho@thu.edu.tw
    AM,RE,SU
Yen, Hong W. 886-04-23590262
    hwyen@thu.edu.tw
    BT,EV,SE

Accredited by: Ministry of Education

Student organization: ChIChE
Advisor: Wang Yeh

Yuan Ze University

Department of Chemical Engineering and Materials Science
Chung-Li, Taoyuan, 32003, Taiwan

University phone 886-3-463-8800
Department phone 886-3-463-6897
Department FAX 886-3-455-9373

Professors

Hong, Shinn G. . . . . ext. 2559 [†]
    cesghong@saturn.yzu.edu.tw
    PO,SU
Hwang, Jenn C. . . . . ext. 2560
    cejhwang@saturn.yzu.edu.tw
    AM,PO
Juang, Ruey S. (Dean) . . . . ext. 2555
    rsjuang@saturnyzu.edu.tw
    SE,SU
Lin, Sheng H. (Emeritus) . . . . ext. 2551
    ceshlin@saturn.yzu.edu.tw
    EN,EV,MO
Wu, Ho S. (Chair) . . . . ext. 2564
    cehswu@saturn.yzu.edu.tw
    BT,EV,MO,RE,SE
Yin, Ken M.
    cekenyin@saturn.yzu.edu.tw
Yu, Tzyy L.
    cetlyu@ce.yzu.edu.tw
    AM,EN,PO

Associate Professors

Chang, Yu C. . . . . ext. 2571
    yjchan@ce.yzu.edu.tw
    NT,SU
Hsuen, Hsiao K. D. . . . . ext. 2569
    skhsuen@saturn.yzu.edu.tw
    RE
Liao, Chien S. . . . . ext. 2567
    csliao@saturn.yzu.edu.tw
    AM,BT,ME,PO
Liau, Chau K. . . . . ext. 2573
    lckliau@saturn.yzu.edu.tw
    AM,EN,MO,PR
Lin, Hsiu L. . . . . ext. 2568
    sherry@saturn.yzu.edu.tw
    AM,EN,PO
Lin, Sheng D. . . . . ext. 2554
    sdlin@saturn.yzu.edu.tw
    AM,EN,NT,RE,SU

Assistant Professors

Lin, Kuen S. . . . . ext. 2574
    kslin@ce.yzu.edu.tw
    AM,EN,EV,NT

[†] Faculty numbers are extensions of 886-3-463-8800.

Accredited by: Ministry of Education

Degrees granted 2005-2006:
    B.S.: 102   M.S.: 44   Ph.D.: 4

Student organization: CEMS student club
Advisor: Dr. Ho-Shing Wu

Department reports to: Ho-Shing Wu(cehswu@saturn.yzu.edu.tw)

Placement service: ceemilie@saturn.yzu.edu.tw

Thailand

Chulalongkorn University

Faculty of Engineering
Department of Chemical Engineering
Phaya Thai Road
Bangkok 10330

University phone . . . . (662) 215-0871-73
Department phone . . . . (662) 218-6878-83
Department FAX . . . . (662) 218-6877

Professors

Praserthdam, Piyasan (Chair) .... 218-6711 [†]
    piyasan.p@chula.ac.th
Tanthapanichakoon, Wiwut 218-6894
    fchwtt@eng.chula.ac.th

Associate Professors

Assabumrungrat, Suttichai 218-6868
    fchsas@kankrow.eng.chula.ac.th
Charinpanitkul, Tawatchai 218-6894
    ctawat@pioneer.chula.ac.th
Kittisupakorn, Paisan 218-6892
    paisan.k@chula.ac.th
Mongkhonsi, Tharathon 218-6867
    tharathon.m@chula.ac.th
Muangnapoh, Chirakarn 218-6874
    chirakarn.m@chula.ac.th
Pancharoen, Ura 218-6891
    ura.p@chula.ac.th
Satayaprasert, Chairit 218-6893
    chairit.s@chula.ac.th
Vanichseni, Sutham 218-6407
    sutham.v@chula.ac.th

Assistant Professors

Boon-Long, Sasithorn 218-6876
    fchsbl@eng.chula.ac.th
Chatsiriwech, Deacha 218-6863
    deacha.c@chula.ac.th
Chongvisal, Vichitra 218-6875
    vichitra.c@chula.ac.th
Damrongsakkul, Siriporn 218-6862
    siriporn.d@chula.ac.th
Duriyabunleng, Hathaichanok 218-6867
    hathaic.d@chula.ac.th
Prechanont, Seeroong 218-6860
    seeroong.p@chula.ac.th
Thongyai, Supakanok 218-6860
    supakanok.t@chula.ac.th

[†] Direct dial with prefix 662.

Accredited by: Engineering Association of Thailand, Ministry of University Affairs

Department reports to: Somsak Panyakeow (218-6310), Dean of Faculty of Engineering

Kasetsart University

Department of Chemical Engineering
Faculty of Engineering
50 Phaholyothin Road, Jatujak
Bankok, Thailand
10900

University phone 66-2942-8500
University FAX 66-2942-8988
Department phone .... 66-2942-8555 ext. 1204
Department FAX 66-2561-4621

Associate Professors

Chareonpanich, Metta .... ext 1215 [†]
    fengmtc@ku.ac.th
    AM,EN,EV,NT,RE
Kongkachuichay, Paisan .... ext 1207
    fengpsk@ku.ac.th
    AM,BT,EN,RE
Limtrakul, Sunun .... ext 1210
    sunun.l@ku.ac.th
    EN,MO,RE
Phanawadee, Phungphai (Associate Chair) .... ext 1212
    fengphi@ku.ac.th
    MO,RE
Srinophakhun, Penjit (Associate Chair) .... ext 1213
    fengpjs@ku.ac.th
    BT,EN
Srinophakhun, Thongchai .... ext 1214
    fengtcs@ku.ac.th
    MO,PR
Suvachitanont, Sirikalaya .... ext 1205
    fengsks@ku.ac.th
    AM,EN,EV,NT

Assistant Professors

Anantawaraskul, Siripon .... ext 1231
    fengsia@ku.ac.th
    AM,MO,PO
Charoenchaitrakool, Manop .... ext 1216
    fengmnc@ku.ac.th
    AM,BM,BT,EN,SE
Chutmanop, Jarun .... ext 1218
    fengjrc@ku.ac.th
    BT,EN
Duangchan, Apinya .... ext 1211
    fengapd@ku.ac.th
    AM,BT,EN,PO,RE
Jaree, Attasak .... ext 1229
    fengasj@ku.ac.th
    BT,EN,EV,RE
Mungcharoen, Thumrongrut (Chair) .... ext 1201
    fengtrm@ku.ac.th
    EN,EV,MO,PR,SE
Sudsakorn, Kandis .... ext 1217
    fengkdsk@ku.ac.th
    BT,EN,RE
Vatanatham, Terdthai .... ext 1208
    terdthai.v@ku.ac.th
    AM,EN,PO,RE,SE

[†] Phone numbers are extensions of 66-2942-8555

Accredited by: Thailand Ministry of University Affairs, Thailand Civil Service Commission

Degrees granted 2005-2006:
    B.S.: 80   M.S.: 36   Ph.D.: 1

Graduate advisor: Assoc. Prof. Dr.Sunun Limtrakul

Undergraduate advisor: Dr. Attasak JAREE

Student organization: KUChE Student Club
Advisor: Dr. Siripon Anantawaraskul

Department reports to: Professor Nontawat Junjaroen, Dean of Engineering

Placement service: Assoc.Prof.Dr. Penjit Srinophakhun

Prince of Songkla University

Chemical Engineering Department
P.O.B. 2, Kohong
Hat-Yai, Songkla 90112

University phone 66 74 211030-49
Department phone 66 74 212896
Department FAX 66 74 212896

Associate Professors

Srisuwan, Galaya (Director) 66 74 287283
    galaya.s@psu.ac.th
Tongurai, Chakrit 66 74 287287
    chakrit.t@psu.ac.th

Assistant Professors

Bunyakan, Charun (Associate Dean) 66 74 287309
    Charun.b@psu.ac.th
Innachitra, Paiboon 66 74 212896
    Paiboon.i@psu.ac.th
Kaewsichan, Lupong 66 74 287292
    Lupong.k@psu.ac.th
Ratanapisit, Juraiwan (Director) 66 74 287286
    juraivan.r@psu.ac.th
Thonglimp, Veerasak 66 74 287293
    veerasak.t@psu.ac.th
Tirawanichakul, Supawan 66 74 212896
    SUPAWANVACHIRAMON@Hotmail.com
Yamsaengsung, Ram 66 74 287291
    ram.y@psu.ac.th

Accredited by:

Degrees granted 2004-2005:
    B.S.: 38   M.S.: 8

Graduate advisor: Galaya Srisuwan

Undergraduate advisor: Juntima Chungsiriporn

Department reports to: Chusak Limsakul, Dean of Engineering

Turkey

Ankara University

Department of Chemical Engineering
Tandoan 06100
Ankara

University phone 90-312-2126040
Department phone 90-312-2126720x1321
Department FAX 90-312-2121546

Professors

Aktas, Zeki .... x1300 [†]
    zaktas@eng.ankara.edu.tr
Alpbaz, Mustafa x1365
    alpbaz@eng.ankara.edu.tr
Berber, Ridvan (Dean) x1351
    berber@eng.ankara.edu.tr
Bilgesu, Ali Y. x1349
    bilgesu@eng.ankara.edu.tr
Calik, Guzide x1367
    calik@eng.ankara.edu.tr
Calimli, Ayla (Chair) x1303
    calimli@eng.ankara.edu.tr
Erol, Murat (Associate Chair) x1375
    erol@eng.ankara.edu.tr
Hapoglu, Hale x1338
    hapoglu@eng.ankara.edu.tr
Mehmetoglu, Ulku x1311
    mehmet@eng.ankara.edu.tr
Oguz, Huseyin x1368
    oguz@eng.ankara.edu.tr
Ozdamar, H. T. x1354
    ozdamar@eng.ankara.edu.tr
Takac, Serpil x1310
    takac@eng.ankara.edu.tr
Togrul, Tamer x1329
    togrul@eng.ankara.edu.tr
Yeniova, Hasip x1381
    yeniova@eng.ankara.edu.tr

Associate Professors

Akay, Bulent x1307
    bakay@eng.ankara.edu.tr
Bayraktar, Emine (Associate Chair) x1373
    bayrakta@eng.ankara.edu.tr
Cicek, Burhanettin x1127
    bcicek@eng.ankara.edu.tr
Guvenc, Afife x1373
    guvenc@eng.ankara.edu.tr
Karacan, Suleyman x1302
    karacan@eng.ankara.edu.tr
Kocak, Cetin x1330
    kocak@eng.ankara.edu.tr
Ozkan, Gulay x1302
    gozkan@eng.ankara.edu.tr
Yildiz, Nuray x1314
    nyildiz@eng.ankara.edu.tr

Assistant Professors

Karaduman, Ali x1385
    akrduman@eng.ankara.edu.tr
Simsek, Emir H. x1374
    simsek@eng.ankara.edu.tr
Suyadal, Yahya x1374
    suyadal@eng.ankara.edu.tr

[†] Phone numbers are extensions of 90-312-2126720.

Accredited by:

Degrees granted 2004-2005:
    B.S.: 51   M.S.: 12   Ph.D.: 3

Graduate advisor: Dr.Murat Erol

Undergraduate advisor: Dr.Emine Bayraktar

Department reports to: Prof.Dr.Ridvan Berber, Dean of Faculty of Engineering

Bogazici University

Chemical Engineering Department
34342 Bebek, Istanbul

University phone .... (90) 212 358 1540
Department phone .... (90)212 358 1540x1471
Department FAX .... (90) 212 287 2460

Professors

Akman, Ugur .... x1867 [†]
    akman@boun.edu.tr
Arikol, Mahir (Chair) x1870
    arikolma@boun.edu.tr
Baysal, Bahattin (Emeritus) x1868
    bmbaysal@hotmail.com
Borak, Fahir (Emeritus) x1470
    borakf@boun.edu.tr
Camurdan, Mehmet C. x1405
    mccamurd@boun.edu.tr

Dincer, Salih x2338
  dincer@yildiz.edu.tr
Haliloglu, Turkan x2003
  turkan@prc.bme.boun.edu.tr
Hortacsu, Amable x1469
  hortacsa@boun.edu.tr
Hortacsu, Oner x1434
  hortacsu@boun.edu.tr
Kirdar, Betul (Associate Dean) x2126
  kirdar@boun.edu.tr
Onsan, Z.Ilsen x1751
  onsan@boun.edu.tr

Associate Professors

Aksoylu, Ahmet E. x2336
  aksoylu@boun.edu.tr
Doruker, Pemra (Associate Chair) x2365
  pemra@prc.bme.boun.edu.tr
Ulgen, Kutlu x1869
  ulgenk@boun.edu.tr

Assistant Professors

Dervisoglu, Murat 1866
  mudervis@boun.edu.tr
Ozer, Ercument x1471
  biyotek@biyotek.com.tr
Yildirim, Ramazan x2248
  yildirra@boun.edu.tr

† Phone numbers are extensions of

Accredited by: ABET

Graduate advisor: Mahir Arikol

Department reports to: Ali Riza Kaylan, Dean of Engineering

Placement service: BUMED (BU Alumni Association)

University of Ege

Faculty of Engineering
Chemical Engineering Department
35100 Bornova-IZMIR

University phone .... (90) 232 339 0204
University FAX .... (90) 232 339 9090
Department phone .... (90) 232 388 7600
Department FAX .... (90) 232 388 77 76

Professors

Atalay, Ferhan S. .... 2284 †
  ferhan.atalay@ege.edu.tr
Atalay, Suheyda (Chair) 2920
  suheyda.atalay@ege.edu.tr
Ayvaz, Zafer 2287
  zafer.ayvaz@ege.edu.tr
Balkan, Firuz 2291
  firuz.balkan@ege.edu.tr
Ballice, Levent 1484
  levent.ballice@ege.edu.tr
Demircioglu, Mustafa 1831
  mustafa.demircioglu@ege.edu.tr
Erbil, Özel (Emeritus)
  ozel.erbil@ege.edu.tr
Gündüz, Gönül 2292
  gonul.gunduz@ege.edu.tr
Helvaci, Serife 1480
  serife.helvaci@ege.edu.tr
Kabay, Nalan 2290
  nalan.kabay@ege.edu.tr
Olgun, Özden 2286
  oolgun@eng.ege.edu.tr
Peker, Sümer 2282
  sumer.peker@ege.edu.tr
Saglam, Mehmet 2456
  mehmet.saglam@ege.edu.tr
Ulutan, Sevgi 1492
  sevgi.ulutan@ege.edu.tr
Yapar, Saadet (Associate Chair) 2459
  saadet.yapar@ege.edu.tr
Yenigül, Mesut 2289
  mesut.yenigul@ege.edu.tr
Yüksel, Mithat 2456
  mithat.yuksel@ege.edu.tr
Övez, Bikem 1483
  bikem.ovez@ege.edu.tr

Associate Professors

Özdemir, Günseli 1485
  gunseli.ozdemir@ege.edu.tr

Assistant Professors

Cesur, Serap 1485
  serap.cesur@ege.edu.tr
Özçelik, Yavuz 1488
  yavuz.ozcelik@ege.edu.tr
Özçelik, Zehra (Associate Chair) 1488
  zehra.ozcelik@ege.edu.tr

† Direct dial with prefix (90) 232 388 4000.

Accredited by:

Degrees granted 2005-2006:
  B.S.: 90   M.S.: 10   Ph.D.: 3

Graduate advisor: Saadet Yapar (90) 232 388 7600

Undergraduate advisor: Zehra Özçelik (90) 232 388 7600

Department reports to: Saadet Yapar (90) 232 388 7600

Placement service: Zehra Özçelik (90) 232 388 7600

Gazi University

Chemical Engineering Department
Faculty of Engineering and Architecture
06570 Maltepe-Ankara

University phone .... (90) 312-2126840
Department phone .... (90) 312-2317400
Department FAX .... (90) 312-2308434

Professors

Alicilar, Ahmet .... x2524 †
  alicilar@gazi.edu.tr
Balci, Suna x2506
  sunabalci@gazi.edu.tr
Bicer, Ahmet x2548
  abicer@gazi.edu.tr
Cabbar, Canan x2557
  hcabbar@gazi.edu.tr
Culfaz, Mujgan x2503
  culfaz@gazi.edu.tr

Dogu, Gulsen x2559
    gdogu@gazi.edu.tr
Erdogan, Sebahat (Head) x2549
    sebaer@gazi.edu.tr
Ergun, Mubeccel x2520
    mubeccel@gazi.edu.tr
Koc, Timur x2553
    timurk@gazi.edu.tr
Murtezaoglu, Kirali x2523
    kirali@gazi.edu.tr
Oksuz, Iskender x2522
    ioksuz@gazi.edu.tr
Pamuk, Vecihi x2554
    vecihi@gazi.edu.tr
Saracoglu, Nurdan x2546
    nsarac@gazi.edu.tr
Uysal, B. Z. x2501
    bzuysal@gazi.edu.tr

Associate Professors
Ar, Irfan x2517
    irfanar@gazi.edu.tr
Dilsiz, Nursel x2547
    ndilsiz@gazi.edu.tr
Dogan, O. M. x2513
    mdogan@gazi.edu.tr
Guldur, Cigdem x2507
    cguldur@gazi.edu.tr
Gunduz Zafer, Ufuk x2505
    ufzafer@gazi.edu.tr
Guru, Metin x2555
    muru@gazi.edu.tr
Murathan, Atilla x2518
    murathan@gazi.edu.tr
Murathan, Ayse x2551
    amurathan@gazi.edu.tr

Assistant Professors
Altÿnten, Ayla x2508
    altinten@gazi.edu.tr
Balbasi, Muzaffer x2512
    balbasi@gazi.edu.tr
Dogan, Meltem x2510
    meltem@gazi.edu.tr
Mutlu, S.Ferda (Associate Head) x2519
    fmutlu@gazi.edu.tr
Oktar, Nuray x2504
    nurayoktar@gazi.edu.tr
Yasyerli, Nail (Associate Head) x2511
    yasyerli@gazi.edu.tr
Yasyerli, Sena x2509
    syasyerli@gazi.edu.tr
Özkan, Göksel x2515
    gozkan@gazi.edu.tr

† Dial (90) 312-2317400 and ask for extension.

Accredited by: MUDEK

Degrees granted 2005-2006:
    B.S.: 60   M.S.: 20   Ph.D.: 1

Department reports to: Prof.Dr.Sebahat Erdogan

Hacettepe University

Chemical Engineering Department
Beytepe-06800 Ankara
University phone .... (90) 312 2977400-1
University FAX .... (90) 2977400
Department phone .... (90) 312 2977400
Department FAX .... (90) 312 2992124

Professors
Aksu, Zumriye .... (90) 3122977434
    zaksu@hacettepe.edu.tr
Alper, Erdogan (Chair) .... (90) 3122977424
    ealper@hacettepe.edu.tr
Beskardes, Oktay .... (90) 3122977474
    obeskard@hacettepe.edu.tr
Caglar, Arif .... (90) 3122977414
    caglar@hacettepe.edu.tr
Celebi, Serdar S. .... (90) 3122977464
    celebi@hacettepe.edu.tr
Durusoy, Tulay .... (90) 3122977494
    tdurusoy@hacettepe.edu.tr
Gumusderelioglu, Menemse .... (90) 3122977447
    menemse@hacettepe.edu.tr
Kutsal, Tulin .... (90) 3122977484
    tkutsal@hacettepe.edu.tr
Ozbas Bozdemir, Tijen .... (90) 3122977455
    bozdemir@hacettepe.edu.tr
Ozdural, Ahmet R. .... (90) 3122977475
    ozdural@hacettepe.edu.tr
Piskin, Erhan .... (90) 3122977473
    piskin@hacettepe.edu.tr
Rzayev, Zakir .... (90) 3122977400
    zakirr@hacettepe.edu.tr
Sag, Yesim .... (90) 3122977444
    yesims@hacettepe.edu.tr
Tanyolac, Abdurrahman .... (90) 3122977404
    tanyolac@hacettepe.edu.tr
Tuncel, S. A. (Associate Chair) .... (90) 3122977433
    tuncel@hacettepe.edu.tr

Associate Professors
Aydogan, Nihal .... (90) 3122976443
    naydogan@hacettepe.edu.tr
Evren, Vural .... (90) 3122977422
    vural@hacettepe.edu.tr
Mutlu, Selma .... (90) 3122977400
    smselma@hacettepe.edu.tr
Tanyolaç, Deniz .... (90) 3122976162
    deniztan@hacettepe.edu.tr

Assistant Professors
Ayhan, Hakan .... (90) 3122977400
    hayhan@hacettepe.edu.tr

Accredited by: Turkish Higher Educational Council

Degrees granted 2004-2005:
    B.S.: 60   M.S.: 15   Ph.D.: 8

Graduate advisor: Erdogan Alper

Department reports to: Selcuk Gecim, Dean

Placement service: Erdogan Alper

Middle East Technical University

Chemical Engineering Department
Inonu Bulvari Ankara, 06531
University phone .... (90) 312 2101000
Department phone .... (90) 312 2102601
Department FAX .... (90) 312 2101264

Professors

Bakir, Ufuk (Associate Chair) .... (90) 312 2102619
    ubakir@metu.edu.tr
Culfaz, Ali .... (90) 312 2102611
    culfaz@metu.edu.tr
Dogu, Timur (Chair) .... (90) 312 2102631
    tdogu@metu.edu.tr
Eroglu, Inci .... (90) 312 2102609
    ieroglu@metu.edu.tr
Gunduz, Gungor .... (90) 312 2102616
    ggunduz@metu.edu.tr
Guruz, Guniz .... (90) 312 2102634
    guniz@metu.edu.tr
Gürkan, Türker .... (90) 312 2102618
    tgurkan@metu.edu.tr
Kincal, Suzan .... (90) 312 2102617
    nskincal@metu.edu.tr
Kisakürek, Bilgin .... (90) 312 2102625
    bilgin@metu.edu.tr
Onal, Isik .... (90) 312 2102639
    ional@metu.edu.tr
Ozbelge, Onder .... (90) 312 2102628
    oozbelge@metu.edu.tr
Ozbelge, Tulay .... (90) 312 2102621
    tozbelge@metu.edu.tr
Ozgen, Canan .... (90) 312 2102605
    cozgen@metu.edu.tr
Selcuk, Nevin .... (90) 312 2102603
    selcuk@metu.edu.tr
Tosun, Ismail .... (90) 312 2102637
    itosun@metu.edu.tr
Yilmaz, Levent .... (90) 312 2102607
    lyilmaz@metu.edu.tr
Yilmazer, Ulku .... (90) 312 2102615
    yilmazer@metu.edu.tr
Yucel, Hayrettin .... (90) 312 2102635
    hyucel@metu.edu.tr

Associate Professors

Calik, Pinar .... (90) 312 2104385
    pcalik@metu.edu.tr
Karakas, Gurkan (Associate Chair) ...... (90) 312 2102630
    gkarakas@metu.edu.tr
Ozbay, Deniz U. .... (90) 312 2104383
    uner@metu.edu.tr

Assistant Professors

Bayram, Goknur .... (90) 3122102632
    gbayram@metu.edu.tr
Kalÿpçÿlar, Halil .... (90) 312 2104357
    hkalipcilar@metu.edu.tr
Uludag, Yusuf .... (90) 3122104374
    yuludag@metu.edu.tr

Accredited by: Recognised by ABET

Graduate advisor: Levent Yilmaz

Student organization: AIChE
Advisor: Deniz Uner

Department reports to: Yildirim Uctug, Dean, Faculty of Engineering

Yildiz Technical University

Faculty of Chemical and Metallurgical Engineering
Chemical Engineering Department
Davutpasa Campus, No.127, 34210 Esenler
Istanbul
University phone 90 212 449 16 67
Department phone 90 212 449 18 94
Department FAX 90 212 449 18 95

Professors

Bolat, Esen
Dincer, Salih
Oner, Mualla
Pala, Mehmet
Piskin, Sabriye (Head)

Associate Professors

Baykara, Sema
Beker, Ulker
Erturan, Seyfettin
Ozbek, Belma

Assistant Professors

Akgun, Mesut
Akgun, Nalan
Gulen, Jale
Ozkan, Semra

Graduate advisor: Sabriye Piskin

Undergraduate advisor: Sabriye Piskin

Department reports to: Nihat Kinikoglu, Dean of Chemical and Metallurgical Engineering

United Arab Emirates

United Arab Emirates University

Department of Chemical and Petroleum Engineering
College of Engineering, UAE University
PO Box 17555, Al-Ain
United Arab Emirates
Department phone .... (971-3) 7133-611
Department FAX .... (971-3) 762-4262

Professors

Abou-Kassem, Jamal 971-3-7133-599
    J.Aboukassem@uaeu.ac.ae
Almehaideb, Reyadh A. (Dean) 971-3-7621-765
    Reyadh@uaeu.ac.ae
Zekri, Abdulrazag 971-3-7133-550
    a.zekri@uaeu.ac.ae

Associate Professors

Al-Attar, Hazim
    Hazim.Alattar@uaeu.ac.ae
Abu-Eishah, Samir 971-3-7133-554
    s.abueishah@uaeu.ac.ae

Castier, Marcelo 971-3-7133-634
    mcastier@uaeu.ac.ae
Ghannam, Mamdouh T. 971-3-7133-635
    Mamdouh.Ghannam@uaeu.ac.ae

Assistant Professors

Abdulkarim, Mohamed (Dean) 971-3-7133-547
    Mohamed.AK@uaeu.ac.ae
Al Matroushi, Eisa A. 971-3-7133-638
    Almatroushi@uaeu.ac.ae
Al-Marzouqi, Ali 971-3-7133-639
    HassanA@uaeu.ac.ae
Al-Marzouqi, Mohamed (Chair) 971-3-7621-695
    MHHassan@uaeu.ac.ae
Al-Muhtaseb, Shaheen 971-3-7133-545
    s.almuhtaseb@uaeu.ac.ae
Al-Zuhair, Sulaiman 971-3-7133-636
    S.Alzuhair@uaeu.ac.ae
El-Naas, Muftah 971-3-7133-637
    muftah@uaeu.ac.ae
Ghasem, Nayef 971-3-7133-546
    nayef@uaeu.ac.ae
Rahman, Motiur 971-3-7133-404
    M.Rahman@uaeu.ac.ae

Accredited by: ABET

Department reports to: Dean of College of Engineering

United Kingdom

Aston University

School of Engineering and Applied Science
Division of Chemical Engineering and Applied Chemistry
Aston Triangle
Birmingham B4 7ET, England

University phone .... (44) 121 359-3611
Department phone .... (44) 121 359 3611 x 4632
Department FAX .... (44) 121 359-4094

Professors

Barker, P. E. (Emeritus)
Bridgwater, A. V. (Head)
Porter, K. E. (Emeritus)

Assistant Professors

Brammer, J
Davies, B.
Drahun, J A
Fletcher, J. P.
Generalis, S.
Titiloye, J

Accredited by: IChemE (UK)

Department reports to: Vice Chancellor

Placement service: Dr J A Drahun

University of Bath

Department of Chemical Engineering
Claverton Down
Bath, BA2 7AY, England

University phone .... (1225) 386826
Department phone .... (1225) 386338
Department FAX .... (1225) 385713

Professors

Chaudhuri, Julian B. .... 386349 [†]
    J.B.Chaudhuri@bath.ac.uk
Crittenden, Barry D. (Head) 386501
    B.D.Crittenden@bath.ac.uk
Greaves, Malcolm (Emeritus) 386624
    M.Greaves@bath.ac.uk
Guy, Keith W. A. (Visiting) 386338
Howell, John A. (Emeritus) 386338
Kolaczkowski, Stan T. 386440
    S.T.Kolaczkowski@bath.ac.uk
Moore, Stephen (Visiting) 386338
Tennison, Steve (Visiting) 386338
    S.Tennison@bath.ac.uk

Associate Professors

Bird, Michael R 386336
    M.R.Bird@bath.ac.uk
England, Richard 386516
    R.England@bath.ac.uk
Hubble, John 386221
    J.Hubble@bath.ac.uk
Lukyanov, Dmitry B 383329
    D.B.Lukyanov@bath.ac.uk
Mays, Tim J 386528
    T.J.Mays@bath.ac.uk
Perera, Semali 386584
    S.Perera@bath.ac.uk
Plucinski, Pawel 386961
    P.Plucinski@bath.ac.uk

Assistant Professors

Arnot, Tom C 386707
    T.C.Arnot@bath.ac.uk
Ellis, Marianne J 384484
    M.J.Ellis@bath.ac.uk
Hicks, David I 386251
    D.I.Hicks@bath.ac.uk
Lapkin, Alexei A. 383369
    A.Lapkin@bath.ac.uk
Rathbone, Richard R. 383961
    R.R.Rathbone@bath.ac.uk
Rigby, Sean P 384978
    S.P.Rigby@bath.ac.uk

[†] Dial area code 1225

Accredited by: IChemE (UK)

Degrees granted 2004-2005:
    B.S.: 9   M.S.: 31   Ph.D.: 7

Graduate advisor: Dr Alexei Lapkin

Undergraduate advisor: Dr. Tim Mays

Student organization: IChemE (UK)
Advisor: Dr. Sean Rigby

Department reports to: Dean, Professor Alan Day (1225) 386469

238 United Kingdom

Placement service: Dr. Mike Bird (1255) 386336

University of Cambridge

Chemical Engineering Department
Pembroke Street
Cambridge CB2 3RA, England
Department phone .... (1223) 334777
Department FAX .... (1223) 334796

Professors
Chase, H. A.
    hac1000@cam.ac.uk
    BT,EV,RE,SE

Davidson, J. F. (Emeritus)

Gladden, L. F. (Head)
    gladden@cheng.cam.ac.uk
    MO,PO,RE,SU,TP

Hayhurst, A. N. (Emeritus)
    allan_hayhurst@cheng.cam.ac.uk

Mackley, M.R.
    malcolm_mackley@cheng.cam.ac.uk

Slater, N K H
    nkhs2@cam.ac.uk
    BT,PO

Associate Professors
Cardoso, S. S. S.
    sssc1@cam.ac.uk
    EN,EV,MO

Dennis, J.S.
    jsd3@cam.ac.uk
    EN,EV,MO,RE

Fisher, A.C.
    acf42@cam.ac.uk
    BT,MO,NT,RE,SU

Johns, M.L.
    mlj21@cheng.cam.ac.uk

Kaminski, C.F.
    cfk23@cam.ac.uk
    BM,EN,RE

Kraft, M
    mk306@cam.ac.uk
    EN,EV,MO,NT,PR,RE

Moggridge, G.M.
    gdm14@cam.ac.uk
    AM,NT,PO

Paterson, W.R.
    wrp1@cheng.cam.ac.uk
    BM,MO,PR,RE,SE

Scott, D.M.
    dms1@cheng.cam.ac.uk
    MO,PR,RE,SE

Vassiliadis, V. S.
    vsv20@cheng.cam.ac.uk
    MO,PR

Wilson, D. I.
    diw11@cam.ac.uk
    AM,EN,SU

Assistant Professors
Barrie, P. J.
    pjb10@cam.ac.uk
    RE

Routh, A.F.
    afr10@cam.ac.uk
    MO,NT,PO,SU

Accredited by: The Institution of Chemical Engineers (UK)

Degrees granted 2005-2006:
    B.S.: 23   M.S.: 9   Ph.D.: 14

Graduate advisor: M L Johns

Undergraduate advisor: P J Barrie

Student organization: CUCES

Placement service: A. J. Raban

University of Edinburgh

Chemical Engineering
The Kenneth Denbigh Chemical Engineering Building
King's Buildings, Mayfield Rd.
Edinburgh EH9 3JL, Scotland
Deliveries: Engineering & Electronics Stores, Sanderson Building, King's Buildings, Gate 3
University phone 0131-650-1000
Department phone 0131-650-4860
Department FAX 0131-650-6551

Professors
Christy, John R. E. (Head) .... 650-4854 [†]
    John.Christy@ed.ac.uk

Glass, Donald H. 650-4870
    Don.Glass@ed.ac.uk

Ponton, Jack W. 650-4858
    J.W.Ponton@ed.ac.uk

Pritchard, Colln L. 650-4852
    colin@chemeng.ed.ac.uk

Seaton, Nigel A. 650-4867
    Nigel.Seaton@ed.ac.uk

Associate Professors
Biggs, Mark J 650-5891
    MBiggs@ed.ac.uk

Sefiane, Khellil 650-4873
    ksefiane@ed.ac.uk

Serghiou, George 650-8553
    George.Serghiou@ed.ac.uk

Skilling, Jennifer M. 650-4863
    J.Skilling@ed.ac.uk

Assistant Professors
Duren, Tina 650-4856
    tina.duren@ed.ac.uk

Duursma, Gail 650-4868
    Gail.Duursma@ed.ac.uk

[†] Direct dial with prefix 0131.

Accredited by: Institute of Chemical Engineers

Degrees granted 2003-2004:
    B.S.: 11   M.S.: 5   Ph.D.: 2

Department reports to: College of Science and Engineering

Heriot-Watt University

Chemical Engineering
School of Engineering & Physical Sciences
Riccarton
Edinburgh EH14 4AS, Scotland

University phone ....(44) 131 449 5111
Department phone ....(44) 131 451 3131
Department FAX ....(44) 131 451 3129

Professors

Docherty, R. (Bob) (Visiting)
Geldart, Derek (Visiting)
Ni, Xiong-Wei
Ocone, Raffaella (Head)
Waldie, Brian (Emeritus)

Associate Professors

Murray, Keith R.
Wilkinson, Derek

Assistant Professors

Bell, George
Boron, Stefan
Bustard, Mark T
Goodwin, Julian A.S.
Pekdemir, Turgay
Thomson, Gillian B
Westacott, Robin E
White, Graeme

Accredited by: Institution of Chemical Engineers

Degrees granted 2003-2004:
    B.S.: 21   M.S.: 7   Ph.D.: 7

Department reports to: J S Archer, Principal

Placement service: R E Westacott

University of Leeds

School of Process, Environmental and Materials Engineering
Leeds LS2 9JT

University phone 44 (0) 113 243 1751
Department phone 44 (0) 113 343 2404
Department FAX 44 (0) 113 343 2405

Professors

Biggs, Simon R .... 343 2790 †
    s.r.biggs@leeds.ac.uk
Ghadiri, Mojtaba 343 2406
    m.ghadiri@leeds.ac.uk
Roberts, Kevin J. 343 2400
    k.j.roberts@leeds.ac.uk
Williams, Richard A 343 2801
    r.a.williams@leeds.ac.uk

† Direct dial with prefix 0113.

Accredited by: IChemE

Graduate advisor: S. Kalliadasis / T. Mahmud

Undergraduate advisor: M.Fairweather

Student organization: Chemical Engineering Society
Advisor: Chair, Chemical Engineering Society

Department reports to: Prof. P.A. Dowd, Head School Process, Environmental and Materials Engineering

Placement service: M.J. Jones

University of London (Imperial College)

Department of Chemical Engineering & Chemical Technology
Imperial College London
South Kensington Campus
London, SW7 2AZ, UK

University phone 44 (0) 20 7589 5111
Department phone 44 (0) 20 7594 5605

Professors

Blackmond, Donna 44 (0) 20 7594 1193
    d.blackmond@imperial.ac.uk
    RE
Briscoe, Brian J. 44 (0) 20 7594 5559
    b.briscoe@imperial.ac.uk
    NT
Chadwick, David 44 (0) 20 7594 5579
    d.chadwick@imperial.ac.uk
    EN,RE
Dugwell, Denis R. 44 (0) 20 7594 5568
    d.dugwell@imperial.ac.uk
    EN,EV
Hewitt, Geoffrey F. (Emeritus) 44 (0) 20 7594 5562
    g.hewitt@imperial.ac.uk
    HT,TP
Higgins, Julia S. 44 (0) 20 7594 5565
    j.higgins@imperial.ac.uk
    PO
Jackson, George 44 (0) 20 7594 5640
    g.jackson@imperial.ac.uk
    EN,MO,NT,PO
Kandiyoti, Rafael 44 (0) 20 7594 5581
    r.kandiyoti@imperial.ac.uk
    EN,EV,RE
Kazarian, Sergei G. 44 (0) 20 7594 5574
    s.kazarian@imperial.ac.uk
    AM,BM,EN,EV,PO,SE
Kelsall, Geoffrey H. 44 (0) 20 7594 5633
    g.kelsall@imperial.ac.uk
    EN,EV,RE
Kershenbaum, Lester S. (Emeritus) ... 44 (0) 20 7594 5566
    l.kershenbaum@imperial.ac.uk
    PR,RE
Lawrence, Christopher J. 44 (0) 20 7594 5622
    c.lawrence@imperial.ac.uk
    EN,MO
Livingston, Andrew 44 (0) 20 7594 5582
    a.livingston@imperial.ac.uk
    BT,EV,SE
Luckham, Paul F. 44 (0) 20 7594 5583
    p.luckham01@imperial.ac.uk
    AM,NT,PO
Macchietto, Sandro 44 (0) 20 7594 6608
    s.macchietto@imperial.ac.uk
    BM,EN,PR
Maitland, Geoffrey C. 44 (0) 20 7594 1830
    g.maitland@imperial.ac.uk
    AM,EN,EV,MO,PO

Pantelides, Costas C. 44 (0) 20 7594 6622
c.pantelides@imperial.ac.uk
PR
Pistikopoulos, Efstratios N. 44 (0) 20 7594 6620
e.pistikopoulos@imperial.ac.uk
PR
Richardson, Stephen M. (Head) 44 (0) 20 7594 5626
s.m.richardson@imperial.ac.uk
EN,MO
Sargent, Roger.W.H. (Emeritus) 44 (0) 20 7594 6613
r.sargent@imperial.ac.uk
MO,PR
Shah, Nilay 44 (0) 20 7594 6621
n.shah@imperial.ac.uk
EN,EV,MO,PR
Stuckey, David C. 44 (0) 20 7594 5591
d.stuckey@imperial.ac.uk
BT,EN,EV
Trusler, J. P. M. 44 (0) 20 7594 5592
m.trusler@imperial.ac.uk
EN,SE
Weinberg, Felix J. (Emeritus) 44 (0) 20 7594 5580
f.weinberg@imperial.ac.uk
EN,EV,RE

Associate Professors

Adjiman, Claire S. 44 (0) 20 7594 6638
c.adjiman@imperial.ac.uk
PR
Bismarck, Alexander 44 (0) 20 7594 5578
a.bismarck@imperial.ac.uk
AM,NT,PO
Galindo, Amparo 44 (0) 20 7594 5606
a.galindo@imperial.ac.uk
MO
Hellgardt, Klaus 44 (0) 20 7594 5577
k.hellgardt@imperial.ac.uk
AM,EN,EV,RE
Kalliadasis, Serafim 44 (0) 20 7594 1373
s.kalliadasis@imperial.ac.uk
MO
Kogelbauer, Andreas 44 (0) 20 7594 5572
a.kogelbauer@imperial.ac.uk
EV,NT,RE
Li, Kang 44 (0) 20 7594 5676
kang.li@imperial.ac.uk
RE,SE
Matar, Omar K. 44 (0) 20 7594 5635
o.matar@imperial.ac.uk
EN,MO,NT
Mueller, Erich A. 44 (0) 20 7494 1569
e.muller@imperial.ac.uk
EN,EV,MO,NT,SE
Ortiz, E. S. (Emeritus) 44 (0) 20 7594 5587
s.ortiz@imperial.ac.uk
SE
Saville, Graham (Emeritus) 44 (0) 20 7594 5578
g.saville@imperial.ac.uk
EN,MO
Stepanek, Frantisek 44 (0) 20 7594 5608
f.stepanek@imperial.ac.uk
MO,RE
Williams, Daryl R. 44 (0) 20 7594 1168
d.r.williams@imperial.ac.uk
BT,NT,PO
Xu, X.Yun 44 (0) 20 7594 5588
yun.xu@imperial.ac.uk
BM,BT,MO

Assistant Professors

Cabral, Joao P.B.T. 44 (0) 20 7594 1662
j.cabral@imperial.ac.uk
AM,BM,NT,PO,RE
Immanuel, Charles D. 44 (0) 20 7594 5594
c.immanuel@imperial.ac.uk
BT,MO,PR
Krishnan, J 44 (0) 20 7594 6633
j.krishnan@imperial.ac.uk
BM,BT,MO,PR
Mantalaris, Athanassios 44 (0) 20 7594 5601
a.mantalaris@imperial.ac.uk
BM
Millan-Agorio, Marcos G. 44 (0) 20 7594 1633
marcos.millan@imperial.ac.uk
EN,RE
Spelt, Peter D.M. 44 (0) 20 7594 1601
p.spelt@imperial.ac.uk
EN,MO

Accredited by: IChemE (UK)

Degrees granted 2005-2006:
    B.S.: 102   M.S.: 26   Ph.D.: 26

Graduate advisor: Professor George Jackson

Undergraduate advisor: Professor Paul F. Luckham

Student organization: Imperial College Chemical Engineering Society
Advisor: Professor Rafael Kandiyoti

Department reports to: The Rector

Placement service: Dr Klaus Hellgardt

University of London (University College London)

Department of Chemical Engineering
Torrington Place
London, WC1E 7JE

University phone 44 020 7679-2000
Department phone 44 020 7679-3825
Department FAX 44 020 7383-2348

Professors

Bogle, I. D. L. 3803
d.bogle@ucl.ac.uk
Brandani, S 2315
s.brandani@ucl.ac.uk
Cornish, A. R. H. (Emeritus) 3824
Gibilaro, L. G. (Emeritus) 3838
Gugan, K. (Visiting) 3824
Jones, A. G. (Head) 3828
a.jones@ucl.ac.uk
Mahgerefteh, H. 3835
h.mahgerefteh@ucl.ac.uk
Mullin, J. W. (Emeritus) 3824
Rowe, P. N. (Emeritus) 3824
Simons, S. J. R. 3805
stefan.simons@ucl.ac.uk
Yates, J. G. 3837
john.yates@ucl.ac.uk

Associate Professors

Elson, T. P. 3821
t.elson@ucl.ac.uk

Fraga, E.S. 3817
  e.fraga@ucl.ac.uk
Gavriilidis, A. 3811
  a.gavriilidis@ucl.ac.uk
Sorensen, E. 3802
  e.sorensen@ucl.ac.uk

Assistant Professors
Angeli, P. 3832
  p.angeli@ucl.ac.uk
Cains, P. 3806
  p.cains@ucl.ac.uk
Lettieri, P. 7867
  p.lettieri@ucl.ac.uk
Manos, G 3810
  g.manos@ucl.ac.uk
Papageorgiou, L 2563
  l.papageorgiou@ucl.ac.uk

Accredited by: IChemE

Graduate advisor: Dr S Brandani

Student organization: Ramsay Society
Advisor: Anna Harrington

Department reports to: Sir Derek Roberts, Provost and President

Placement service: Dr. G. Manos

Loughborough University

Department of Chemical Engineering
Ashby Road
Loughborough, Leicestershire
LE11 3TU, England

University phone 01509-263171
Department phone 01509-222532
Department FAX 01509-223923

Professors
Brooks, Brian W. (Emeritus) 01509-222510
  B.W.Brooks@lboro.ac.uk
Buffham, Bryan A. (Emeritus) 01509-222503
  B.A.Buffham@lboro.ac.uk
Hall, Alan (Visiting)
Hankinson, Geoff 01509-222540
  G.Hankinson@lboro.ac.uk
Kletz, Trevor A. (Visiting)
  T.A.Kletz@lboro.ac.uk
Lydon, Richard (Visiting)
Mason, Geoff 01509-222509
  G.Mason@lboro.ac.uk
Nassehi, Vahid 01509-222522
  V.Nassehi@lboro.ac.uk.
Neville, Brewis (Visiting)
Preston, Malcolm L (Visiting) 01509-222515
  M.L.Preston@lboro.ac.uk
Rielly, Chris D 01509-222504
  C.D.Rielly@lboro.ac.uk
Starov, Victor M 01509-222508
  V.M.Starov@lboro.ac.uk
Streat, Michael (Emeritus) 01509-222506
  M.Streat@lboro.ac.uk
Theliander, Hans (Visiting)
Wakeman, Richard J. (Head) 01509-222500
  R.J.Wakeman@lboro.ac.uk

Associate Professors
Cumming, Iain W. 01509-222523
  I.W.Cumming@lboro.ac.uk
Edwards, David W. (Visiting) 01509-222515
  D.W.Edwards@lboro.ac.uk
Holdich, Richard G. 01509-222519
  R.G.Holdich@lboro.ac.uk
Klaus, Hellgardt 01509-222518
Pinder, Colin C. (Visiting)
Saha, Basu 01509-222505
  B.Saha@lboro.ac.uk
Shama, Gilbert 01509-222514
  G.Shama@lboro.ac.uk
Tarleton, E. S. 01509-222535
  E.S.Tarleton@lboro.ac.uk

Assistant Professors
Drott, David W. 01509-222516
  D.W.Drott@lboro.ac.uk
Malik, Danish J. 01509-222507
  D.J.Malik@lboro.ac.uk
Nagy, Zoltan 01509-222516
Stapley, Andy G.F. 01509-222525
  A.G.F.Stapley@lboro.ac.uk
Wilcockson, Robin W. 01509-222521
  R.B.Wilcockson@lboro.ac.uk

Accredited by: IChemE

Degrees granted 2004-2005:
  B.S.: 49   M.S.: 15   Ph.D.: 5

Undergraduate advisor: ES Tarleton

Student organization: IChemE

Department reports to: Vice Chancellor

Placement service: Wendy Llewellyn, Appts. Officer

University of Manchester (formerly UMIST)

School of Chemical Engineering and Analytical Science,
P.O. Box 88
Sackville Street
Manchester, M60 1QD, England

University phone 0161 306 6000
Department phone 0161 306 9320
Department FAX 0161 306 9321

Professors
Alder, John F .... 44 (0)161 306 4885
  fred.alder@manchester.ac.uk
  SE
Azpagic, Adisa .... 44 (0)161 306 4340
  adisa.azpagic@manchester.ac.uk
Baker, John G (Visiting)
  j.baker@manchester.ac.uk
Davey, Roger J .... 44 (0)161 306 4409
  roger.davey@manchester.ac.uk
  AM,SU,TH
Dewhurst, Richard J .... 44 (0)161 306 4886
  richard.dewhurst@manchester.ac.uk
  BM
Fielden, Peter R .... 44 (0)161 306 4889
  peter.fielden@manchester.ac.uk

242 United Kingdom

Garrett, Peter (Visiting) ....44 (0)161 306 4558
p.garrett@manchester.ac.uk
SU
Goddard, Nicholas ....44 (0)161 306 4895
nick.goddard@manchester.ac.uk
Griffiths, Richard F (Emeritus) ....44 (0)161 306 9320
enquiries-ceas@manchester.ac.uk
Heggs, Peter J ....44 (0)161 306 4370
peter.heggs@manchester.ac.uk
EN,EV,HT,MO,PR,RE,SE,SU,TH,TP
Hulse, Joseph (Visiting)
Mann, Reg (Emeritus) ....44 (0)161 306 4378
reginald.mann@manchester.ac.uk
BT,PR,RE
Mavituna, Ferda ....44 (0)161 306 4372
ferda.mavituna@manchester.ac.uk
BM,BT,EV,PR,RE
McCarthy, John ....44 (0)161 306 8916
john.mccarthy@manchester.ac.uk
Munro, Andrew W ....44 (0)161 306 5151
andrew.munro@manchester.ac.uk
Persaud, Krishna C ....44 (0)161 306 4892
krishna.persaud@manchester.ac.uk
Rankin, Hamish (Visiting) ....44 (0)161 306 3982
hamish.rankin@manchester.ac.uk
AM,EN,EV,SU,TP
Sharratt, Paul N ....44 (0)161 306 4367
paul.sharratt@manchester.ac.uk
EV,MO,PR,RE,TP
Smith, Robin ....44 (0)161 306 4382
robin.smith@manchester.ac.uk
EN,EV,PR,RE,SE
Snoep, Jacob .... 44(0)161 306 4340
Jacob.Snoep@manchester.ac.uk
Snook, Richard D ....44 (0)161 306 4893
richard.snook@manchester.ac.uk
Sutcliffe, Michael ....44 (01)161 306 5153
Michael.Sutcliffe@manchester.ac.uk
BT,MO
Tiddy, Gordon J ....44 (0)161 306 8865
gordon.tiddy@manchester.ac.uk
Vickerman, John ....44 (0)161 306 4544
john.vickerman@manchester.ac.uk
Webb, Colin (Head) ....44 (0)161 306 4379
colin.webb@manchester.ac.uk
BT,EV
Westerhoff, Hans ....44 (0)161 306 4407
hans.westerhoff@manchester.ac.uk
BT,EN,EV,MO,PR,TH
Woodcock, Les .... 44(0)161 306 4481
les.woodcock@manchester.ac.uk
EN,MO,TH,TP

Associate Professors

Campbell, Grant M ....44 (0)161 306 4472
grant.campbell@manchester.ac.uk
BT,EN,PR
Gardner, Peter ....44 (0)161 306 4483
peter.gardner@manchester.ac.uk
BM,BT,RE,SU
Garforth, Arthur ....44 (0)161 306 8850
arthur.garforth@manchester.ac.uk
AM,EN,EV,NT,RE
Grassia, Paul ....44 (0)161 306 8851
paul.s.grassia@manchester.ac.uk
MO,SE,SU,TP
Holmes, Stuart ....44 (0)161 306 4376
stuart.holmes@manchester.ac.uk
AM,EN,EV,NT,RE,SE,SU

Jobson, Megan ....44 (0)161 306 4381
megan.jobson@manchester.ac.uk
PR,SE
Lue, Leo ....44 (0)161 306 4867
leo.lue@manchester.ac.uk
AM,MO,NT,PO,TH,TP
Markx, Gerard H ....44 (0)161 306 4394
gerard.markx@manchester.ac.uk
AM,BM,BT,EV,ME,MO,NT,RE,SE,SU,TP
Martin, Philip ....44 (0)161 306 5779
Philip.Martin@manchester.ac.uk
AM,EV,ME,RE
Masters, Andrew J ....44 (0)161 275 4679
andrew.masters@manchester.ac.uk
MO,PO,TH,TP
Narayanaswamy, Ramaier ....44 (0)161 306 4891
ramaier.narayanaswamy@manchester.ac.uk
AM,EV,ME,NT,PO,PR
Pandiella, Severino S ....44 (0)161 306 4429
severino.s.pandiella@manchester.ac.uk
BM,BT,MO,RE,TP
Reader, Andrew J ....44 (0)161 306 4900
andrew.j.reader@manchester.ac.uk
MO
Roberts, Edward P L ....44 (0)161 306 8849
edward.roberts@manchester.ac.uk
EN,EV,MO,NT,RE,TP
Schroeder, Sven ....44 (0)161 306 5780
S.Schroeder@manchester.ac.uk
AM,EV,NT,RE,SU
Scully, Patricia J ....44 (0)161 306 8923
patricia.scully@manchester.ac.uk
BM,BT,EV,NT,PO,PR
Senior, Peter R ....44 (0)161 306 4403
peter.senior@manchester.ac.uk
HT,MO,PR
Stephens, Gill M ....44 (0)161 306 4377
gill.stephens@manchester.ac.uk
Theodoropoulos, Constantinos ....44 (0)161 306 4386
K.Theodoropoulos@manchester.ac.uk
ME,MO,NT,PR,RE,TP
Thomas, Paul ....44 (0)161 306 4910
paul.thomas@manchester.ac.uk
EV,PR,SE
Thorniley, Maureen S ....44 (0)161 306 8932
maureen.s.thorniley@manchester.ac.uk
Yuan, Xue-feng ....44 (0)161 306 4887
Xue-feng.Yuan@manchester.ac.uk
AM,BM,MO,NT,PO,TP
Zhang, Nan ....44 (0)161 306 4384
nan.zhang@manchester.ac.uk
Zweit, Jamal ....44 (0)161 446 3152
jamal.zweit@manchester.ac.uk

Assistant Professors

Bennett, Michael ....44 (0)161 306 4388
mike.bennett@manchester.ac.uk
EV
Bruggeman, Frans .... 44(0)161 306 5178
frans.bruggeman@manchester.ac.uk
BT,MO,PR,RE
Curtis, Robin ....44 (0)161 306 4401
R.Curtis@manchester.ac.uk
de Visser, Samuel ....44 (0)161 306 4882
sam.devisser@manchester.ac.uk
MO,RE,TH
Horsch, Mark .... 44(0)161 306 4340
mark.horsch@manchester.ac.uk

James, Alec ....44 (0)161 306 4368
    alec.e.james@manchester.ac.uk
Kim, Jin-Kuk ....44 (0)161 306 8755
    J.Kim-2@manchester.ac.uk
    EN,EV,MO,PR
Lockyer, Nick ....44 (0)161 306 4479
    nick.lockyer@manchester.ac.uk
Martin, Alastair ....44 (0)161 306 4395
    alastair.martin@manchester.ac.uk
    EN,EV,MO,SE,TH
Miller, Aline ....44 (0)161 306 5781
    Aline.Miller@manchester.ac.uk
    AM,BM,BT,PO,SU
Sadhukhan, Jhuma ....44 (0)161 306 4396
    jhuma.sadhukhan@manchester.ac.uk
Saiani, Alberto ....44 (0)161 306 8861
    A.Saiani@manchester.ac.uk
    AM,PO
Siperstein, Flor .... 44(0)161 306 4340
    flor.siperstein@manchester.ac.uk
Ventura-Medina, Esther .... 44(0)161 306 4346
    e.ventura-medina@manchester.ac.uk

Accredited by: IChemE

Degrees granted 2005-2006:
    B.S.: 60   M.S.: 56   Ph.D.: 34

Graduate advisor: Professor F Mavituna

Undergraduate advisor: Dr A Garforth

Student organization: Ms A Marron

Department reports to: Prof J Perkins (Dean of Faculty, Engineering and Physical Sciences)

Placement service: Careers Services, 0161 306 4330

University of Newcastle Upon Tyne

Chemical Engineering and Advanced Materials
University of Newcastle Upon Tyne
Merz Court
Newcastle Upon Tyne
NE1 7RU, England

University phone..........................0191-222-6000
Department phone 0191-222-7266
Department FAX 0191-222-5292

Professors

Akay, Galip 0191-222-7269
    galip.akay@ncl.ac.uk
Bull, Steve 0191 2227913
    s.j.bull@ncl.ac.uk
Martin, Elaine B 0191 2226231
    e.b.martin@ncl.ac.uk
Montague, Gary 0191-222-7265
    gary.montague@ncl.ac.uk
Morris, Julian (Head) 0191-222-7342
    julian.morris@ncl.ac.uk
Page, Trevor 0191 2227201
    t.f.page@ncl.ac.uk
Ramshaw, Colin 0191-222-7270
    colin.ramshaw@ncl.ac.uk
Scott, Keith 0191-222-8771
    k.scott@ncl.ac.uk
Snowden, Ken
    k.j.snowden
Thompson, Derek 0191 2227202
    d.p.thompson@ncl.ac.uk
White, Jim 0191 2227906
    jim.white@ncl.ac.uk

Associate Professors

Egerton, Terry 0191 2225618
    t.a.egerton@ncl.ac.uk
Norman, Peter W. 0191-222-7278
    peter.norman@ncl.ac.uk
Tham, Ming T. 0191-222-7285
    ming.tham@ncl.ac.uk
Willis, Mark J. 0191-222-7242
    mark.willis@ncl.ac.uk

Assistant Professors

Boodhoo, Kamelia 0191 222 7243
    k.v.k.boodhoo@ncl.ac.uk
Charles, Alastair 0191 2227900
    e.a.charles@ncl.ac.uk
Glassey, Jarka 0191-222-7275
    jarka.glassey@ncl.ac.uk
Haile, Sue 0191-222-7279
    s.m.haile@ncl.ac.uk
Jachuck, Roshan 0191-222-5202
    r.j.j.jachuck@ncl.ac.uk
Jaros, Milan 0191 2227379
    milan.jaros@ncl.ac.uk
Lee, Jon 0191-222-7241
    j.g.m.lee@ncl.ac.uk
Roy, Sudipta 0191-222-7274
    s.roy@ncl.ac.uk
Siller, Lidija 0191 2227323
    lidija.siller@ncl.ac.uk
Zhang, Jie 0191-222-7240
    jie.zhang@ncl.ac.uk

Accredited by: IChemE (UK), Institute of Energy (London)

Department reports to: Prof Peter Hills, Dean

Placement service: Cathryn Harvey, Chief Careers Advisor

University of Nottingham

School of Chemical, Environmental and Mining Engineering (SChEME)
University of Nottingham
University Park
Nottingham, NG7 2RD, England

University phone....................... 44 115 951 5151
Department phone 44 115 951 4163
Department FAX 44 115 951 4115

Professors

Azzopardi, B. J. 44 115 951 4160
    barry.azzopardi@nottingham.ac.uk
    EN,SE,TP
Miles, N. J. (Head) 44 115 951 4085
    nick.miles@nottingham.ac.uk
    EN,EV,SE
Patrick, J. W. 44 115 951 4175
    john.patrick@nottingham.ac.uk
Snape, C. E. 44 115 951 4166
    colin.snape@nottingham.ac.uk

Associate Professors

Chen, George Z 44 115 951 4171
    george.chen@nottingham.ac.uk
Hilal, N 44 115 951 4168
    nidal.hilal@nottingham.ac.uk
    AM,NT,PR,SE
Kingman, S. W. 44 115 951 4165
    sam.kingman@nottingham.ac.uk
    EN,EV,MO,PR,RE,SE
Lester, E 44 115 951 4974
    edward.lester@nottingham.ac.uk
    AM,BM,EN,NT,RE
Lowndes, I. S 44 115 951 4086
    ian.lowndes@nottingham.ac.uk
Maroto-Valer, M. 44 115 846 6893
    mercedes.maroto-valer@nottingham.ac.uk
Wilson, J. A. 44 115 951 4179
    j.a.wilson@nottingham.ac.uk

Assistant Professors

Andresen, J 44 115 951 4640
    john.andresen@nottingham.ac.uk
    AM,BT,EN,EV,MO,NT,RE,SE
Drage, T (Research) 44 115 951 4099
    trevor.drage@nottingham.ac.uk
    EN,PO,SE,SU
Haw, M. 44 115 951 4178
    mark.haw@nottingham.ac.uk
    AM,NT,PO,SU
Irvine, D.
    PO
Jones, W. E. 44 115 951 4172
    warren.jones@nottingham.ac.uk
Langston, P. A. 44 115 951 4177
    p.langston@nottingham.ac.uk
    MO,PR
Large, D. J. 44 115 951 4114
    david.large@nottingham.ac.uk
Li Puma, G 44 115 951 4170
    gianluca.li.puma@nottingham.ac.uk
    RE
Licence, P. 44 115 846 6176
    peter.licence@nottingham.ac.uk
Robinson, J.P. 44 115 951 4092
    john.p.robinson@nottingham.ac.uk
    AM,EN,EV,NT,PO,PR,SE,TP
Steel, K 44 115 951 4078
    karen.steel@nottingham.ac.uk
    AM,EN,EV,PO,SE,TP
Thielemens, W.
    NT
Waller, M. D. 44 115 951 4096
    martin.waller@nottingham.ac.uk

Accredited by: IChemE (UK)

Degrees granted 2005-2006:
    B.S.: 63   M.S.: 54

Graduate advisor: Professor B J Azzopardi

Undergraduate advisor: Dr W E Jones

Department reports to: University Senate

Placement service: Careers Advisory Service

Queen's University of Belfast

School of Chemistry & Chemical Engineering
David Keir Building
Stranmillis Road
Belfast BT9 5AG, Northern Ireland

University phone 028 90245133
University FAX 02890975137
Department phone 028 90974253
Department FAX 028 90381753

Professors

Allen, Stephen J. 028 90974295
    s.allen@qub.ac.uk
    EV
Magee, T. R. A. 028 90274255
    r.magee@qub.ac.uk
    HT
Richardson, Philip (Visiting)

Associate Professors

Ahmad, Mohammad N. 028 90974389
    mnm.ahmad@qub.ac.uk
    RE,TH
Gan, Quan 028 90274463
    q.gan@qub.ac.uk
    BT,RE
Holland, Clive 028 90974256
    c.r.holland@qub.ac.uk
McNally, Gerry M. 028 90974705
    g.mcnally@qub.ac.uk
    AM,PO
Rooney, David 028 90974050
    d.rooney.@qub.ac.uk
    RE
Walker, Gavin 028 90974172
    g.walker@qub.ac.uk
    BM,PR

Assistant Professors

Aiouache, Farid 02890974253
    f.aiouahe@qub.ac.uk
Sen Gupta, Bhaskar 028 90974554
    b.sengupta@qub.ac.uk
    BT

Accredited by: IChemE (UK)

Degrees granted 2005-2006:
    M.S.: 12   Ph.D.: 6

Graduate advisor: Q. Gan

Undergraduate advisor: C.R. Holland

Department reports to: The Vice-Chancellor

Placement service: Appointments and Careers, 14 Malone Road, Belfast

University of Sheffield

Department of Chemical and Process Engineering
Mappin Street
Sheffield, S1 3JD, England

University phone 44 114 222 2000
Department phone 44 114 222 7500
Department FAX 44 114 222 7501

Professors

Allen, R.W.K. ....44-(0)114-222 7601
r.w.k.allen@shef.ac.uk
Hounslow, M.J. (Head) ....44-(0)114-222 7565
m.j.hounslow@shef.ac.uk
James, D. ....44-(0)114-222 7505
BM,BT
Sharifi, V.N. ....44-(0)114-222 7518
v.n.sharifi@shef.ac.uk
EN,EV
Swithenbank, J. (Emeritus) ....44-(0)114-222 7502
j.swithenbank@shef.ac.uk
EN,EV
Wright, P.C. ....44-(0)114-222 7577
p.c.wright@shef.ac.uk
BM,BT
Zimmerman, W.B.J. ....44-(0)114-222 7517
w.zimmerman@shef.ac.uk

Associate Professors

Edyvean, R.G.J. (Associate Head) ......44-(0)114-222 7506
r.edyvean@shef.ac.uk
MacInnes, J.M. ....44-(0)114-222 7511
j.m.macinnes@shef.ac.uk
Pitt, M.J. ....44-(0)114-222 7513
m.j.pitt@shef.ac.uk
Priestman, G.H. ....44-(0)114-222 7512
g.priestman@shef.ac.uk
Ristic, R.I. ....44-(0)114-222 7516
r.i.ristic@shef.ac.uk
Salman, A.D. ....44-(0)114-222 7560
a.d.salman@shef.ac.uk
Styring, P. ....44-(0)114-222 7571
p.styring@shef.ac.uk

Assistant Professors

Biggs, C.A. ....44-(0)114-222 7510
c.biggs@shef.ac.uk
BM,BT
Dickman, M.J. ....44-(0)114-222 7541
m.dickman@shef.ac.uk
BM,BT
Ewan, B.C.R. ....44-(0)114-222 7504
b.c.ewan@shef.ac.uk
Russell, N.V. ....44-(0)114-222 7521
n.russell@shef.ac.uk
EN
Wu, Y. ....44-(0)114-222 7514
y.wu@shef.ac.uk

Accredited by: IChemE, Energy Institute

Degrees granted 2005-2006:
  B.S.: 25   M.S.: 48   Ph.D.: 14

Department reports to: The Vice-Chancellor of the University

University of Strathclyde

Department of Chemical and Process Engineering
James Weir Building
Montrose St.
Glasgow, G1 1XJ, Scotland

University phone........................ 141 552 4400
Department phone 141 548 2361
Department FAX 141 552 2302

Professors

Grant, Colin (Dean)
Hall, Peter
Magee, Ronnie (Visiting) .... 44 (0)141 548 2361

Associate Professors

Cousins, Rod
Crawley, Frank
Larsen, Vidar
Muir, David
Postlethwaite, Bruce
Schaschke, Carl (Head)
Sefcik, Jan
Shilton, Simon

Assistant Professors

Bull, Laura-Anne
Campbell, Linda
Dickson, Brian
Heslop, Mark
Morrison, Karen
Sweatman, Martin

Accredited by: Institution of Chemical Engineers

Degrees granted 2005-2006:
  B.S.: 53   M.S.: 7   Ph.D.: 4

Graduate advisor: Dr Jan Sefcik

Undergraduate advisor: Dr Laura-Anne Bull (Yr1), Dr Simon Shilton(Yrs 2&3), Mr Rod Cousins (Yr4), Dr Carl Schaschke(Yr5,MEng)

Student organization: Chemical Engineering Society

Department reports to: Dean of the Faculty of Engineering

Placement service: University Careers Service

University of Surrey

Department of Chemical and Process Engineering,
School of Engineering, Guildford, Surrey GU2 7XH, England

University phone........................ 44 1483 300800
University FAX 44 1483 300803
Department phone 44 1483 686580
Department FAX 44 1483 686581

Professors

Koenders, M.A. (Visiting)
koenders@kingston.ac.uk
Kokossis, A.C. .... 44(0)1483 686573
A.Kokossis@surrey.ac.uk
Lawson, G. (Visiting)
Smith, J. M. (Emeritus) .... 44(0)1483 689276
J.Smith@surrey.ac.uk
Taylor, K. (Visiting)
K.Taylor@surrey.ac.uk
Thorpe, R.B. .... 44(0)1483 689270
Rex.Thorpe@surrey.ac.uk
Tuzun, U. (Head) .... 44(0)1483 686587
U.Tuzun@surrey.ac.uk

Associate Professors

Kirkby, N. F. .... 44(0)1483 686577
N.Kirkby@surrey.ac.uk
Millington, C. A. .... 44(0)1483 686582
A.Millington@surrey.ac.uk

Assistant Professors

Cleaver, J. A. S. .... 44(0)1483 686598
    J.Cleaver@surrey.ac.uk
Linke, P .... 44(0)1483 689116
    P.Linke@surrey.ac.uk
Sharif, A.O. .... 44(0)1483 686584
    A.Sharif@surrey.ac.uk

Accredited by: IChemE (UK)

Degrees granted 2004-2005:
    B.S.: 15   M.S.: 36   Ph.D.: 9

Graduate advisor: Prof U Tuzun

Undergraduate advisor: Dr N Kirkby

Student organization: Chemical Engineering Society
Advisor: Dr JAS Cleaver

Department reports to: Prof Paul Smith, Head of School of Engineering

Placement service: Dr JAS Cleaver

University of Teesside

School of Science and Technology
University of Teesside,
Middlesbrough, TS1 3BA, England

University phone 44 (0) 1642 218121
Department phone 44 (0) 1642-342499
Department FAX 44 (0) 1642-342401

Professors

Matchett, A. J. .... 44 (0) 1642-342438 [†]
    a.j.matchett@tees.ac.uk
    MO

Associate Professors

Gerrard, A. M. 44 (0) 1642-342444
    a.m.gerrard@tees.ac.uk
    MO,PR

Assistant Professors

Martin, G. F. 44 (0) 1642-342447
    g.f.martin@tees.ac.uk
    EV
Nithiamanthan, V. 44 (0) 1642-342445
    v.nthiananthan@tees.ac.uk
    EN
O'Reilly, A. J. 44 (0) 1642-342447
    a.j.oreilly@tees.ac.uk
    MO,PR
Peel, C 44 (0) 1642 342436
    c.peel@tees.ac.uk
    MO,PR
Russell, Paul A. 44 (0) 1642 342488
    p.russell@tees.ac.uk
    NT,RE,SE

[†] Phone numbers are preceded by country code 44.

Accredited by: IChemE (UK)

Degrees granted 2005-2006:
    B.S.: 12   M.S.: 15   Ph.D.: 1

Graduate advisor: A J Matchett

Undergraduate advisor: C Peel

Department reports to: Dean of School of Science and Technology

Placement service: Careers Office

Venezuela

Universidad Central de Venezuela

Escuela de Ing. Quimica
Apartado 50656-50361
Sabana Grande
Caracas 1040-A

University phone 58-212-6054050
University FAX 58-212-6053178
Department phone 58-212-6053293
Department FAX 58-212-6053178

Professors

Albano, Carmen 58-212-6051698
    albanoc@ucv.ve
    PO
Martin, Rafael
Marzuka, Samir 58-212-6051717
    marzukas@ucv.ve
    EN
Papa, Jose 58-212-6051693
    papaj@ucv.ve
    PO,RE

Associate Professors

Alonso, Mary L. 58-212-6051713
    mlalonso@cantv.net
    SE
Garcia, Luis 58-212-6051692
    lberfon@cantv.net
    RE
Guzmán, Nólides (Head) 58-212-6051706
    guzmann@ucv.ve
    EN,SE
Hernandez, Jose D. 58-212-6051696
    jodhersu@cantv.net
    RE
Sorrentino, José A. (Chair) 58-212-6053292
    sorrentj@ucv.ve
    SE

Assistant Professors

Blanco, Berenice 58-212-6051704
    bereniceblanco@cantv.net
    PR
Fernández, José F. 58-212-6053081
    fernandj@ucv.ve
    AM,SE
Kum, Humberto 58-212-6051701
    hkum@cantv.net
    RE
Morales, Carlos 58-212-6051721
    cj_morales@hotmail.com
    SE,SU
Perez, Anubis 58-212-6051703
    pereza@ucv.ve
    SE

Vizcaya, Armando 58-212-6051699
vizcayaa@ucv.ve
EN

Yánez, Francisco 58-212-6051697
fyanez@hotmail.com
RE

Accredited by:

Degrees granted 2005-2006:
    B.S.: 59   M.S.: 3

Student organization: Centro de Estudiantes de Ingeniería Química

Omega Chi Epsilon

Omega Chi Epsilon promotes high scholarship, encourages original investigation in all areas of chemical engineering, and recognizes the valuable traits of character, integrity and leadership. The society renders service to chemical engineering department and its students and fosters meaningful faculty-student dialogue within forty-nine chapters.

ALPHA	University of Illinois, 1931 (inactive)
BETA	Iowa State University, 1932, reactivated 1966
GAMMA	University of Minnesota, 1934 (inactive)
DELTA	Clarkson College of Technology, 1941
EPSILON	The University of Texas, 1941
ZETA	Purdue University, 1943
ETA	New Jersey Institute of Technology, 1957
THETA	West Virginia University, 1958
IOTA	University of Pittsburgh, 1959
KAPPA	Polytechnic Institute of New York, 1964
LAMBDA	City University of New York, 1964
MU	Oklahoma State University, 1964
NU	University of Detroit, 1965
XI	Northeastern University, 1965
OMICRON	Lamar University, 1967
PI	Rose-Hulman Institute of Technology, 1969
RHO	Texas A&M University, 1970
SIGMA	University of Arkansas, 1970
TAU	University of Alabama, 1971
UPSILON	University of Kentucky, 1971
PHI	Louisiana Tech University, 1971
CHI	University of Maryland, 1973
PSI	University of Southern California, 1974
OMEGA	University of Missouri-Rolls, 1974
ALPHA ALPHA	Auburn University, 1975
ALPHA BETA	Northwestern University, 1975
ALPHA GAMMA	University of Lowell, 1975
ALPHA DELTA	University of Houston, 1975
ALPHA EPSILON	Kansas State University, 1976
ALPHA ZETA	Michigan State University, 1976
ALPHA ETA	Mississippi State University, 1978
ALPHA THETA	Youngstown State University, 1978
ALPHA IOTA	Tulane University, 1978
ALPHA KAPPA	University of South Carolina, 1979
ALPHA LAMBDA	University of Southwestern Louisiana, 1979
ALPHA MU	New Mexico State University, 1979
ALPHA NU	Tri-State University, 1980
ALPHA XI	University of Colorado, 1980
ALPHA OMICRON	Texas Tech University, 1981
ALPHA PI	Howard University, 1982
ALPHA RHO	Georgia Institute of Technology, 1983
ALPHA SIGMA	University of Connecticut, 1983
ALPHA TAU	Colorado State University, 1984
ALPHA UPSILON	California State Polytechnic University, 1984
ALPHA PHI	University of Iowa, 1984
ALPHA CHI	Virginia Polytechnic Institute & State University, 1984
ALPHA PSI	Manhattan College, 1985
ALPHA OMEGA	Louisiana State University, 1988
BETA ALPHA	Pennsylvania State University, 1988
BETA BETA	University of Missouri-Columbia, 1991
BETA GAMMA	North Carolina A&T State University, 1992
BETA DELTA	Tuskegee University, 1992
BETA EPSILON	University of Cincinnati, 1995
BETA ZETA	University of Minnesota-Duluth, 1995
BETA ETA	Texas A&M University-Kingsville, 1995
BETA THETA	University of Michigan, 1995
BETA IOTA	Prairie View A&M University, 1996
BETA KAPPA	University of South Florida, 1996
BETA LAMBDA	Washington University, 1996
BETA MU	Widener University, 1996

Index of U.S. Faculty

A

Abbasian, Javad 33
Abbott, James 125
Abbott, Nicholas L. 136
Abbud-Madrid, Angel 19
Abegg, Carl F. 37
Abraham, Martin A. 95
Abrams, Cameron F. 103
Abu-Lail, Nehal 133
Abubakr, Said 60
Acevedo Nazario, Aldo 109
Achenie, Luke E. K. 21
Ackerman, John 137
Ackerson, Michael D. 7
Acrivos, Andreas 8, 78
Adams, Craig 66
Aderangi, Nader 33
Adewuyi, Yusuf G. 86
Adeyiga, Adeyinka A. 128
Adidharma, Hertanto 137
Adler, Stuart 131
Admassu, Wudneh 30
Adomaitis, Raymond A. 49
Advincula, Rigoberto C. 119
Agarwal, Sumit 19
Aghara, Sukesh 120
Agrawal, Pradeep K. 29
Agrawal, Rakesh 36
Aguayo, Guillermo 109
Ahleman, Larry 60
Ahmed, Zikri 79
Akers, William W. 121
Akins, Richard G. 41
Aksay, Ilhan A. 72
Aksoy, Burak 3
Akyurtlu, Ates 128
Akyurtlu, Jale F. 128
Al-Dahhan, Muthanna 67
Al-Hallaj, Said 33
Al-Saadoon, Faleh T. 123
Alamo, Rufina 27
Albright, Lyle F. 36
Alcantar, Norma 28
Aldrich, J. Winthrop 15
Alexander, Keith 8
Alexandridis, Paschalis 84
Alkire, Richard C. 32
Allan, G. Graham 131
Allbritton, Nancy L. 10
Allen, David T. 123
Allen, Emily 16
Allen, Jonathan 5
Allen, Mark G. 29
Altman, Eric 22
Altwicker, Elmar R. 82
Aluko, Mobolaji E. 24
Alvarado, Vladimir 137
Amaral, Luis A. N. 34
Amiridis, Michael D. 113
Amundson, Neal R. 119
Andersen, Paul 77
Anderson, Brian J. 134

Anderson, Erik 55
Anderson, Frank 63
Anderson, Harold M. 76
Anderson, John L. 90, 102
Anderson, Kimberly 42
Anderson, Paul 33
Anderson, Thomas F. 21
Anderson, Timothy J. 25
Andino, Jean M. 5
Andreadis, Stelios T. 84
Andres, Ronald P. 36
Androulakis, Ioannis (Yannis) P. ..73
Angenent, Lars 67
Angus, John C. 90
Anisimov, Mikhail A. 49
Anklam, Mark R. 37
Annapragada, Ananth 119
Anseth, Kristi S. 18
Anthamatten, Mitchell 83
Anthony, Jennifer 41
Anthony, Rayford G. 122
Anton, A. Brad 80
Antonio, Vincitore 81
Appleby, Anthony J. 122
April, Gary C. 1
Aranda-Espinoza, Helim 49
Aranovich, Gregory 48
Arastoopour, Hamid 33
Aravamuthan, Raja 60
Arce, Pedro E. 117
Archer, Lynden A. 80
Argyle, Morris 137
Armaou, Antonios 106
Armenante, Piero 71
Armeniades, Constantine 121
Armstrong, Robert C. 53
Arnold, David W. 1
Arnold, Frances H. 14
Arnold, Robert 6
Aronson, Mark 128
Artigue, Ronald S. 37
Artyushkova, Kateryna 76
Ashbaugh, Henry S. 46
Ashurst, William (Bob) 3
Ashworth, Sharon 47
Askew, William C. 4
Assaf-Anid, Nada 81
Asta, Mark 9
Asthagiri, Anand R. 14
Asthagiri, Aravind 25
Asthagiri, Dilip 48
Aston, Eric 30
Ataai, Mohammad M. 107
Atanassov, Plamen 76
Atwood, Glenn A. 89
Auerbach, Scott M. 51
Aurand, Gary 38
Aven, Russell E. 63
Avilés-Molina, Misael O. 109
Aydil, Eray 61
Azad, Abdul-Majeed 95
Azbel, David S. 66

B

Babb, Albert L. 131
Babcock, Robert E. 7
Babu, S. V. 78
Baertsch, Chelsey D. 36
Bagajewicz, Miguel J. 97
Baglin, Frank G. 69
Bailey, Travis 20
Bailie, Richard C. 134
Baird, Donald G. 130
Bajpai, Rakesh K. 65
Baker, Ian 70
Bakshi, Bhavik 94
Balakotaiah, Vemuri 119
Balazs, Anna C. 107
Balbuena, Perla 122
Balcarcel, R. Robert 118
Baldwin, John T. 122
Baldwin, Robert M. 19
Balk, John 42
Balsara, Nitash P. 8
Baltus, Ruth E. 78
Baltzis, Basil 71
Banaszak, Michael 137
Banerjee, Sanjoy 13
Banerjee, Sujit 29
Baneyx, François 131
Bang, Sookie 114
Banish, Michael 2
Bankoff, S. George 34
Banta, Scott 79
Bao, Zhenan 17
Barabino, Gilda A. 54
Barat, Robert 71
Baratuci, William B. 132
Barbari, Timothy A. 49
Barile, Ron 26
Barkat, Omar 45
Barkel, Barry 57
Barker, Dee H. 126
Barkey, Dale P. 70
Barna, Bruce A. 59
Barnett, Stanley M. 111
Barron, Annelise E. 34
Barron, Charles H. 112
Bart, Ernest 71
Barteau, Mark A. 23
Bartholomew, Calvin H. 126
Barton, Paul I. 53
Basaran, Osman 36
Baskaran, Harihara 90
Bates, Frank S. 61
Bauer, Larry G. 114
Baumann, Melissa 58
Baumgart, Tobias 105
Baxter, Gregory 80
Baxter, Larry L. 126
Baygents, James 6
Bayles, Taryn 50
Bayuzick, Robert J. 118
Beér, János M. 53
Beard, John N. 112

Index of U.S. Faculty

Beatty, Kenneth O. 87
Beaucage, Gregory 91
Beaudoin, Stephen 36
Beckman, Eric J. 107
Beckman, James R. 5
Beckwith, William F. 112
Beddow, John K. 38
Beers, Kenneth J. 53
Behbahani, Ahmed 45
Behrens, Sven 29
Beitle, Robert 7
Belfiore, Laurence A. 20
Belfort, Georges 82
Bell, Alexis T. 8
Bell, David A. 137
Bell, James P. 21
Bell, Kenneth J. 98
Bell, Richard L. 9
Belovich, Joanne M. 92
Ben-Jebria, Aziz 106
Benítez, Jaime 109
Bender, Jonathan W. 113
Bender, Timothy 58
Bennett, Carroll O. 21
Bennett, Gary F. 95
Benoit, Gaboury 22
Bent, Stacey 17
Bentley, William E. 49
Benziger, Jay B. 72
Bequette, B. Wayne 82
Berg, John C. 131
Berglund, Kris A. 58
Beris, Antony N. 23
Berkland, Cory J. 40
Berman, Neil S. 5
Bernstein, Barry 33
Beroes, Charles 107
Berson, R. Eric 43
Besser, Ronald 74
Betenbaugh, Michael J. 48
Bethea, Robert 125
Bevan, Michael 122
Beyenal, Haluk 133
Bhatia, Surita R. 51
Bhattacharyya, Dibakar 42
Bhethanabotla, Venkat R. 28
Bidani, Akhil 119
Bidstrup Allen, Sue Ann 29
Biegler, Lorenz T. 102
Bieler, Thomas 58
Bienkowski, Paul R. 116
Bier, Milan 6
Biernacki, Joseph J. 117
Bird, R. Byron 136
Bischoff, Kenneth B. 23
Bishop, Kenneth A. 40
Biswal, Sibani Lisa 121
Biswas, Pratim 67
Blanch, Harvey W. 8
Blankschtein, E. Daniel 53
Blau, Gary E. 36
Block, David E. 9
Block, Robert J. 97
Block, Seymour S. 25
Blowers, Paul 6
Blythe, Philip A. 104
Boder, Eric T. 105
Boehlert, Carl 58

Boerio, James F. 91
Bogere, Moses N. 109
Bohn, Paul W. 35
Bommarius, Andreas 29
Bonnecaze, Roger T. 123
Bonner, Francis J. 52
Book, Neil L. 66
Borhan, Ali 106
Borovetz, Harvey S. 107
Bose, Arijit 111
Bothwell, Michelle K. 100
Botsaris, Gregory B. 55
Botte, Gerardine G. 95
Boudart, Michel 17
Boulton, Roger 9
Bousfield, Douglas W. 47
Bowden, Warren W. 37
Bowen, J. Ray 131
Bowman, Christopher N. 18
Bowman, Frank 88
Braatz, Richard D. 32
Brace, John 95
Bradshaw, Jerry L. 122
Brady, John F. 14
Brand, Jennifer I. 68
Braun, Walter G. 106
Bray, Robert S. 15
Brazel, Chris 1
Brazinsky, Irv 79
Breedveld, L. Victor 30
Brennecke, Joan F. 35
Brenner, Howard 53
Brenner, James R. 26
Bretz, Robert E. 75
Breuer, Kenneth 110
Brevnov, Dimitri 76
Brewer, Lawrence 133
Briano, Julio G. 109
Bricka, R. Mark 63
Briedis, Daina M. 58
Briggs, Dale E. 57
Briggs, James M. 119
Brinker, C. Jeffrey 76
Broadbelt, Linda J. 34
Brock, James R. 123
Brodkey, Robert 94
Brown, George M. 34
Brown, Gilbert Jay 52
Brown, Richard 111
Brown, Robert C. 39
Browning, Charles E. 93
Browning, Nigel 9
Bruce, David A. 112
Bruns, Duane D. 116
Bryant, Stephanie 18
Bryers, James D. 131
Buchanan, Relva C. 91
Buettner, Helen M. 73
Bukur, Dragomir B. 122
Bullard, Lisa G. 87
Bullin, Jerry A. 122
Bungay, Henry R. III 82
Bunge, Annette L. 19
Buonopane, Ralph A. 54
Burghardt, Wesley R. 34
Burkey, Daniel 54
Burnet, George 39
Burns, Mark A. 57

Burris, Conrad T. 81
Burrows, Veronica 5
Busch, Robert D. 76
Busot, J. Carlos 28
Butera, Robert J. 23
Buthod, Paul 99
Butler, Audrey 38
Butler, Jason 25
Butt, John B. 34
Buttrey, Douglas J. 23
Byrne, Mark E. 3

C

Córdova-Figueroa, Ubaldo M. ... 109
Caenepeel, Christopher L. 15
Cagin, Tahir 122
Cahela, Don 3
Cairncross, Richard A. 103
Cairns, Elton J. 8
Cal, Mark 75
Calabrese Barton, Scott 58
Calabrese, Richard V. 49
Cale, Timothy S. 82
Calo, J.M. 110
Camarda, Kyle V. 40
Cameron, John 60
Cameron, Michael R. 95
Camesano, Terri A. 56
Campbell, Charles T. 131
Campbell, Gregory A. 78
Campbell, Kenneth 133
Campbell, Scott W. 28
Canavan, Heather 76
Caneba, Gerard T. 59
Cannon, Joseph N. 24
Caram, Hugo S. 104
Carbonell, Ruben G. 87
Cardona-Martínez, Nelson 109
Caretta, Raul 61
Carlson, Alfred 37
Carlson, Eric S. 1
Carlson, Ross P. 68
Carmichael, Gregory R. 38
Carnahan, Brice 57
Carney, Laurel 85
Carpen, Ileana 117
Carr, Robert W. Jr. 61
Carr, Russell T. 70
Carr, Stephen H. 34
Carrier, Rebecca 54
Carstensen, Hans-Heinrich 19
Carta, Giorgio 128
Carter, C. Barry 61
Caruthers, James M. 36
Case, Eldon 58
Caskey, Jerry A. 37
Castaldi, Marco 81
Castellanos, Mariajose 50
Castner, David G. 131
Castro, Sigifredo 41
Caswell, Bruce 110
Cavanagh, Daniel P. 101
Cecchi, Joseph L. 76
Center, Alfred M. 80
Cerro, Ramon L. 2
Cha, Chang Yul 137
Chaikof, Elliot L. 29
Chakraborty, Arup 53

Challa, Sivakumar 76
Chalmers, Jeffrey 94
Chambers, Robert P. 3
Chan, Christina 58
Chan, Paul C. H. 65
Chance, Ronald 30
Chang, Chih-hung (Alex) 100
Chang, Hsueh-Chia 35
Chang, Jane P. 11
Chang, W. Victor 16
Chao, K. C. 36
Chapman, Thomas W. 136
Chapman, Walter G. 121
Charles, Marvin 104
Chase, George G. 89
Chatzimavroudis, George P. 92
Chau, Pao C. 12
Chaudhury, Manoj K. 104
Chauhan, Anuj 25
Chawla, Ramesh C. 24
Chegini, Amir 128
Chelikowsky, James R. 123
Chella, Ravindran 27
Chellam, Shankar 119
Chen, Chien-Pin 2
Chen, Daniel H. 120
Chen, David H. T. 108
Chen, Jingguang 23
Chen, John C. 104
Chen, Kevin 27
Chen, Rachel 29
Chen, Shaw H. 83
Chen, Wei-Yin 63
Chen, Wen-Jia 95
Chen, Wilfred 12
Chen, Zilin 35
Cheng, Shang-I 79
Cheng, Zhengdong 122
Cheung, H. Michael 89
Chiang, Shiao-Hung 107
Chiew, Yee C. 73
Chimowitz, Eldred H. 83
Chin, Der-Tau 78
Chirdon, William 46
Chisholm, John 123
Chitra, Surya 108
Chittur, Krishnan K. 2
Chmelka, Bradley F. 13
Chmielewski, Donald 33
Cho, Eung H. 134
Choi, Kyu Yong 49
Christofides, Panagiotis D. 11
Chu, Jhih-Wei 8
Chuang, Steven 89
Churchill, Stuart W. 105
Cilento, Eugene V. 134
Cinar, Ali 33
Ciric, Amy 93
Cirino, Patrick C. 106
Clancey, M. Sean 59
Clancy, Paulette 80
Clapp, Aaron 39
Clark, Douglas S. 8
Clark, Peter E. 1
Clark, William M. 56
Clark, Yvette 117
Clarson, Stephen J. 91
Clausen, E. C. 7

Clement, Gilles 95
Clements, L. Davis 68
Clements, Wm. C. Jr. 1
Clough, David E. 18
Co, Albert 47
Co, Carlos 91
Co, Tomas B. 59
Cobb, James T. Jr. 107
Cochran, Eric W. 39
Cocke, David L. 120
Cocker, David R. 12
Cococcioni, Matteo 62
Cohen, Claude 80
Cohen, Robert E. 53
Cohen, William C. 34
Cohen, Yoram 11
Cokelet, Giles 68
Colón, Guillermo 109
Cole, Robert 78
Coleman, Maria R. 95
Collier, John 70
Collins, Dermot J. 43
Collins, William E. 24
Collura, Michael A. 22
Colton, Clark K. 53
Colucci-Rios, José A. 109
Colvin, Vicki 121
Composto, Russell J. 105
Conner, W. Curtis Jr. 51
Constantinides, Alkis 73
Converse, Alvin O. 70
Cooney, Charles L. 53
Cooper, Douglas J. 21
Cooper, Gary W. 76
Cooper, Stuart 94
Coppens, Marc-Olivier 82
Corn, John 94
Cornell, David 63
Coronell, Daniel G. 37
Coronella, Charles J. 69
Corripio, Armando B. 44
Corti, David S. 36
Cosgrove, Stanley 91
Couchman, Peter 73
Coughlin, Robert W. 21
Coulman, George A. 92
Counce, Robert M. 116
Couper, James R. 7
Couzis, Alexander 78
Cox, Chris D. 116
Cox, David F. 130
Cox, Kenneth 121
Cramer, Steven M. 82
Crimp, Martin 58
Crisalle, Oscar D. 25
Crisman, Everett 111
Crist, Buckley Jr. 34
Crist, Kevin C. 95
Crocker, John C. 105
Cross, Robert 7
Crosser, Orrin K. 66
Crowl, Daniel A. 59
Crum, Edward 135
Crunkleton, Daniel 99
Crynes, Billy L. 97
Csernica, Jeffrey 101
Cui, Shengting 116
Cullinan, Harry T. 3

Cummings, Peter T. 118
Cunningham, James 115
Curl, Rane 57
Curro, John 76
Curtis, Christine W. 3
Curtis, Jennifer 25
Curtis, Wayne R. 106
Cushman-Roisin, Benoit 70
Cussler, Edward L. 61
Cutlip, Michael 21
Cygan, Margaret 44
Czermak, -Ing. Peter 41

D

d'Itri, Julie L. 107
da Rocha, Sandro 60
Dadyburjor, Dady B. 134
Dahl, Kris Noel 102
Dahler, John S. 61
Dahm, Kevin D. 73
Dai, Lenore 125
Dai, Liming 93
Dale, Bruce E. 58
Dalzell, William H. 53
Dan, Nily R. 103
Dandy, David S. 20
Daneshy, Ali 119
Daniels, Raymond D. 97
Danner, Ronald P. 106
Daoutidis, Prodromos 61
Darby, Ronald 122
Darcy, Patricia 103
DaSilva, Nancy A. 10
Datta, Ravindra 56
Datye, Abhaya 76
Daubert, Thomas E. 106
Daugherty, Patrick 13
Dave, Rajesh 71
David, Morton M. 131
Davidson, Burton Z. 73
Davis, Denny C. 133
Davis, E. James 131
Davis, H. Ted 61
Davis, James 11
Davis, Jeffrey M. 51
Davis, Mark E. 14
Davis, Richard 61
Davis, Richey M. 130
Davis, Robert H. 18
Davis, Robert J. 128
Davis, Sam H. 121
Davis, Virginia A. 3
Davison, Brian 116
Davison, Richard R. 122
Daw, C. Stuart 116
De Kee, Daniel 46
de Pablo, Juan J. 136
Dean, Anthony M. 19
Deans, Harry A. 137
Debelak, Kenneth A. 118
Debenedetti, Pablo G. 72
Deen, William A. 53
deGrazia, Janet 18
Deibert, Max C. 68
Del Valle, Eddie 22
Del Valle, Francisco 77
DeLancey, George B. 74
Delcamp, Robert M. 91

Delgass, W. Nicholas 36
Delhommelle, Jerome P. 113
DeLisa, Matthew P. 80
Demirel, Yasar 68
Deng, Shuguang 77
Deng, Yulin 29
Denn, Morton M. 78
Deo, Milind D. 127
DePaoli, David 116
Derby, Jeffrey J. 61
Deshpande, Pradeep B. 43
Deshusses, Marc A. 12
DeSimone, Joseph M. 87
Desio, Peter J. 22
DeSisto, William 47
Detamore, Michael S. 40
Devine, Kevin 81
DeWitt, Kenneth J. 95
Dewitt, Matthew 93
Dhawan, Jagdish C. 4
Dhurjati, Prasad 23
Diamond, Scott L. 105
DiBenedetto, Anthony T. 21
DiBiasio, David 56
Dickinson, Richard 25
Diemer, Russell 23
DiMarzio, Edmund A. 49
Dimitrakopoulos, Panagiotis 49
Dirk, Elizabeth LeBleu 76
Discher, Dennis 105
Dismukes, John P. 95
Dittman, F. W. 73
Dixon, Anthony G. 56
Dixon, David J. 114
Dobbins, Richard A. 110
Doherty, Michael F. 13
Domach, Michael M. 102
Donahue, Darrell 47
Donahue, Francis 57
Donahue, Neil M. 102
Donatelli, Alfred Anthony 52
Doner, David 135
Dong, Winny 15
Donglu, Shi 91
Donnelly, Vincent M. 119
Donohue, Marc D. 48
Dooley, Kerry M. 44
Doraiswamy, Deepak 134
Doraiswamy, L. K. 39
Dordick, Jonathan S. 82
Dorfman, Kevin 62
Dorgan, John R. 19
Dorland, Dianne 73
Doskocil, Eric 65
Dougherty, Elmer L. 16
Douglas, James M. 51
Doyle, Frank 13
Doyle, Patrick S. 53
Dranoff, Joshua S. 34
Drazer, German M. 48
Drown, David C. 30
Drzal, Lawrence T. 58
Du, Henry 74
Duarte, Horacio 123
Duda, J. Larry 106
Dudukovic, Milorad 67
Duffy, James E. 68
Duke, Steve R. 3

Dumesic, James A. 136
Dunbar, Paul 42
Duncan, T. Michael 80
Dungan, Stephanie R. 9
Dunston, Doug 75
Durbin, Leonel D. 122
Durning, C. J. 79
Durrill, Preston L. 130
Dutta, Subhash 26
Dyson, Derek C. 121
Dziubla, Thomas 42

E
Earthman, James C. 10
Ebenhack, Ben 83
Eckert, Charles A. 29
Economides, Michael J. 119
Economou, Demetre J. 119
Eddings, Eric 127
Eden, Mario 3
Edgar, James H. 41
Edgar, Thomas F. 123
Edie, Danny D. 112
Edwards, Aurelie 55
Edwards, Brian J. 116
Edwards, Jeremy S. 76
Edwards, Louis L. Jr. 30
Edwards, Robert V. 90
Ehrman, Sheryl H. 49
Eisenberg, Richard F. 83
Eitel, Richard 42
Ekechukwu, Kenneth 24
Ekerdt, John G. 123
El-Aasser, Mohamed S. 104
El-Farra, Nael 9
El-Genk, Mohamed S. 76
El-Halwagi, Mahmoud M. 122
Ela, Wendell 6
Elabd, Yossef 103
Eldridge, R. Bruce 124
Eliassen, John D. 117
Elimelech, Menachem 22
Elliott, J. Richard 89
Elsass, Michael 93
Ely, James F. 19
Emery, Alden H. 36
Empie, Howard L. 29
Engbretson, Gustav A 85
Engel, Alfred J. 106
Engstrom, James R. 80
Enick, Robert M. 107
Eniola-Adefeso, Omolola 57
Epps, Thomas H. 23
Erickson, Larry E. 41
Erkey, Can 21
Erlebacher, Jonah 48
Ernst, William R. 29
Errington, Jeffrey R. 84
Ershaghi, Iraj 16
Escobar, Isabel C. 95
Escobedo, Fernando 80
Espino, Ramon 128
Estévez, L. Antonio 109
Etchells, Arthur W. III 23
Eubank, Philip T. 122
Evans, Edward A. 89
Evans, Robert 19
Evers, John F. 137

Eylon, Daniel 93

F
Fabiano, Leonard A. 105
Fair, James R. 123
Falconer, John L. 18
Faller, Roland 9
Famularo, Jack 81
Fan, L. S. 94
Fan, L. T. 41
Fan, Stephen 70
Farag, Ihab H. 70
Farmer, Richard 69
Farrell, James 6
Farrell, Stephanie 73
Farshad, Fred F. 46
Federspiel, William 107
Fedkiw, Peter S. 87
Feintuch, Howard 81
Feke, Donald L. 90
Felder, Richard M. 87
Fenn, John B. 22
Fernandez, Erik J. 128
Ferri, James 103
Fichthorn, Kristen A. 106
Field, James 6
Finlayson, Bruce A. 131
Finley, David 37
Finn, Robert K. 80
Finney, Wright C. 27
Firoozabadi, Abbas 22
Fisher, Edward R. 59
Fisher, John P. 49
Flach, Lawrance 93
Flagan, Richard C. 14
Flake, John C. 44
Fleischman, Marvin 43
Fleming, Dan 60
Fletcher, Thomas H. 126
Flood, H. William 52
Floudas, Christodoulos A. 72
Floyd, Tamara 4
Flumerfelt, Raymond W. 119
Flynn, Ann Marie 81
Flynn, George W. 79
Flytzani-Stephanopoulos, Maria .. 55
Fogler, H. Scott 57
Foley, Henry C. 106
Fong, Stephen 129
Fontijn, Arthur 82
Forbes, Neil S. 51
Forciniti, Daniel 66
Ford, David M. 51
Ford, Laura P. 99
Ford, Roseanne M. 128
Forney, Larry J. 29
Fort, Tomlinson 118
Foster, David 83
Fotouh, Kamel H. 120
Foutch, Gary L. 98
Fox, George E. 119
Fox, Rodney O. 39
Francis, Lorraine F. 61
Frank, Curtis W. 17
Franses, Elias I. 36
Frechet, Jean M. J. 8
Frechette, Joelle 48
Frederick, William 29

Index of U.S. Faculty

Fredrickson, Arnold G. 61
Fredrickson, Glenn H. 13
Freeman, Benny D. 123
French, W. Todd 63
Frey, Douglas D. 50
Fried, Joel R. 91
Friedlander, Sheldon K. 11
Frisbie, C. Daniel 61
Froment, Gilbert 122
Frost, Harold 70
Frymier, Paul 116
Fuchs, Alan 69
Fulghum, Julia E. 76
Fuller, Gerald G. 17
Fuller, Thomas 29
Funk, Edward 31
Furst, Eric M. 23

G

Gabitto, Jorge F. 120
Gadala-Maria, Francis A. 113
Gaddy, Glen 48
Gaden, Elmer L. Jr. 128
Gainer, John L. 128
Galkin, Oleg N. 119
Galli, Alfred F. 134
Gallivan, Martha 30
Gallois, Bernard 74
Ganesan, Venkat 124
Ganley, Jason 24
Gao, Di 107
Gappa-Fahlenkamp, Heather 98
Garber, James D. 46
Garde, Shekhar 82
Gardner, Tracy Q. 19
Garlid, Kermit L. 131
Garr, Jeanette 96
Gary, James H. 19
Gasem, Khaled A. M. 98
Gast, Alice P. 53
Gates, Bruce C. 9
Gatica, Jorge E. 92
Gatzke, Edward P. 113
Gavalas, George R. 14
Gecol, Hatice 69
Gehrke, Stevin H. 40
Gellman, Andrew J. 102
Genco, Joseph M. 47
Gentry, James W. 49
Genzer, Jan 87
Georgakis, Christos 55
George, Clifford E. 63
George, Steven 18
George, Steven C. 10
Georgiou, George 123
Gephardt, Zenaida Otero 73
Gerberich, William W. 61
Gerhard, Earl R. 43
Gerngross, Tillman 70
Ghorashi, Bahman 92
Giapis, Konstantinos P. 14
Gibbons, Joseph H. 113
Gibeling, Jeffery C. 9
Gibson, Urusla 70
Gidalevitz, David 33
Gidaspow, Dimitri 33
Gilbert, Jeremy 85
Gilbert, Richard 28

Gilbert, Richard E. 68
Gilchrist, James F. 104
Gilchrist, M. Lane 78
Gilcrease, Patrick 114
Gill, Ryan 18
Gill, William N. 82
Gillham, John K. 72
Gillis, Peter P. 42
Gin, Douglas 18
Glandt, Eduardo D. 105
Glasgow, Larry A. 41
Glasser, Benjamin J. 73
Glatz, Charles E. 39
Gleason, Karen K. 53
Gleaves, John T. 67
Glotzer, Sharon C. 57
Glover, Charles J. 122
Godbey, W T. 46
Godbold, Thomas M. 118
Godleski, Edward S. 92
Goetz, Douglas J. 95
Gogos, Costas 71
Gold, Scott 45
Golden, John O. 19
Goldstein, Aaron 130
Goldstick, Thomas K. 34
Gomezplata, Albert 49
Gonzalez, Ramon 121
Gonzalez, Richard 46
Goo, Edward 16
Good, Robert J. 84
Good, Theresa 50
Gooding, Charles H. 112
Goodwin, Bernard M. 54
Goodwin, James G. 112
Goren, Simon L. 8
Gorte, Raymond J. 105
Gossage, John L. 120
Goswami, Yogi 28
Gould, Ronald 73
Govind, Rakesh 91
Gow, Arthur S. III 22
Gracias, David 48
Grady, Brian P. 97
Grady, Michael 73
Graedel, Thomas E. 22
Graessley, William W. 34, 72
Graff, Robert A. 78
Graham, Michael D. 136
Grant, Christine S. 87
Grant, James J. III 23
Grant, Samuel 27
Grant, Stanley 10
Grasselli, Robert 23
Graves, David B. 8
Graves, David J. 105
Gray, Donald J. 111
Gray, Jeffrey 48
Green, David L. 128
Green, Don W. 40
Green, William H. 53
Greenberg, David B. 91
Greene, Howard 89
Greener, Jehuda 83
Greenfield, Michael L. 111
Greenkorn, Robert A. 36
Greenstein, Teddy 71
Greer, Sandra C. 49

Gregory, Otto J. 111
Grenville, Richard K. 23
Griffin, Gregory L. 44
Griffith, D. John 45
Grossberg, A. L. 8
Grossmann, Elihu D. 103
Grossmann, Ignacio E. 102
Groves, Frank R. 44
Groza, Joanna R. 9
Grulke, Eric A. 42
Grummon, David 58
Gryte, Carl C. 79
Grzybowski, Bartosz A. 34
Gu, Tingyue 95
Gubbins, Keith E. 80, 87
Guelcher, Scott A. 118
Guin, James A. 3
Guiseppi-Elie, Anthony 112
Gulari, Erdogan 57
Gulari, Esin 60
Guliants, Vadim 91
Gulino, Daniel A. 95
Gupta, Arunava 1
Gupta, Nivedita 70
Gupta, Rakesh K. 134
Gupta, Ram 3
Gustafson, Richard 131
Gutoff, Edgar B. 54
Guymon, C. Allan 38
Guzman, Javier 40
Guzman, Roberto 6
Gyamerah, Michael 120
Gyure, Dale C. 55

H

Ha, Su 133
Hacker, Barbara 15
Hackleman, David 100
Haddon, Robert C. 12
Haghgooie, Ramin 55
Hahm, Jong-in 106
Hahn, Juergen 122
Hahn, Mariah 122
Haile, Sossina 14
Hales, Hugh B. 126
Hall, Carol K. 87
Hall, Kenneth R. 122
Hall, William B. 63
Haller, Gary L. 22
Halligan, James E. 98
Hammack, William S. 32
Hammer, Daniel A. 105
Hammond, Paula T. 53
Hamrin, Charles E. Jr. 42
Han, Sang 76
Handegama, Naresh 116
Hanes, Justin 48
Hanesian, Deran 71
Hanley, Thomas 3
Hanneman, Larry F. 39
Hannemann, Robert 36
Hanratty, Thomas J. 32
Hanson, Donald N. 8
Hanson, Francis V. 127
Hanyak, Michael E. Jr. 101
Hara, Masanori 73
Harb, John N. 126
Harding, David R. 83

Harding, W. David 22
Hariri, M. Hossein 37
Harold, Michael P. 119
Harper, Dean O. 43
Harriott, Peter 80
Harris, H. Gordon 137
Harris, Michael T. 36
Harris, Robert E. 90
Harris, Sandra L. 78
Harris, William M. 15
Harrison, B. Keith 4
Harrison, Douglas P. 44
Harrison, Graham M. 112
Harrison, Roger G. 97
Hart, Elizabeth 93
Harwell, Jeffrey H. 97
Hasan, A. Rashid 61
Hasenwinkel, Julie 85
Hassler, John 130
Hatfield, G. Wesley 10
Hatton, T. Alan 53
Hatziavramidis, Dimitri 33
Hatzimanikatis, Vassily 34, 80
Hauan, Steinar 102
Haug, Warren 34
Haugh, Jason M. 87
Havens, Jerry A. 7
Hawley, Martin C. 58
Hayes, Douglas G. 116
Haynes, Henry W. 137
Hazelwood, Vikki 74
He, Qinghua 4
Hebert, Kurt R. 39
Hecker, William C. 126
Heenan, William A. 123
Heichelheim, H. R. 125
Heideger, William J. 131
Heist, Richard 81
Heller, Adam 123
Hellums, J. David 121
Hendela, Art 71
Henderson, Clifford 29
Henderson, Louis 108
Hendrix, James L. 68
Henley, Ernest J. 119
Henry, James 44
Henry, Jim 115
Henry, Joseph D. 134
Henson, Michael A. 51
Henthorn, David 66
Henthorn, Kimberly 66
Hernández-Maldonado, Arturo J. 109
Hernandez, Rafael 63
Herring, Andy 19
Hersel, Allen 37
Hershey, Daniel 91
Hershey, Harry C. 94
Herz, Richard K. 12
Hesketh, Robert P. 73
Hess, Dennis W. 29
Heydweiller, John C. 85
Heys, Jeffrey 5
Hicks, Robert F. 11
Higdon, Jonathan J. L. 32
Higgins, Brian G. 9
Higgins, Charles J. 52
High, Karen A. 98
High, Martin S. 98

Hightower, Joe W. 121
Hiiemae, Karen 85
Hilborn, David 83
Hile, Lloyd 15
Hill, Charles G. Jr. 136
Hill, Davide A. 35
Hill, Donald O. 63
Hill, James C. 39
Hill, Priscilla J. 63
Hillhouse, Hugh W. 36
Hillier, Andrew 39
Hilt, J. Zachary 42
Himmelblau, David M. 123
Hinds, Bruce 42
Hirasaki, George J. 121
Hirt, Douglas E. 112
Hirth, Leo J. 3
Hirtzel, Cynthia 96
Hjortso, Martin 44
Hladky, Harold 60
Hlavacek, Vladimir 84
Ho, Chia-Chi 91
Ho, Thomas C. 120
Ho, W.S. Winston 94
Hoagland, David A. 51
Hodges, Russel 3
Hodgson, Kevin 131
Hoffman, Allan S. 131
Hoffman, Robert 73
Hoflund, Gar B. 25
Hoh, Jan 48
Hohmann, Edward C. 15
Hohn, Keith L. 41
Holder, Gerald D. 107
Holland, Charles D. 122
Holland, Nolan B. 92
Holland, William D. 117
Hollein, Helen C. 81
Holles, Joseph H. 59
Holmes, Russell 62
Holste, James C. 122
Holt, Bradley R. 131
Holtzapple, Mark T. 122
Hong, Juan 10
Hoo, Karlene 125
Hoopes, John W. Jr. 108
Hopfenberg, Harold B. 87
Hopke, Philip K. 78
Hopper, Jack R. 120
Horabik, Carol 61
Horbett, Thomas A. 131
Houze, R. Neal 36
Howard, G. Michael 21
Howard, Jack B. 53
Howat, Colin S. 40
Howat, Julie 40
Howitt, David G. 9
Hrenya, Christine M. 18
Hsieh, Jeffery 29
Hsu, James T. 104
Hu, Wei-Shou 61
Huang, Chien-Yueh 71
Huang, Ching-Rong 71
Huang, Yinlun 60
Huber, George W. 51
Hudson, J. L. 128
Huff, Marylin C. 105
Huffman, Gerald 42

Huggins, Frank 42
Hurt, Robert H. 110
Husson, Scott M. 112
Huvard, Gary 129
Hwalek, John J. 47
Hwang, Gyeong S. 124
Hwang, Sun-Tak 91
Hyun, Kun Sup 71

I

Ierapetritou, Marianthi G. 73
Iglesia, Enrique 8
Ignarro, L. J. 11
Iisa, M. Kristina 30
Ilias, Shamsuddin 86
Inan, Mehmet 68
Iroh, Jude O. 91
Isaacs, Leslie L. 78
Islam, Mohammad 75, 102
Israelachvili, Jacob 13
Ivnitski, Dmitri 76
Ivory, Cornelius F. 133

J

Jabarin, Saleh A. 95
Jabbari, Esmaiel 113
Jablonski, Erin L. 101
Jachuck, Roshan J.J. 78
Jackman, Alan P. 9
Jackson, Roy 72
Jacobs, Stephen J. 83
Jacobson, Allan J. 119
Jacobson, Annette 102
Jacoby, William A. 65
Jaffe, Charles M. 134
Jagota, Anand 104
Jameson, Cynthia J. 31
Jamison, Russell 129
Jang, Larry 15
Jaroszeski, Mark 28
Jayaraman, Arul 122
Jayaraman, Krishnamurthy 58
Jechura, John 19
Jefcoat, Irvin 63
Jenekhe, Samson A. 131
Jennings, G. Kane 118
Jennings, Michael B. 16
Jennings, Paul A. 26
Jensen, Klavs F. 53
Jeon, Noo Li 10
Jessen, Kristian 16
Jessop, Julie 38
Jhon, Myung S. 102
Jiang, Peng 25
Jiang, Shaoyi 131
Johannes, Arland H. 98
Johanson, Lennart N. 131
John, Vijay T. 46
Johns, Lewis E. 25
Johnson, Charles L. 77
Johnson, Duane T. 1
Johnson, Ernest F. 72
Johnson, J. Karl 107
Johnson, Jay 93
Johnson, Patrick 137
Johnston, Barry S. 53
Johnston, Keith P. 123
Jolls, Kenneth R. 39

Jones, Christopher 30
Jones, Frank 115
Joo, Yong L. 80
Jorne', Jacob 83
Jose-Yacaman, Miguel 123
Joseph, Babu 28
Josephson, Bill 3
Josephson, William 4
Jovanovic, Goran 100
Joyce, Margaret 60
Joyce, Tom 60
Joye, Donald D. 108
Ju, Jingyue 79
Ju, Lu-Kwang 89
Juda, Walter 55

K

Kabadi, Vinayak N. 86
Kabel, Robert L. 106
Kadlec, Robert H. 57
Kaler, Eric W. 23
Kalia, Rajiv 16
Kalika, Douglass 42
Kalu, Eric 27
Kalyon, Dilhan 74
Kamat, Prashant 35
Kane, Ravi 82
Kang, Choon-Hyoung 3
Kang, Kyung A. 43
Kannan, Rangaramanujam 60
Kantor, Jeffrey C. 35
Kao, Camilla 17
Kao, Y. K. 91
Kaplan, David 55
Kardos, John L. 67
Karel, Marcus 53
Karim, M. Nazmul 125
Karlsson, Sture K. F. 110
Karweit, Michael 48
Katz, Alexander 8
Katz, Joseph L. 48
Kauffman, David 76
Kavianian, Hamid 15
Kawatra, S. Komar 59
Kazantzis, Nikolaos K. 56
Kaznessis, Yiannis 62
Keasling, Jay D. 8
Keffer, David 116
Keith, Jason M. 59
Keller, George 134
Keller, Kenneth H. 61
Kelley, Brian 55
Kelly, C. Michael 108
Kelly, Christine 100
Kelly, Robert M. 87
Kelly, William J. 108
Kenis, Paul J. A. 32
Kennedy, Francis 70
Kermode, Richard I. 42
Kerr, Clayton P. 117
Kevrekidis, Yannis G. 72
Kezirian, Michael 16
Khalil, Yehia F. 22
Khan, Saad A. 87
Khang, Soon-Jai 91
Khare, Rajesh 125
Khinast, Johannes G. 73
Khomami, Bamin 67, 116

Khosla, Chaitan S. 17
Khoury, Fouad M. 119
Kidnay, Arthur J. 19
Kiefer, John 31
Kilbey, S. Michael 112
Kilpatrick, Peter K. 87
Kim, Dong-Shik 95
Kim, Jin Ryoun 81
Kim, Jingsang 57
Kim, Nam K. 59
Kim, Sangtae 36
Kim, Seong Han 106
Kimmel, Howard 71
Kimura, Shoichi 100
King, C. Judson 8
King, Franklin G. 86
King, Jerry 7
King, Julia A. 59
King, Michael 83
King, Terry S. 41
King, William E. Jr. 101
Kipper, Matt J. 20
Kiran, Erdogan 130
Kirwan, Donald J. 128
Kiselev, Sergei B. 19
Kitchens, Christopher 112
Klapa, Maria I. 49
Klein, Andrew 104
Klein, Michael T. 73
Klein, Tonya M. 1
Kline, Andy 60
Klingenberg, Daniel J. 136
Klinke, David J. 134
Klinzing, George E. 107
Klosterman, Donald 93
Knecht, Robert D. 19
Knickle, Harold N. 111
Knopf, F. Carl 44
Knotts, Thomas A. 126
Knox, Dana 71
Knudsen, James G. 100
Knuth, Eldon L. 11
Knutson, Barbara L. 42
Kobayashi, Riki 121
Koberstein, J. 79
Koch, Donald L. 80
Koelling, Harold A. 63
Koelling, Kurt W. 94
Koffas, Mattheos 84
Kofinas, Peter 49
Kofke, David A. 84
Kofke, Tamara G. 84
Koh, Carolyn 19
Kohl, Paul 29
Kokkoli, Efrosini 62
Kolodka, Edward B. 88
Komives, Claire 16
Kompala, Dhinakar S. 18
Konkar, Atul 16
Kono, Hisashi O. 134
Konstantopoulos, Konstantinos ... 48
Kopatsis, Alexander 108
Kopelevich, Dmitry 25
Koretsky, Milo D. 100
Korgel, Brian A. 124
Kornfield, Julia A. 14
Koros, William J. 29
Korus, Roger A. 30

Kostin, Morton D. 72
Kothare, Mayuresh V. 104
Kotov, Nicholas 57
Kovalchuk, Vladimir 107
Kovenklioglu, Suphan 74
Krahenbuhl, Melinda 127
Krambeck, Fredrick 48
Kramer, Edward J. 13
Krantz, William 91
Kretzschmar, Ilona 78
Krieger-Brockett, Barbara 131
Krishnagopalan, Gopal A. 3
Krishnagopalan, Jaya 4
Krishnamoorti, Ramanan 119
Kroenke, William 76
Kuech, Thomas F. 136
Kugler, Edwin L. 134
Kuhl, Tonya L. 9
Kumar, Sanat 79, 82
Kumar, Satish 62
Kummler, Ralph H. 60
Kung, Harold H. 34
Kung, Mayfair 34
Kunz, H. Russell 21
Kuo, Chiang-Hai 63
Kuo, Yue 122
Kusaka, Isamu 94
Kushner, Mark 39
Kwon, K. C. 4
Kwon, Soonjo 27
Kyle, Benjamin G. 41

L

Ladd, Anthony J.C. 25
LaForce, Tara 137
Lahann, Joerg 57
Lai, Eva 48
Laibinis, Paul E. 118
Lam-Anderson, Marca 79
Lamb, H. Henry 87
Lamm, Monica 39
Landau, Uziel 90
Lane, Alan M. 1
Langer, Robert S. 53
Langer, Stanley H. 136
LaRoche, Richard K. 23
Larsen, Gustavo 68
Larsen, Ronald W. 68
Larson, Ronald G. 57
Lashover, Jacob 44
Lauderback, Lee L. 68
Lauffenburger, Douglas A. 53
Laukhuf, W. L. S. 43
Laurence, Jennifer S. 40
Laurence, Robert L. 51
Laurencin, Cato 128
Laurenzi, Ian J. 104
Lauterbach, Jochen A. 23
Lavernia, Enrique 9
Law, Victor J. 46
Lawal, Adeniyi 74
Lawton, Carl W. 52
Le, Lloyd 15
Leach, Jennie 50
Leal, L. Gary 13
LeBlanc, Steven E. 95
Leckband, Deborah E. 32
Lee, Andre 58

Lee, C. Ted 16
Lee, C. William 93
Lee, Carolyn W. T. 54
Lee, Gil U. 36
Lee, Ilsoon 58
Lee, Jae W. 78
Lee, James Ly 94
Lee, James M. 133
Lee, Jay 29
Lee, Kelvin H. 80
Lee, Kyongbum 55
Lee, Lloyd L. 97
Lee, Robert L. 75
Lee, Sang-Yong 123
Lee, Stephen 94
Lee, Sunggyu 66
Lee, William E. III 28
Lee, Woo Y. 74
Lee, Y. Y. 3
Lee-DeSautels, Rhonda 42
Lefebvre, Brian G. 73
Leggoe, Jeremy 125
Lei, Yu 21
Leighton, Christopher 62
Leighton, David T. Jr. 35
Leipziger, Stuart 37
Lele, Tanmay 25
Lemlich, Robert 91
Lemon, Lois 60
Lenhoff, Abraham M. 23
Leonard, Edward F. 79
Leong, Kam W. 48
LeVan, M. Douglas 118
Levenspiel, Octave 100
Levicky, Rastislav 81
Levien, Keith L. 100
Lewandowski, Gordon 71
Lewis, Randy S. 126
Li, Ku-Yen 120
Li, Kun 102
Li, Qiming 76
Liang, Jenn-Tai 40
Liang, Ruifeng 134
Liao, James C. 11
Liapis, Athanasios I. 66
Libera, Matthew R. 74
Liberatore, Matthew W. 19
Licht, William 91
Liddell, KNona C. 133
Lidstrom, Mary E. 131
Lightfoot, Edwin N. 136
Lighty, JoAnn S. 127
Lilleleht, Lembit 128
Lim, Henry C. 10
Lim, Phooi K. 87
Lim, Soon-Sik 96
Lin, David 133
Lin, Jerry Y. S. 5
Lin, Jerry Y.S. 91
Lin, Sidney 120
Lindahl, Harold 33
Linden, Henry 33
Linden, James C. 20
Linderman, Jennifer J. 57
Lindt, Jan T. 107
Linhardt, Robert 82
Linic, Suljo 57
Linninger, Andreas 31

Liotta, Charles L. 29
Lipscomb, G. Glenn 95
Lira, Carl T. 58
Little, Steven 107
Littman, Howard 82
Litvinov, Dmitri 119
Liu, Chung-Chiun 90
Liu, Joseph T. C. 110
Liu, Julie 36
Liu, Juncheng 3
Liu, Y. A. 130
Llewellyn, J. Anthony 28
Lloyd, Douglas R. 123
Lobban, Lance 97
Lobo, Raul F. 23
Locke, Bruce R. 27
Locke, Carl E. Jr. 40
Lodge, Keith B. 61
Lodge, Timothy P. 62
Loehman, Ronald E. 76
Loewenberg, Michael 22
Lombardo, Stephen J. 65
Loney, Norman 71
Long, Richard L. Jr. 77
Longo, Marjorie L. 9
Loo, Yueh Lin 124
Lopez, Gabriel P. 76
Lopina, Stephanie T. 89
Lotero, Edgar 112
Lou, Helen H. 120
Lou, Jianzhong 86
Louie, Beverly 18
Louvar, Joeseph 60
Loveland, Stephanie 39
Lowman, Anthony M. 103
Loyalka, Sudarshan 65
Lu, Christopher C. 93
Lu, Hang 30
Lu, Jia Grace 10
Lu, Yunfeng 46
Lucas, Glenn E. 13
Lucas, James 58
Lucas, Robert 81
Lucia, Angelo 111
Ludlow, Douglas K. 66
Ludovice, Peter 30
Luecke, Richard 65
Luk, Yan Yeung 85
Luks, Kraemer D. 99
Lund, Carl R. F. 84
Luss, Dan 119
Lustig, Rolf 92
Lutz, Robert J. 24
Luyben, William L. 104
Luzik, Eddie 22
Lynd, Lee R. 70
Lynn, David M. 136
Lynn, Scott 8
Lyon, David N. 8
Lyons, James E. 23

M

Ma, Teng 27
Ma, Yi Hua 56
Maboudian, Roya 8
Mackay, Michael 58
Macosko, Christopher W. 62
Maddox, Robert N. 98

Madhukar, Anupam 16
Madihally, Sundararajan V. 98
Madou, Marc 10
Maffia, Gennaro J. 108
Magda, Jules J. 127
Maginn, Edward J. 35
Mahoney, Melissa 18
Mainardi, Daniela 45
Maiorella, Brian 8
Maldarelli, Charles 78
Mallapragada, Surya K. 39
Mallinson, Richard G. 97
Malone, Michael F. 51
Maloney, James O. 40
Mandavilli, Satya N. 109
Mandell, John F. 68
Maneval, James E. 101
Manke, Charles 60
Manley, David B. 66
Mann, J. Adin Jr. 90
Mann, Michael D. 88
Mann, Uzi 125
Mannan, Sam 122
Manning, Francis S. 99
Manning, Michael 54
Manogue, William H. 23
Manousiouthakis, Vasilios 11
Mansfeld, Florian 16
Mansoori, G. Ali 31
Mantzaris, Nikolaos 121
Mao, Guangzhao 60
Maples, Glennon 3
Maranas, Costas 106
Maranas, Janna K. 106
Marand, Eva 130
Maravelias, Christos T. 136
Marnell, Paul 81
Maroudas, Dimitrios 51
Marple, Stanley Jr. 119
Marr, David W. M. 19
Marrero, Thomas R. 65
Marshall, John 75
Martínez-Iñesta, María M. 109
Marten, Mark 50
Martin, George C. 85
Martin, Heidi B. 90
Martin, J. Ronald 103
Martin, Lealon 82
Martin, Stephen 130
Martirosyan, Karen S. 119
Masel, Richard I. 32
Mason, Geoffrey 137
Mason, Michael 47
Mathur, Virendra K. 70
Matsoukas, Themistoklis 106
Matsumoto, Mark R. 12
Matthew, Howard 60
Matthews, John C. 41
Matthews, Michael A. 113
Mavrikakis, Manos 136
Mayer, Michael 57
Maynard, Jennifer 62
McAvoy, Thomas J. 49
McCabe, Clare 118
McCandless, Frank P. 68
McCarthy, Joseph 107
McClellan, Scott J. 37
McCluskey, Richard J. 78

Index of U.S. Faculty

McCormick, Alon V. 62
McCoy, Benjamin J. 9
McCoy, John 75
McCready, Mark J. 35
McDonald, Karen A. 9
McFarland, Eric 13
McFarlane, Joanna 116
McFetridge, Peter S. 97
McGee, Henry 129
McGee, John C. 117
McGinn, Paul J. 35
McGuire, Joseph 100
McHugh, Anthony J. 32, 104
McHugh, Mark 129
McIntire, Larry 29
McIntosh, Steven 128
McKean, Wm. T. 131
McKelvey, James M. 67
McKenna, Greg 125
McKetta, John J. 123
McKinley, Marvin D. 1
McKinnon, J. Thomas 19
McLaughlin, Edward 44
McLaughlin, John B. 78
McMicking, James H. 60
McNeil, Kenneth M. 108
McNeil, Melanie 16
McRae, Gregory J. 53
McWilliams, Thomas G. Jr. 108
Mead, Richard W. 76
Meagher, Michael M. 68
Mecartney, Martha 10
Medlin, J. William 18
Mehta, Narinder K. 109
Meldon, Jerry 55
Mellichamp, Duncan A. 13
Melsheimer, Stephen S. 112
Menkhaus, Todd 114
Meredith, J. Carson 30
Merrill, Edward W. 53
Mesler, Russell B. 40
Metters, Andrew 112
Meyer, Randall 31
Meyer, Stephen P. 103
Michelsen, Donald L. 130
Mijovic, Jovan 81
Mikos, Antonios G. 121
Miletic, Marina 32
Millard, Paul 47
Miller, Albert E. 35
Miller, Clarence A. 121
Miller, David C. 37
Miller, David R. 12
Miller, Dennis J. 58
Miller, James B 102
Miller, Kelly 19
Miller, Reid C. 133
Miller, Ronald L. 19
Miller, William M. 34
Mills, David R. 3
Mills, Patrick 123
Minerick, Adrienne R. 63
Minnick, Michael 135
Mischke, Roland A. 130
Misra, Devesh 46
Misra, Manish 4
Mitch, William 22
Mitchell, Brian S. 46

Mitchell, James 24
Mitchell, Martha C. 77
Mitragotri, Samir 13
Mockros, Lyle F. 34
Moghe, Prabhas V. 73
Mohalley-Snedeker, Jacqueline ...30
Mohamed, Farghalli 10
Mohanty, Kishore K. 119
Monbouquette, Harold G. 11
Monson, Peter A. 51
Montgomery, Susan M. 57
Moore, Noel E. 37
Moore, O. C. 3
Moreira, Antonio R. 50
Morel, Jim 76
Morgan, John A. 36
Morgan, Morris H. III 128
Morisani, Stephen J. 4
Morosoff, Nicholas 66
Morris, James G. 42
Morris, Jeffrey 78
Morrison, Faith A. 59
Morrow, Norman R. 137
Morse, David C. 62
Morsi, Badie I. 107
Morton, Samuel 103
Moschandreas, Demetrios 33
Moser, William 56
Moshfeghian, Mahmood 98
Moss, Melissa A. 113
Mountziaris, T. J. 84
Mountziaris, T.J. 51
Mowrey, Daniel B. 44
Muggli, Darrin S. 88
Mukasyan, Alexander 35
Mukherjee, Amiya K. 9
Mulchandani, Ashok 12
Mulholland, Kenneth L. 23
Muller, Susan J. 8
Mullins, C. Buddie 123
Mullins, Joseph C. 112
Mullins, Michael E. 59
Mumm, Daniel 10
Mumme, Kenneth 47
Munir, Zuhair A. 9
Munro, James M. 114
Munson-McGee, Stuart 77
Murad, Sohail 31
Murhammer, David W. 38
Murphy, David 123
Murphy, Regina M. 136
Murphy, Vincent G. 20
Murray, Glen 19
Murthy, Shashi 54
Muscat, Anthony 6
Musgrave, Charles 17
Muske, Kenneth R. 108
Mutharasan, Raj 103
Muzzio, Fernando J. 73
Muzzy, John D. 29
Myers, Alan L. 105
Myers, John A. 108
Myers, Kevin J. 93
Myers, Wm. A. 7
Myerson, Allan 33
Myung, Nosang V. 12

N
Nadarajah, Arunan 95

Nagarajan, R. 106
Naimpally, Ashok 15
Nair, Sankar 30
Narang, Atul 25
Narasimhan, Balaji 39
Narayan, Ramani 58
Narayanan, Ranga 25
Naser, Jamil 4
Natoli, John 73
Nauman, E. Bruce 82
Nave, Felecia M. 120
Navrotsky, Alexandra 9
Nealey, Paul F. 136
Nedwick, Robert 106
Neelamegham, Sriram 84
Neivandt, David 47
Nemeth, Laszlo 31
Nenes, Athanasios 30
Neogi, Partho 66
Nerem, Robert M. 29
Nesbitt, Carl 59
Nesic, Srdjan 95
Neuman, Ronald D. 3
Neurock, Matthew 128
Newby, Bi-min Zhang 89
Newell, James A. 73
Newman, John S. 8
Ng, K. Y. Simon 60
Nguyen, Thuan K. 15
Nguyen, Trung Van 40
Nicholson, Bruce J. 84
Nickelson, Robert L. 68
Nikolaou, Michael 119
Nikolov, Alex 33
Nippert, Charles R. 108
Nitsche, Johannes M. 84
Nitsche, Ludwig 31
Nobe, Ken 11
Noble, Richard D. 18
Nohe, Anja 47
Noll, Kenneth 33
Nollert, Matthias U. 97
Norbeck, Joseph M. 12
Nordheden, Karen J. 40
Norris, David 62
Nosa, Egiebor 4
Noureddini, Hossein 68
Nowick, Henry 56
Nutt, Steven 16
Nuttall, H. Eric 76
Nychka, John 42

O
O'Connell, John P. 128
O'Connor, Kim C. 46
O'Haver, John Howard 63
O'Rear, Edgar A. III 97
O'Shaughnessy, B. 79
Oakey, John 19
Oberhauser, James P. 128
Ofoli, Robert Y. 58
Ogale, Amod A. 112
Ogden, Kimberly 6
Ogunnaike, Babatunde A. 23
Okazaki, Kenji 42
Okorator, Charles 79
Okos, M. 36
Olbricht, William L. 80

Ollis, David F. 87
Olsen, Thor 83
Olson, David H. 105
Olson, Fred 134
Olson, Jon H. 23
Ontiveros, Cordelia 15
Orazem, Mark E. 25
Oriani, Richard A. 62
Orkoulas, Gerassimos 11
Oroskar, Anil 31
Osborne-Lee, Irvin W. 120
Oscarson, John L. 126
Ostermann, Russell D. 40
Ostermeier, Marc 48
Ottino, Julio M. 34
Overcash, Michael R. 87
Overney, Rene M. 131
Owens, Thomas C. 88
Oyama, S. Ted 130
Ozkan, Umit 94

P

Pack, Daniel W. 32
Padhye, Nisha 68
Pagilla, Krishna 33
Palanki, Srinivas 4, 27
Palazoglu, Ahmet N. 9
Palecek, Sean P. 136
Palmer, Andre F. 35
Palmer, James 45
Palmese, Giuseppe R. 103
Palmstrom, Christopher J. 62
Panagiotopoulos, Athanassios Z. ..72
Pandis, Spyros 102
Pang, K. Hing 15
Pant, Kamal 3
Papadopoulos, Kyriakos D. 46
Papathanasiou, Thanasis D. 113
Papavassiliou, Dimitrios 97
Papoutsakis, E. Terry 34
Park, Chang-Won 25
Park, Eugene 111
Park, Jin Y. 30
Park, Sheldon 84
Parker, Peter 60
Parker, Robert S. 107
Parnas, Richard B. 21
Parsons, Gregory N. 87
Parulekar, Satish J. 33
Pasquali, Matteo 121
Patel, Navin 55
Patterson, Gary K. 66
Patton, Christi L. 99
Patzer, John F. II 107
Paul, Donald R. 123
Paulaitis, Michael 48, 94
Payne, Gregory F. 50
Peccia, Jordan 22
Pedersen, Henrik 73
Peeples, Tonya 38
Pekarovicova, Alexandra 60
Pekny, Joseph F. 36
Pendse, Hemant 47
Penney, W. Roy 7
Peppas, Nicholas A. 124
Peretti, Steven W. 87
Perez, Joseph M. 106
Perez-Luna, Victor 33
Perkins, Tracy 20
Perlmutter, Daniel D. 105
Perna, Angelo 71
Perschetti, John 19
Pershing, David W. 127
Peters, Michael 129
Petersen, Eugene E. 8
Petersen, James N. 133
Peterson, david 60
Peterson, Thomas W. 6
Petrenko, Victor 70
Petrou, Athos 84
Petrovan, Simioan 116
Petsev, Dimiter 76
Petty, Charles A. 58
Peyton, Brent M. 68
Pfeffer, Robert 71
Pfefferle, Lisa D. 22
Pfeifer, Blaine 55
Pfromm, Peter 41
Pham, Hien 76
Phelps, James Parkhurst 52
Philipossian, Ara 6
Phillips, Paul 91
Phillips, Ronald J. 9
Piergiovanni, Polly 103
Pignatello, Joseph 22
Pike, Ralph W. 44
Pilat, Michael J. 131
Pilehvari, Ali 123
Pintar, Anton J. 59
Pintauro, Peter N. 90
Pinto, Jose M. 81
Pinto, Neville G. 91
Pipes, Byron 36
Pishko, Michael V. 106
Pitt, William G. 126
Placek, Timothy D. 3
Plank, Charles A. 43
Plawsky, Joel L. 82
Ploehn, Harry J. 113
Plouffe, Paul 8
Podlaha, Elizabeth J. 44
Poehl, Michael 124
Poling, Bruce E. 95
Pommersheim, James M. 101
Ponter, Anthony P. 46
Popov, Branko N. 113
Porter, Marc D. 39
Potoff, Jeffrey J. 60
Pourhashemi, Ali 115
Pourki, Forouza 108
Powell, Robert L. 9
Powers, Gary J. 102
Powers, John E. 57
Pozo de Fernandez, Maria E. 26
Pozrikidis, Constantine 12
Prados, John W. 116
Prakash, Jai 33
Prather, Kristala J. 53
Prausnitz, John M. 8
Prausnitz, Mark R. 29
Preston, Floyd W. 40
Price, Douglas 96
Price, Geoffrey L. 99
Price, Randel M. 115
Prieve, Dennis C. 102
Prince, Michael J. 101
Prinja, Anil K. 76
Proehl, Jeffrey A. 70
Prokop, Ales 118
Prud'homme, Robert K. 72
Prudich, Michael E. 95
Przybycien, Todd 102
Pugmire, Ronald J. 127
Pun, Suzie 132
Punzi, Vito L. 108
Puszynski, Jan A. 114
Putatunda, Susil K. 60
Putnam, Andrew 10
Putnam, David A. 80

Q

Qammar, Helen C. 89
Qi, Dewie 60
Qin, S. Joe 124
Queneau, Paul 70
Quinn, John A. 105
Qutubuddin, Syed 90

R

Radhakrishnan, Ravi 105
Radke, Clayton J. 8
Radosz, Maciej 137
Ragan, Regina 10
Raghavan, Srinivasa R. 49
Rajagopalan, Padma 104
Ralston, Patricia A. S. 43
Ramírez, Carlos A. 109
Ramachandran, P. A. 67
Ramakrishnan, Subramanian 27
Ramanarayanan, Trikur A. 105
Ramani, Vijay 33
Ramirez, W. Fred 18
Ramkrishna, D. 36
Ranade, Madhav B. 49
Randall, D'Arcy 124
Randolph, Theodore W. 18
Rangel, Roger H. 10
Rankin, Stephen 42
Rao, Balaji 87
Rao, Christopher V. 32
Rao, Govind 50
Rao, M. Gopala 24
Rao, Raj 129
Raper, Judy A. 66
Rase, Howard F. 124
Rasmussen, Don H. 78
Rathman, James F. 94
Ratner, Buddy D. 131
Raupp, Gregory B. 5
Ravi, Vilupanur 15
Rawlings, James B. 136
Ray, Asit K. 42, 115
Ray, W. Harmon 136
Raymond, Timothy M. 101
Realff, Matthew J. 30
Reardon, Kenneth F. 20
Reed, X B Jr. 66
Regalbuto, John 31
Regan, Thomas M. 49
Rege, Kaushal 5
Register, Richard A. 72
Reible, Danny 44
Reid, Robert C. 53
Reilly, Peter J. 39

Index of U.S. Faculty

Reimer, Jeffrey A. 8
Reinhardt, James R. 46
Reisfeld, Brad 20
Rekhson, Simon 92
Reklaitis, Gintaras V. 36
Ren, Dacheng 85
Ren, Fan 25
Rengaswamy, Raghunathan 78
Resasco, Daniel E. 97
Rethwisch, David G. 38
Retzloff, David G. 65
Reucroft, Philip 42
Reynolds, Joseph 81
Rezac, Mary 41
Rhinehart, R. Russell 98
Rials, Timothy G. 116
Ribeiro, Fabio 36
Rice, Philip A. 85
Rice, Richard W. 112
Rice, William J. 108
Richards, John 23
Richardson, James T. 119
Richardson, Peter D. 110
Richter, Horst 70
Ricker, N. Lawrence 131
Ridgway, Darin 95
Riedel, Edward 19
Riggs, James B. 125
Rinaldi-Ramos, Carlos Manuel ...109
Rinard, Irven H. 78
Ring, Terry A. 127
Rinker, Robert G. 13
Risbud, Subhash 9
Ritchie, Stephen M. 1
Ritter, Arthur 74
Ritter, Edward R. 108
Ritter, James A. 113
Rivera, Daniel 5
Rivero-Hudec, Mercedes 111
Robert, Marc J. 121
Roberts, Christopher 3
Roberts, Christopher J. 23
Roberts, George W. 87
Roberts, Kenneth L. 86
Roberts, Richard 16
Roberts, Robert W. 89
Roberts, Ronnie S. 30
Roberts, Susan C. 51
Robertson, Channing R. 17
Robinson, Anne Skaja 23
Robinson, Richard L. 45
Robinson, Robert L. Jr. 98
Rochefort, Willie E. 100
Rochelle, Gary T. 124
Rockstraw, David A. 77
Roderick, Norman F. 76
Rodríguez-Ramírez, Abraham ... 109
Rodriguez, Ferdinand 80
Rodriguez, Harold V. 4
Rogers, Bridget R. 118
Rogers, Rudy E. 63
Rogers, Tony N. 59
Rollins, Derrick K. 39
Romagnoli, Jose A. 44
Romonchuk, Wayne 19
Rony, Peter R. 130
Root, Thatcher W. 136
Roper, D. Keith 127

Rorrer, Gregory L. 100
Rose, Vincent C. 111
Roseman, Rodney D. 91
Rosen, Stephen L. 66
Rosner, Daniel E. 22
Ross, Julia M. 50
Rossky, Peter J. 124
Rosson, Harold F. 40
Roth, Charles M. 73
Roth, John A. 118
Rothberg, Gerald 74
Rothberg, Lewis 83
Rothe, Erhard W. 60
Rother, Michael 61
Rousseau, Ronald W. 29
Rowell, George 103, 108
Rowley, Richard L. 126
Ruckenstein, Eli 84
Rumschitzki, David 78
Russel, William B. 72
Russell, Alan J. 107
Russell, T. W. Fraser 23
Ruthven, Douglas M. 47
Rutkowski, Gregory 61
Rutledge, Gregory C. 53
Ryan, Michael E. 84
Ryder, Daniel 55
Ryu, Dewey 9

S

Sacco Jr., Al 54
Sachs, Frederick 84
Sadana, Ajit 63
Saddawi, Salma 35
Saez, Eduardo 6
Sahimi, Muhammad 16
Sahinidis, Nikolaos V. 32
Saiers, James E. 22
Saliba, Tony E. 93
Saliby, Michael J. 22
Saliceti-Piazza, Lorenzo 109
Salim, Majid 37
Salley, Steven O. 60
Salovey, Ronald 16
Saltsburg, Howard 55
Saltzman, Mark 22
Sama, Dominick Anthony 52
Sambanis, Athanassios 29
Sampson, Kendree J. 95
Samuels, Robert J. 29
San, Ka-Yiu 121
Sanchez, Isaac C. 124
Sandall, Orville C. 13
Sandell, John 59
Sandhu, Sarwan S. 93
Sandvig, Robert L. 114
Sangani, Ashok 85
Sani, Rajesh 114
Sani, Robert L. 18
Sankaran, R. Mohan 90
Santanam, Suresh 85
Santore, Maria M. 51
Saraf, Ravi 68
Sarikaya, Mehmet 131
Sarkar, Casim A. 105
Sarofim, Adel 127
Sater, Vernon E. 5

Sather, Glenn A. 136
Satterfield, Charles N. 53
Satvat, Behrooz 54
Sauer, Sharon G. 37
Savage, Nancy O. 22
Savage, Phillip E. 57
Savelski, Mariano J. 73
Saville, Dudley A. 72
Savinell, Robert F. 90
Sawin, Herbert H. 53
Saxena, Satish C. 31
Scamehorn, John F. 97
Scarrah, Warren P. 68
Schaefer, Dale W. 91
Schaffer, David 8
Schaffer, James P. 103
Schall, Constance A. 95
Schechter, Robert S. 124
Scheinbeim, Jerry I. 73
Scheldorf, Jay J. 30
Schieber, Jay D. 33
Schiesser, William 104
Schimmel, Keith A. 86
Schlup, John R. 41
Schmidt, Christine 124
Schmidt, Lanny D. 62
Schmidtke, David W. 97
Schmitz, Roger A. 35
Schneider, James 102
Schneider, William F. 35
Schnelle, Karl B. Jr. 118
Schoenung, Julie M. 9
Schork, F. Joseph 29
Schott, Jeffrey 62
Schowalter, William 72
Schrader, Glenn L. 6, 39
Schrieber, Loren 27
Schroder, Klaus 85
Schrodt, J. Thomas 42
Schulson, Erland 70
Schultz, Jerold M. 23
Schulz, Kirk H. 63
Schure, Mark R. 23
Schwaber, James S. 23
Schwank, Johannes W. 57
Schwartz, Daniel 18
Schwartz, Daniel T. 131
Schwartz, Pauline M. 22
Sciance, Carroll T. 124
Scott, Garland E. 15
Scott, Susannah L. 13
Scovazzo, Paul 63
Scranton, Alec 38
Scriven, L. E. 62
Scurto, Aaron M. 40
Seagrave, Richard C. 39
Seames, Wayne S. 88
Searson, Peter 48
Seborg, Dale E. 13
Seebauer, Edmund G. 32
Segalman, Rachel A. 8
Segatori, Laura 121
Segura, Tatiana 11
Seidel, Robert W. 62
Seider, Warren D. 105
Seinfeld, John H. 14
Sekhar, Jainagesh A. 91
Selman, J. Robert 33

Scminario, Jorge 122
Sengers, Jan V. 49
Sengupta, Arup K. 104
Senkan, Selim M. 11
Serbezov, Atanas 37
Serth, Robert W. 123
Seshadri, K. 12
Setzer, C. John 87
Seymour, Joseph 68
Shackelford, James F. 9
Shadman, Farhang 6
Shaeiwitz, Joseph A. 134
Shaffer, Daniel L. 68
Shah, Dhananjai B. 92
Shah, Dinesh O. 25
Shah, Naresh 42
Shambaugh, Robert L. 97
Shanks, Brent H. 39
Shanks, Jackie V. 39
Shantz, Daniel F. 122
Shapir, Yonthan 83
Shapley, Nina 79
Shaqfeh, Eric S. G. 17
Sharfstein, Susan 82
Shariat, Ahmad 98
Sharma, Mrityunjai P. 137
Sharma, Rajendra 75
Sharp, Julie E. 118
Shaw, Montgomery T. 21
Shea, Kenneth 10
Shea, Lonnie D. 34
Sheff, James Robert 52
Sheldon, Brian W. 110
Shell, Scott 13
Shen, Hong 132
Shen, Youqing 137
Sheng, Henry P. 15
Sheppard, Keith 74
Sheth, Atul 116
Shi, Frank 10
Shieh, Wen K. 105
Shin, Eun-Jae 19
Shin, Jong-shik 125
Shine, Annette D. 23
Shing, Katherine S. 16
Shinnar, Reuel 78
Sholl, David 102
Shonnard, David R. 59
Shores, David A. 62
Short, David 23
Shreve, Gina S. 60
Shuler, Michael L. 80
Shusta, Eric V. 136
Shvartsman, Stanislav Y. 72
Sibbett, Scott S. 76
Siberio-Pérez, Diana Y. 109
Sibulkin, Merwin 110
Sidebotham, George 79
Sides, Paul 102
Sierka, Raymond 6
Sierks, Michael 5
Sierra Alvarez, Maria Reyes 6
Siirola, Jeffrey 116
Sikavitsas, Vassilios I. 97
Silano, A. A. 7
Silas, James 122
Silcox, Geoffrey D. 127
Silebi, Cesar A. 104

Silverstein, David 42
Simon, Laurent 71
Simon, Sindee 125
Sinha, Jayanta 68
Sinno, Talid R. 105
Sircar, Shivaji 104
Sirkar, Kamalesh 71
Sitton, Oliver C. 66
Sivertsen, John M. 62
Siviniah, Easan 125
Skaf, Dorothy W. 108
Skelland, A. H. Peter 29
Skliar, Mikhail 127
Slater, C. Stewart 73
Sleicher, Charles A. 131
Slider, H. C. 94
Sliepcevich, C. M. 97
Sloan, E. Dendy Jr. 19
Slonaker, Robert E. Jr. 101
Smart, Jimmy 42
Smay, James E. 98
Smirniotis, Panagiotis 91
Smith, Carlos A. 28
Smith, Edwin E. 94
Smith, J.M. 9
Smith, James E. Jr. 2
Smith, Julian C. 80
Smith, Kenneth A. 53
Smith, Philip J. 127
Smith, Robert L. 85
Smith, T. Gordon 20
Smith, Theodore G. 49
Smolke, Christina D. 14
Smoot, L. Douglas 126
Smyrl, William H. 62
Snide, James A. 93
Snurr, Randall Q. 34
Snyder, William J. 101
Solen, Kenneth A. 126
Solomon, Michael J. 57
Sommerfeld, Jude T. 29
Sonnerup, Bengt 70
Soroush, Masoud 103
Southard, Marylee Z. 40
Spakowitz, Andrew 17
Sparrow, Charles A. 63
Spencer, Barry B. 116
Spencer, Hugh T. 43
Spencer, Jordan L. 79
Sperling, Leslie H. 104
Spicer, Thomas O. 7
Spivey, James J. 44
Spontak, Richard J. 87
Squires, Arthur M. 130
Squires, Robert G. 36
Squires, Todd 13
Sridhar, Lakshmi N. 109
Srienc, Friedrich 62
Srivastava, Ranjan 21
Stadtherr, Mark A. 35
Stancell, Arnold 29
Stanford, Thomas G. 113
Stanier, Charles 38
Stansbury, Jeff 18
Starling, Kenneth E. 97
Starr, Thomas 56
Starr, Thomas L. 43
Stasaik, Raymond 63

Stebe, Kathleen J. 48
Steele, William V. 116
Steen, Paul H. 80
Stein, Fred P. 104
Steiner, Carol 78
Steinfink, Hugo 124
Stempfer, Berthe 19
Stephanopoulos, George 53
Stephanopoulos, Gregory 53
Sterling, Arthur 44
Sternberg, Moshe 8
Sternberg, Steven P.K. 61
Stevens, William F. 34
Stewart, Phillip 68
Stewart, Richard R. 54
Stice, James E. 124
Stiehl, Cory 39
Stiel, Leonard I. 81
Stiller, Alfred H. 134
Stinespring, Charter D. 134
Stock, Richard 79
Stoll, A. George 15
Storvick, Truman 65
Strano, Michael S. 32
Strasser, Peter 119
Streett, William B. 80
Stretz, Holly A. 117
Strieder, William C. 35
Striolo, Alberto 97
Stroeve, Pieter 9
Stroock, Abraham D. 80
Strunk, Mailand R. 66
Stuve, Eric M. 131
Sublette, Kerry L. 99
Subramaniam, Bala 40
Subramanian, Anu 68
Subramanian, K.N. 58
Subramanian, R. Shankar 78
Subramanian, Ravi 69
Subramanian, Venkat R. 117
Subramanian, Venkiteswaran 38
Sukanek, Peter C. 63
Suleiman-Rosado, David 109
Sullivan, Jonathan 45
Sullivan, Joseph 130
Sullivan, Millicent M. O. 23
Sullivan, Samuel L. Jr. 46
Sum, Amadeu 130
Sun, Lianhong 51
Sun, Lizhi 10
Sundaresan, Sankaran 72
Sundstrom, Donald W. 21
Sung, Nakho 55
Suni, Ian I. 78
Sunkara, Mahendra K. 43
Sunol, Aydin K. 28
Suppes, Galen J. 65
Sureshkumar, R. 67
Sutherland, James C. 127
Suuberg, Eric M. 110
Svoronos, Spyros A. 25
Swaney, Ross E. 136
Swanson, Todd 69
Swartz, James 17
Sweeney, Thomas L. 94
Sweeny, Robert F. 108
Swift, George W. 40
Swihart, Mark T. 84

Swinnea, J. Steven 124
Sylvester, Nicholas 4
Szepe, Stephen 31

T

Taconi, Katherine 2
Tadmor, Rafael 120
Takamura, Koichi 137
Takoudis, Christos 31
Talbot, Jan B. 12
Talu, Orhan 92
Tan, Jinglu 65
Tande, Brian 88
Tang, Ching W. 83
Tang, Yi 11
Tanner, Robert D. 118
Tao, Luh C. 68
Tardos, Gabriel I. 78
Tarrer, A. Ray 3
Tassoney, Joseph P. 15
Tatarchuk, Bruce J. 3
Tatterson, Gary B. 86
Tavakoli, Javad 103
Tavlarides, Lawrence L. 85
Taylor, Ross 78
Tedder, Dan W. 30
Teja, Amyn 29
Telotte, John C. 27
Terry, Ronald E. 126
Tester, Jefferson W. 53
Tewari, Surendra N. 92
Teymour, Fouad A. 33
Tharakan, John P. 24
Thatcher, Charles M. 7
Theodore, Louis 81
Theofanous, Theofanis 13
Thibodeaux, Louis J. 44
Thies, Curt 67
Thies, Mark C. 112
Thoen, Paul 19
Thoma, Greg 7
Thomas, Aaron 30
Thomas, Garth 135
Thompson, Edward V. 47
Thompson, Karsten E. 44
Thompson, Levi T. 57
Thompson, Richard E. 99
Thompson, Robert W. 56
Thomson, Kendall T. 36
Thomson, William J. 133
Thorgerson, Eric 54
Tien, Chi 85
Tierney, John W. 107
Tilton, James N. 23
Tilton, Robert 102
Timm, Delmar C. 68
Tirrell, David A. 14
Tirrell, Matthew 13
Tock, R. W. 125
Toghiani, Hossein 63
Toghiani, Rebecca K. 63
Tomadakis, Manolis M. 26
Tomasko, David 94
Tomassone, M. Silvina 73
Tomkins, Reginald 71
Tomlinson, John L. 15
Toomey, Ryan 28
Toor, Herbert L. 102

Torres, Walter 112
Torres-Lugo, Madeline 109
Toups, Harry 44
Tournier, Jean-Michel 76
Towler, Brian F. 137
Tranquillo, Robert T. 62
Tree, D. Alan 98
Tripathi, Anubhav 110
Troian, Sandra M. 72
Trout, Bernhardt L. 53
Trujillo, Edward M. 127
Truskett, Thomas M. 124
Tsai, Shirley 15
Tsang, Tate T. H. 42
Tsao, George T. 36
Tsapatsis, Michael 62
Tschoegl, Nicholas W. 14
Tseng, Yiider 25
Tsotsis, Theodore T. 16
Tsoulfanidis, Nicholas 69
Tu, Raymond 78
Turian, Raffi 31
Turner, C. Heath 1
Turner, Jay R. 67
Turpin, Jim L. 7
Turro, Nicholas J. 79
Turton, Richard 134
Tuzla, Kemal 104
Tyler, Bonnie 127
Tzanakakis, Emmanouhl 84

U

Uebler, E. Alan 23
Ueki, Taro 76
Ugaz, Victor 122
Uitenham, Leonard 86
Ulrich, Gael D. 70
Ulrich, Richard K. 7
Ulrichson, Dean L. 39
Ultman, James S. 106
Ungarala, Sridhar 92
Utgikar, Vivek 30
Uz, Mehmet 103

V

Vaeth, Kathleen 80
Vahdat, Nader 4
Valsaraj, Kalliat T. 44
Van Brunt, Vincent 113
Van Cott, Kevin 68
van Heiningen, Adriaan 47
Van Ness, Hendrick C. 82
Van Ooij, Wim J. 91
van Oss, Carel J. 84
Van Tassel, Paul 22
Van Vorst, W. D. 11
Van Wie, Bernard J. 133
Van Zee, John W. 113
VanAuker, Michael 28
Vanderlick, T. Kyle 72
VanKirk, Jesse 73
Vannice, M. Albert 106
VanWormer, Kenneth A. 55
Varanasi, Sasidhar 95
Varma, Arvind 36
Varner, Jeffrey D. 80
Vasavada, Anita 133
Vasenkov, Sergey 25

Vashishta, Priya 16
Vasilos, Thomas 52
Vasquez, Victor R. 69
Vasudevan, Palligarnai T. 70
Vasudevan, Vijay K. 91
Vaughn, Mark 125
Vedula, Krishna 52
Vekilov, Peter G. 119
Velázquez-Figueroa, Carlos 109
Velander, William H 68
Velankar, Sachin 107
Velegol, Darrell 106
Velev, Orlin D. 87
Venerus, David C. 33
Venkatasubramanian, Venkat 36
Venugopalan, Vasan 10
Veser, Goetz 107
Vestal, Charles 19
Vigeant, Margot A.S. 101
Vigil, R. Dennis 39
Viljoen, Hendrik J. 68
Villafañe-Ruiz, Gilberto 109
Vinals, Jorge 27
Violi, Angela 57
Virk, Preetinder S. 53
Visco, Donald P. 116, 117
Viswanath, Dabir S. 65
Vives, Donald L. 3
Vlachos, Dionisios 23
Vogt, Bryan 5
Vohs, John M. 105
von Braun, Margrit C. 30
VonBerg, Robert L. 80
Vossoughi, Shapour 40
Vrentas, James S. 106

W

Wachs, Israel E. 104
Waggoner, Raymond C. 66
Wagner, Eric S. 14
Wagner, Jan 98
Wagner, John E. 37
Wagner, Norman J. 23
Wagner, Robert 56
Wagner, William R. 107
Walas, Stanley M. 40
Walawender, Walter P. 41
Walker, Lynn 102
Walker, Sharon L. 12
Walkinshaw, John W. 52
Wallace, William E. 134
Wallis, Graham B. 70
Wallman, P. Henrik 8
Walters, Keisha 63
Walton, Krista 41
Walton, S. Patrick 58
Walz, John 130
Wan, Kai-Tak 66
Wang, Chunsheng 117
Wang, Daniel I. C. 53
Wang, Henry Y. 57
Wang, Hongjun 74
Wang, Jee-Ching 66
Wang, Jin 3, 135
Wang, Joseph 5
Wang, Linda Nien-Hwa 36
Wang, Nam Sun 49
Wang, Pin 16

Index of U.S. Faculty 261

Wang, Ping 89
Wang, Qiang (David) 20
Wang, Shaw S. 73
Wang, Szu-Wen 10
Wang, Tse-Wei 116
Wang, Yong 21
Wang, Zhen-Gang 14
Wankat, Phillip C. 36
Ward, Timothy L. 76
Wasan, Darsh T. 33
Watkins, James J. 51
Watson, A. Ted 20
Watson, John 88
Watters, James C. 43
Way, J. Douglas 19
Wayner, Peter C. 82
Wazzan, A. R. 11
Weatherley, Laurence R. 40
Weaver, Harry 76
Weaver, Jason 25
Weaver, Mark 1
Weber, Frederick E. 116
Weber, James H. 68
Weber, Thomas W. 84
Weber, Walter J. Jr. 57
Weeks, Brandon 125
Wei, James 72
Weidner, John W. 113
Weigand, William A. 49
Weimer, Alan W. 18
Weimer, Jeffrey 2
Weinberger, Charles B. 103
Weinkauf, Donald H. 75
Weinstein, Herbert 78
Weinstein, Michael 83
Weinstein, Randy D. 108
Weinstein, Steven 80, 83
Weiss, Alvin 56
Weiss, Robert A. 21
Welker, J. Reed 7
Weller, Sol W. 84
Wender, Irving 107
Wendt, Jost O. L. 127
Wentzcovitch, Renata 62
Wesson, Rosemarie D. 49
West, A. C. 79
West, Jennifer 121
Westenberg, David 66
Westerberg, Arthur W. 102
Westmoreland, Phillip R. 51
Wetzel, David M. 44
Wheeler, Clayton 47
Wheeler, Dean 126
Wheeler, George L. 22
Wheelock, Thomas D. 39
Whitaker, Stephen 9
White, Don H. 6
White, John Robert 52
White, Lee 102
White, Mark G. 63
White, Ralph E. 113
Whiteley, James R. 98
Whiting, Wallace B. 69
Whitlow, Jonathan E. 26
Whitten, David G. 76
Whitty, Kevin 127

Wickramasinghe, Ranil 20
Wicks, Charles E. 100
Wiegandt, H. F. 80
Wiencek, John 38
Wiesner, Theodore F. 125
Wiest, John M. 1
Wilcox, Jennifer L. 56
Wilcox, William R. 78
Wildi, Robert 134
Wilding, W. Vincent 126
Wilhite, Benjamin 21
Wilkens, Robert 93
Wilkes, Garth L. 130
Wilkes, James O. 57
Willey, Ronald J. 54
Willhite, G. Paul 40
Williams, Christopher T. 113
Williams, John A. 54
Williams, Michael C. 8
Williams, Susan M. 40
Williamson, Kenneth 100
Williford, Clint W. Jr. 63
Willing, Gerold A. 43
Willis, Brian G. 23
Wills, George B. 130
Willson, C. Grant 124
Willson, Richard C. 119
Winey, Karen I. 105
Winnick, Jack 29
Winter, H. Henning 51
Winter, Jessica 94
Winter, Robb M. 114
Wirtz, Denis 48
Wise, Donald L. 54
Wisecarver, Keith D. 99
Wissler, Eugene H. 124
Wittrup, K. Dane 53
Wnek, Gary 90
Wo, Shaochang 137
Wolden, Colin A. 19
Wolf, Eduardo E. 35
Won, You-Yeon 36
Wong, Michael 121
Wood, David W. 72
Wood, Thomas 122
Woodruff, Gene L. 131
Wool, Richard P. 23
Woolf, Peter 57
Worden, R. Mark 58
Wornat, Mary J. 44
Wrenn, Steven 103
Wu, David T. 19
Wu, J. H. David 83
Wu, Jianzhong 12
Wu, Jing 71
Wyman, Charles E. 12
Wynne, Kenneth 129
Wyslouzil, Barbara 94

X

Xanthos, Marino 71
Xia, Younan 131
Xing, Yangchuan 66
Xomeritakis, George 76
Xu, Qiang 120

Y

Yablonsky, Gregory 67
Yager, Paul 131
Yan, Yushan 12
Yang, Arthur J.-M. 49
Yang, Fuqian 42
Yang, Hong 83
Yang, Ralph T. 57
Yang, Ray Y. K. 134
Yang, Shang-Tian 94
Yang, Shu 105
Yarbrough, David W. 117
Yarmush, Martin L. 73
Yasuda, H. K. 65
Yates, Matthew 83
Yaws, Carl L. 120
Ydstie, B. Erik 102
Yee, Albert F. 10
Yesavage, Victor F. 19
Yi, Hyunmin 55
Yin, John 136
Yoganathan, Ajit P. 29
Yokochi, Alexandre 100
York, Otto 71
Yortsos, Yannis C. 16
You, Seong 3
Young, Edwin H. 57
Young, Gregory L. 16
Young, Ming-Wan 71
Young, Valerie L. 95
Youngquist, Gordon R. 39
Yu, Qingsong 65
Yu, Xiaojun 74
Yurttas, Lale 122

Z

Zafiriou, Evanghelos 49
Zakin, Jacques L. 94
Zappi, Mark E. 46
Zasadzinski, Joseph 13
Zawodzinski, Thomas 90
Zeimer, Katherine S. 54
Zenz, Frederick A. 81
Zhai, Tongguang 42
Zhang, Fuming 82
Zhang, Wu 134
Zhao, Huimin 32
Zhao, Jian 3
Zhao, Shihuai 3
Zhou, H. Susan 56
Zhou, Yaoqi 84
Zhou, Youngquan 115
Zhu, Lei 21
Zhu, Wen-Hua 3
Zhu, Yingxi Elaine 35
Ziegler, Edward N. 81
Ziegler, Kirk 25
Ziff, Robert M. 57
Zilm, Kurt 22
Zollars, Richard L. 133
Zondlo, John W. 134
Zones, Stacey 8
Zukoski, Charles F. 32
Zurawsky, Walter P. 81
Zydney, Andrew L. 106
Zygourakis, Kyriacos 121

Index of Institutions

A
Aachen Technical University ... 168
Abo Akademi University ... 166
Adelaide, University of ... 140
Akron, University of ... 89
Alabama in Huntsville, University of ... 2
Alabama, University of ... 1
Alberta, University of ... 151
Alcala de Henares Ingenieria Quimica, University of ... 214
Alicante, University of ... 214
Amirkabir University of Technology(Tehran Polytechnic) 182
Amsterdam, University of ... 200
Ankara University ... 233
Annamalai University ... 174
Aristotle University ... 172
Arizona State University ... 5
Arizona, University of ... 6
Arkansas, University of ... 7
Aston University ... 237
Auburn University ... 3
Auckland, University of ... 202
Autonoma University of Barcelona ... 214

B
Bandung, Institut Teknologi ... 181
Bangladesh University of Engineering and Technology .. 144
Bath, University of ... 237
Beijing University of Chemical Technology ... 163
Ben-Gurion University of the Negev ... 184
Bicol University ... 204
Birla Institute of Technology & Science (BITS) - Pilani . 175
Bogazici University ... 233
Bologna, Universita di ... 186
Braunschweig, Technische Universitaet ... 168
Brigham Young University ... 126
British Columbia, University of ... 150
Brown University ... 110
Bucknell University ... 101
Budapest University of Technology and Economics ... 174

C
Cagliari, Universita di ... 186
Calcutta University ... 175
Calgary, University of ... 152
California Institute of Technology ... 14
California State Polytechnic University, Pomona ... 15
California State University, Long Beach ... 15
California, Berkeley, University of ... 8
California, Davis, University of ... 9
California, Irvine, University of ... 10
California, Los Angeles, University of ... 11
California, Riverside, University of ... 12
California, San Diego, University of ... 12
California, Santa Barbara, University of ... 13
Cambridge, University of ... 238
Campinas, Universidade Estadual de ... 146
Cantabria, University of ... 215
Canterbury, University of ... 203
Cape Town, University of ... 211
Carnegie Mellon University ... 102
Case Western Reserve University ... 90
Castilla Universidad ... 216
Chang Guag University ... 226
Chonnam National University ... 195
Christian Brothers University ... 115
Chulalongkorn University ... 231
Cincinnati, University of ... 91
City College of The City University of New York ... 78
Clarkson University ... 78
Clausthal, Technical University of ... 168
Clemson University ... 112
Cleveland State University ... 92
Colorado School of Mines ... 19
Colorado State University ... 20
Colorado, University of ... 18
Columbia University ... 79
Complutense University of Madrid ... 216
Connecticut, University of ... 21
Cooper Union ... 79
Cork Institute of Technology ... 183
Cornell University ... 80
Curtin University of Technology ... 141

D
Dalhousie University ... 151
Darmstadt University of Technology ... 168
Dartmouth College ... 70
Dayton, University of ... 93
De La Salle University ... 204
Delaware, University of ... 23
Delft University of Technology ... 201
Dortmund, University of ... 169
Drexel University ... 103

E
Ecole Polytechnique-University of Montreal ... 153
Edinburgh, University of ... 238
Ege, University of ... 234
Eindhoven University of Technology ... 201
Erlangen-Nurnberg, Universitat ... 169
Estudios Superiores de Monterrey, Instituto Tecnologico y de ... 199

F
Fachhochschule Mannheim, University of Applied Sciences ... 172
Feng-Chia University ... 226
Florida Institute of Technology ... 26
Florida State University/Florida A&M University ... 27
Florida, University of ... 25
Fukuoka University ... 189

G
Gazi University ... 234
Gent, University of ... 145
Georg-Simon-Ohm Fachhochschule Nuernberg ... 171
Georgia Institute of Technology ... 29
Granada, University of ... 217

H
Hacettepe University ... 235
Hampton University ... 128
Hannover, University of ... 170

Helsinki University of Technology 166
Heriot-Watt University 239
Hiroshima University 189
Hong Kong University of Science and Technology 173
Houston, University of 119
Howard University 24

I

Iberoamericana, Universidad 199
Idaho, University of 30
Illes Balears, University of 218
Illinois at Chicago, University of 31
Illinois at Urbana-Champaign, University of 32
Illinois Institute of Technology 33
Indian Institute of Science 175
Indian Institute of Technology, Delhi 177
Indian Institute of Technology, Kanpur 176
Indian Institute of Technology, Madras 176
Indian Institute of Technology, Roorkee 179
Industries Chimiques Institut, Ecole Nationale
    Superieure des 167
Instituto Tecnologico de Buenos Aires 138
Iowa State University of Science and Technology 39
Iowa, University of 38
Isfahan University of Technology 182

J

Johns Hopkins University 48

K

Kaiserslautern, Universitat 171
Kanazawa University 189
Kansai University 190
Kansas State University 41
Kansas, University of 40
Karlsruhe, Universitat 171
Kasetsart University 232
Katholieke Uiniversiteit Leuven 145
Keimyung University 195
Kentucky, University of 42
King Fahd University of Petroleum & Minerals 207
King Saud University 208
Korea Advanced Institute of Science and Technology 195
Korea University 196
Kuwait University 197
KwaZulu-Natal, University of 212
Kyoto University 190
Kyushu University 191

L

L'Aquila, Universita de 187
la Plata, Universidad Nacional de 138
Lafayette College 103
Lakehead University 154
Lamar University 120
Las Americas, Puebla, Universidad de 199
Laval, Universite 154
Laxminarayan Institute of Technology, Nagpur
    University 177
Leeds, University of 239
Lehigh University 104
Liege, University of 145
Litoral, Universidad Nacional del 138
Ljubljana, University of 210
Lodz, Technical University of 205
London (Imperial College), University of 239
London (University College London), University of ... 240
Loughborough University 241

Louisiana at Lafayette, University of 46
Louisiana State University 44
Louisiana Tech University 45
Louisville, J. B. Speed School of Engineering,
    University of 43
Lund, University of 224

M

Madrid-ETSI Industriales, Universidad Politecnica de ... 218
Maine, University of 47
Malaviya National Institute of Technology, Jaipur 178
Malaya, University of 198
Malaysia, Universiti Sains 198
Manchester, University of 241
Manhattan College 81
Manipal Institute of Technology 178
Maribor, University of 211
Maringa, Universidade Estadual de 147
Maryland Baltimore County, University of 50
Maryland, College Park, University of 49
Massachusetts Institute of Technology 53
Massachusetts-Amherst, University of 51
Massachusetts-Lowell, University of 52
McGill University 154
McMaster University 155
McNeese State University 45
Melbourne, University of 141
Metropolitana-Iztapalapa, Universidad Autonoma 200
Michigan State University 58
Michigan Technological University 59
Michigan, University of 57
Middle East Technical University 236
Minnesota at Duluth, University of 61
Minnesota at Minneapolis St. Paul, University of 61
Mississippi State University 63
Mississippi, University of 63
Missouri-Columbia, University of 65
Missouri-Rolla, University of 66
Monash University 142
Mons, Faculté Polytechnique de 146
Montana State University 68
Mumbai, University of 178
Murcia, University of 218

N

Nagoya University 191
National Central University 227
National Cheng Kung University 227
National Chung Cheng University 228
National Institute of Technology Warangal 179
National Taiwan University 229
National Taiwan University of Science and Technology .. 228
National Tsing Hua University 229
Nebraska, University of 68
Nevada, Reno, University of 69
New Brunswick, University of 156
New Hampshire, University of 70
New Haven, University of 22
New Jersey Institute of Technology 71
New Mexico Institute of Mining and Technology 75
New Mexico State University 77
New Mexico, University of 76
Newcastle Upon Tyne, University of 243
Newcastle, University of 143
Niigata University 192
North Carolina Agricultural and Technical State
    University 86
North Carolina State University 87

Index of Institutions

North Dakota, University of 88
North-West University, Potchefstroom Campus 212
Northeastern University 54
Northwestern University 34
Norwegian University of Science and Technology 203
Notre Dame, University of 35
Nottingham, University of 243

O

Ohio State University 94
Ohio University 95
Oklahoma State University 98
Oklahoma, University of 97
Oregon State University 100
Osaka University 192
Oulu, University of 166

P

Pais Vasco, University of 219
Pardubice, University of 165
Patras, University of 173
Pennsylvania State University 106
Pennsylvania, University of 105
Pisa, Universita di 187
Pittsburgh, University of 107
Pohang University of Science and Technology 196
Polytechnic University 81
Pontificia Universidad Catolica de Chile 162
Porto, Universidade do 206
Prairie View A&M University 120
Pretoria, University of 212
Prince of Songkla University 232
Princeton University 72
Puerto Rico, University of 109
Purdue University 36

Q

Qatar, University of 207
Queen's University 156
Queen's University of Belfast 244
Queensland, University of 143

R

Regional Engineering College, Affiliated with Bharathidasan University 179
Rensselaer Polytechnic Institute 82
Rhode Island, University of 111
Rice University 121
Rio De Janeiro, Universidade Federal Do 148
Rochester, University of 83
Rose-Hulman Institute of Technology 37
Rovira i Virgili University 219
Rowan University 73
Royal Institute of Technology 224
Royal Melbourne Institute of Technology 144
Royal Military College of Canada 157
Rutgers, The State University of New Jersey 73
Ryerson University 157

S

S. V. National Institute of Technology 180
Salamanca, University of 220
Salerno, University of 188
Salta, Universidad Nacional del 139
San Jose State University 16
Santander, Universidad Industrial de 164
Santiago de Chile, Universidad de 162
Santiago de Compostela, Universidade de 220
Santo Tomas, University of 204
Sao Paulo, Universidade de 149
Saskatchewan, University of 158
SDM College of Engineering and Technology 180
Seikei University 192
Seoul National University 197
Sevilla, Escuela Superior de Ingenieros Industriales de .. 221
Sevilla, Universidad de 221
Sharif University of Technology 183
Sheffield, University of 244
Sherbrooke, Universite de 158
Siddaganga Institute of Technology, Affiliated with Viveswaraiah University 180
Singapore, National University of 209
South Alabama, University of 4
South Carolina, University of 113
South Dakota School of Mines and Technology 114
South Florida, University of 28
Southern California, University of 16
Stanford University 17
State University of New York at Buffalo 84
Stevens Institute of Technology 74
Strathclyde, University of 245
Sur, Universidad Nacional del 140
Surrey, University of 245
Swiss Federal Institute of Technlolgy 225
Swiss Federal Institute of Technology, Ecole Polytechnique Federale de Lausanne 225
Sydney, University of 144
Syracuse University 85
Szczecin, Technical University of 205

T

Tamkang University 230
Technion-Israel Institute of Technology 185
Teesside, University of 246
Tennessee at Chattanooga, University of 115
Tennessee at Knoxville, University of 116
Tennessee Technological University 117
Texas A&M University 122
Texas A&M University-Kingsville 123
Texas at Austin, University of 123
Texas Tech University 125
Thapar Institute of Engineering and Technology 181
Tokyo Institute of Technology 193
Tokyo, University of 193
Toledo, University of 95
Toronto, University of 159
Toyama University 194
Tri-State University 37
Tsinghua University 163
Tufts University 55
Tulane University 46
Tulsa, University of 99
Tunghai University 230
Tuskegee University 4

U

Uberlandia, Universidade Federal de 150
United Arab Emirates University 236
University College Cork 184
University College Dublin 184
Utah, University of 127

V

Valencia, Universitat de 222
Valladolid, University of 223
Vanderbilt University 118

Vellore Institute of Technology 181
Venezuela, Universidad Central de 246
Villanova University 108
Virginia Commonwealth University 129
Virginia Tech 130
Virginia, University of 128

W
Warsaw University of Technology 205
Washington State University 133
Washington University 67
Washington, University of 131
Waterloo, University of 160
Wayne State University 60
West Virginia University 134
West Virginia University Institute of Technology 135
Western Michigan University 60
Western Ontario, University of 161

Widener University 108
Wisconsin-Madison, University of 136
Witwatersrand, Johannesburg, University of the 213
Worcester Polytechnic Institute 56
Wyoming, University of 137

Y
Yale University 22
Yamagata University 194
Yildiz Technical University 236
Yokohama National University 194
Youngstown State University 96
Yuan Ze University 231

Z
Zagreb, University of 165
Zaragoza, University of 223